# FREE THO
# FAITH AND SCIENCE

*Finding Unity Through
Seeking Truth*

Roger Pullin

outskirtspress
DENVER, COLORADO

The opinions expressed in this manuscript are solely the opinions of the author and do not represent the opinions or thoughts of the publisher. The author has represented and warranted full ownership and/or legal right to publish all the materials in this book.

Free Thought, Faith, and Science
Finding Unity Through Seeking Truth
All Rights Reserved.
Copyright © 2014 Roger Pullin
v7.0

Cover Image © 2014 Roger Pullin. All rights reserved - used with permission.

This book may not be reproduced, transmitted, or stored in whole or in part by any means, including graphic, electronic, or mechanical without the express written consent of the publisher except in the case of brief quotations embodied in critical articles and reviews.

Outskirts Press, Inc.
http://www.outskirtspress.com

ISBN: 978-1-4787-3570-0

Outskirts Press and the "OP" logo are trademarks belonging to Outskirts Press, Inc.

PRINTED IN THE UNITED STATES OF AMERICA

# Author's Acknowledgements

I thank all the wonderful people who helped me to complete this work, especially the following: Jim Atkinson and Gill and John Vance, who spent much time reviewing and editing my efforts, while holding views that sometimes differed from mine, a little or a lot; Aque Atanacio for turning my scruffy diagrams into polished images, and Emma del Rosario for keeping track of a manuscript that went through many changes. Above all I thank God for His gift of Free Thought and His constant guidance in the spiritual warfare that rages during the writing of books like this and, whether you believe it or not, in the reading of them.

# Table of Contents

Author's Acknowledgements ................................................... i
PREFACE ............................................................................... ix

**CHAPTER 1**
**INTRODUCTION** ................................................................. 1
   Purposes and Perspectives ............................................... 1
   Free Will ............................................................................ 4
   Free Thought and the Mind-Soul Interface ..................... 6
   Divine Interventions? ....................................................... 7
   Faith and Organized Religion .......................................... 8
   Science ............................................................................. 10
   The False Divide Between Faith and Science ................ 12
   Expanding the Faith-Science Quest for Truth ............... 13

**CHAPTER 2**
**THE HUMAN CONDITION** ............................................. 19
   Existence ......................................................................... 19
   Humans and God ........................................................... 25
   Original Sin? ................................................................... 30
   The Self ........................................................................... 33
   Happiness and Morality ................................................. 37
   Consciousness and the Brain ......................................... 42
   The Soul .......................................................................... 49

**CHAPTER 3**
**FREE THOUGHT: SCOPE AND FOUNDATIONS** ........ 59
   Free Thought Territory ................................................... 59
   Free Thought and free thought/freethinking ................ 61
   Information .................................................................... 66
   The Free Will Debate ..................................................... 69
      *The Debate in Organized Religion* ........................... 74
      *The Debate in Science* ............................................... 78

## CHAPTER 4
### FREE THOUGHT: MODEL AND MECHANISMS ................. 82
- Combining Mental Reasoning and Soul Processing .............. 82
- Turning to Quantum Theory ..................................................... 85
- Model and Mechanisms ............................................................. 88
- Comparisons with Other Perspectives and Findings ............... 94
  - *The Quantum Observer* ........................................................ 94
  - *Dual-Aspect Monism* ............................................................ 97
  - *Consciousness and the Big Picture* ...................................... 99
- Postscript .................................................................................. 101

## CHAPTER 5
### FAITH ............................................................................................ 103
- What Is Faith? .......................................................................... 103
- Belief Without Doubt .............................................................. 106
- Faith and Organized Religion ................................................ 109
- Faith in the Quest for Truth ................................................... 112
- Soul States: Baselines, Shifts and Leaps ................................ 115
- Some Personal Experiences of Faith ..................................... 123
- Postscript .................................................................................. 129

## CHAPTER 6
### SCIENCE ....................................................................................... 132
- God and Science ...................................................................... 132
- The Philosophy of Science ..................................................... 133
- Creativity in Science ............................................................... 137
- The Conduct of Science .......................................................... 140
- Scientific Method .................................................................... 141
- Indicators of Correctness ....................................................... 145
- Undervalued Science and Persistent Pseudoscience ........... 146
- A Theory of Everything .......................................................... 150
- Postscript .................................................................................. 152

## CHAPTER 7
## BATTLEFIELDS ............................................................. 155
- No Neutral Ground ................................................. 155
- Faith Versus Unbelief .............................................. 155
- Evil Versus Good..................................................... 157
- Organized Religion Versus Science ........................ 163
- Evolution ................................................................ 165
- Intelligent Design................................................... 175
- Interpersonal, Group and Institutional Battles ..... 183
- A Few Battles At Church........................................ 186

## CHAPTER 8
## TRUTH AND MORALITY ......................................... 189
- Truth, Morality and Us........................................... 189
- Perspectives on Truth............................................. 193
- Postmodernism ...................................................... 197
- Truth About the Material Realm ........................... 201
- Truth About the Spiritual Realm ........................... 204
- Sources of Morality................................................ 205
- Altruism ................................................................. 212
- The Universal Moral Code..................................... 219

## CHAPTER 9
## IN THE MIDST OF LIES AND NONSENSE .............. 225
- Lies and Nonsense from Organized Religion ....... 225
- Sacred Texts ........................................................... 228
- Interpretations....................................................... 232
- Literal Impossibilities ............................................ 238
- Angels .................................................................... 242
- Faith Healing ......................................................... 244
- Heaven.................................................................... 246
- Homosexuality ....................................................... 248
- The Rapture ........................................................... 251
- Research on Prayer ................................................ 254
- Miscellaneous Superstitions .................................. 261

## CHAPTER 10
## REFORMATIONS AND REVOLUTIONS ..................................263
Concepts, Definitions and Prospects ................................. 263
The Protestant Reformation ............................................... 270
Revolutions in Science ........................................................ 271
Christianity in Flux.............................................................. 273
Reformed Epistemology....................................................... 276
Christian Apologetics........................................................... 280
The Second Vatican Council ............................................... 290
Postscript............................................................................... 298

## CHAPTER 11
## UNITY ........................................................................................301
Drawing Threads Together ................................................. 301
Expanding the Faith-Science Quest for Truth ...................... 304
*Prerequisites*................................................................... *304*
*Perspectives* ................................................................... *307*
*Prospects* ........................................................................ *313*
Research on Free Thought and Further Development
of Theory............................................................................... 314
*Brain and Consciousness Research* ............................. *326*
*Towards Combined 'M and S' Theory* ......................... *333*

APPENDIX I: DEFINITIONS .................................................336
KEY TERMS......................................................................... 336
OTHER TERMS................................................................... 340

APPENDIX II: CREDO .............................................................347
Church Creeds and Personal Creeds.................................. 347
The Apostles' Creed and the Nicene Creed ...................... 348
The Athanasian Creed ......................................................... 356
Miscellaneous Topics........................................................... 357
*Angels*............................................................................. *357*
*Demons*........................................................................... *359*
*Evolution*........................................................................ *361*
*Heaven and Hell* ........................................................... *362*
*Miracles*.......................................................................... *365*

 *Prayer*......368
 *Predestination*......370
 *Souls*......371
 *The Spiritual Force for Evil*......372
 *The Universal Moral Code*......373

APPENDIX III: ABOUT ME......375
 Origins......375
 School and Church......376
 University......378
 The Isle of Man......380
 The Philippines......382
 Lessons Learned and Looking Ahead......387
 Epilogue: 'The Most Important Thing in Life'......388

APPENDIX IV: BATTLEFIELD LITERATURE......390
 PRO-FAITH......390
 PRO-UNBELIEF......395

APPENDIX V: GATHERING SUBJECTIVE EVIDENCE FROM WITHIN CONCERNING FAITH AND UNBELIEF..420

INDEX......432

# PREFACE

Throughout history, individual humans have felt the need to connect with something spiritual. Believers explain that need as coming from our souls as we seek God and are sought by Him. Nonbelievers explain it as a product of the evolution of the human brain and as delusions encouraged by organized religion.

Our bodies, brains and minds are made of the same materials as those of other forms of life, but we are very different. Human consciousness operates at a level far above anything seen in animals. We are highly self-aware and potentially God-aware. Our bodies, brains and minds are parts of the material realm, but each of us also has a soul, which is part of the spiritual realm.

We make our choices about the practicalities of life, such as what to eat and what to wear, by using what I call basic thought. We use what I call Free Thought to make choices about the higher things of life: creativity; faith; justice and morality. Faith is defined here as personal belief and trust in God, not as an organized religion or a religious affiliation.

Free Thought is not the same as freethinking in the conventional sense of rebelling against some religious or other form of orthodoxy. A believer has made a Free Thought choice for faith. Choosing faith does not mean accepting everything in the doctrines, dogma and books of an organized religion. In faith or unbelief, everyone has a private and personal creed, derived from Free Thought.

I take the human condition to be a state of combined existence in the material and spiritual realms. There is plenty of objective evidence for our material realm existence, but our spiritual realm existence is entirely a matter of subjective experience. From that dualistic perspective of reality, with its inevitable mixing of objective and subjective evidence, my model for Free Thought has the material realm components of self (body, brain and mind) and the spiritual core of self (soul) working as an integrated whole.

Science is based upon objective evidence, repeatability of findings, and the explanatory and predictive power of theory. As theories get better, we approach truths about the material realm. Strictly speaking, all truthful disclosures through science are verisimilitudes, not absolute truths. For all practical purposes, however, we can take as true the existence of energy, electromagnetism, gravity, the genetic code, matter etc.

Nonbelievers, especially scientists, hold that the question of whether anything spiritual truly exists can be answered only by rigorous scientific investigation. Probing the human brain with the instruments and methods of science is not likely to detect the soul or spiritual revelations and responses. The only clearly researchable evidence for our choices of faith or unbelief is our subjective evidence. Nevertheless, that subjective evidence can be gathered systematically and analysed rigorously, as part of science.

Free Thought is the processing of information from the material and spiritual realms about the higher things of life. The mind reasons about what is sensed from the material realm. The results of that reasoning pass through the mind-soul interface to the soul. The soul processes revelations from God and/or the spiritual force for evil. Information about the state of the soul passes though the mind-soul interface to the mind. Free Thought is the whole of that interactive and iterative, spiritual and mental process, together with responses from the body-mind to the surrounding material realm and from the soul to the spiritual realm.

Information passing through the mind-soul interface must be in one or more common non-material formats, so that it can be combined and appraised to produce the outcomes of Free Thought. Our individual mind-soul interfaces are the *only* connections between the material and spiritual realms. Free Thought takes place in an open system - an individual human who is always part of and connected to the material and spiritual realms.

Believers have diverse perspectives on the spiritual realm and the extents to which God intervenes in the material realm. Deists believe in a non-interventionist God. Some theists believe that God intervenes, either a little or a lot, in general or specifically, on behalf of believers who ask for His help. Some believe that God has predestined everything to proceed only according to His fixed and perfect plan, in His perfect timing. That would prevent God and humans from doing anything beyond playing out His fully scripted drama.

I am a theist, but I believe that God's interventions in the material realm are made *only* through His spiritual revelations to individual souls. I believe that the outcomes of an individual's Free Thought change her/his behaviour and can thereby change the behaviour of others and the state of the world.

I believe that we all live free-willed lives in a free process material realm and must accept the attendant risks. I do not believe that God choreographs the physics, chemistry and biology of accidents, diseases and natural disasters. God allows nature to help us or harm us. I do not believe that God invites or grants requests for any miracles that would require contravening the laws of nature, by which He made this world to work.

I believe that good and evil are at war, in us and around us. God does not direct acts of evil. The rebel soldiers who chop limbs off children and the jihadists who commit mass murder are not fulfilling God's perfect plan in His 'perfect timing.' They are acting for the spiritual force for evil, which engineers all human acts of evil: betrayal; corruption; cruelty; discrimination; greed; injustice and lying. God allows evil to remain a potent force.

I believe that the Big Bang, which led to the evolution of our observable material realm, was the initial mega-miracle in the material realm. I believe that the Incarnation of God, as Jesus Christ, was the only other opportunity in the material realm for some God-performed miracles that overrode the God-given laws of nature. I have seen no other convincing reports of any supernatural miracles, signs and wonders. Believers who cling to such notions are hindering the quest for truth and turning honest seekers away from exploring faith.

I believe that God created the means by which everything in the material and spiritual realms came into being. Some nonbelievers, especially scientists, argue that explanations of reality involving God and the spiritual realm are unnecessarily complex and therefore almost certainly wrong. But explanations based solely on the material realm are also highly complex. Scientific theory has posited the parallel existence of 10 to the power 500 universes and a material realm made of tiny strings with up to 11 dimensions. What matters is the *truth*, regardless of its complexity or simplicity.

We are all on the same side when we refuse to make compromises over truth and recognize that an inconvenient truth is still a truth. Faith and science are complementary paths to truth. Believers find truth through spiritual revelations from God. I believe that all truthful disclosures about the material realm through mathematics and science also come from God, as believers and nonbelievers develop and use their God-given creativity in Free Thought.

Exploring faith requires no compromise in the rigorous methods of science. Embracing science requires no compromise in faith. The faith-science quest for truth can be expanded through science-friendly reformations in organized religion and faith-friendly revolutions across science. The reformations will require believers, especially religious authorities, to embrace science. The revolutions will require nonbelievers, especially scientists, to admit the validity of subjective evidence from within concerning faith and unbelief.

I am a biologist, working mostly on fish and other aquatic biodiversity. I am also a Christian and a member of an international church, where I have served as an Elder and in the music ministry. My faith and my enthusiasm for science are fully in tune. This book is sprinkled with stories from my life and work in the Isle of Man and the Philippines and my various travels. I hope that it will contribute to bringing more believers and nonbelievers together in an expansion of the faith-science quest for truth. In this book, 'we' means everybody.

Roger Pullin                    Makati City, Philippines; May 6, 2014

# CHAPTER 1

# INTRODUCTION

*"Whatsoever things are true, whatsoever things are honest...if there be any virtue, and if there be any praise, think on these things"* [1]

## PURPOSES AND PERSPECTIVES

The Apostle Paul gave good advice for finding the best way through this world of wonders and horrors, pleasures and pain - *think* and focus on whatever is *true*. All the truth that we can glimpse, whether through faith, where faith is defined as personal belief and trust in God, or through science, or through a combination of both, comes from one whole body of truth. Truth cannot contradict itself.

Richard Dawkins[2] stated as follows the purpose of his most famous book against religion: *"If this book works as I intend, religious readers who open it will be atheists when they put it down."* My purpose here is to encourage believers who distrust science and nonbelievers who regard faith as a delusion to reassess their positions and consider participation in an expanded faith-science quest for truth.

---

1    Philippians 4:8
2    Dawkins, R. 2006. *The God Delusion*. Bantam Press: London.406p.p.5.

# Free Thought, Faith, and Science

This book is an attempted contribution in a complex field of study to which I am a relative newcomer. It is also a personal testimony. Writing in the first person singular is generally frowned upon in science, but I had no other option here in order to describe some of my experiences concerning faith and science, my present beliefs and my journey through life so far.

I begin with general explanations about the human condition, human thought processes, faith and science. I move on through reviews of the battlefields on which we fight about the nature of truth and morality, in the midst of lies and nonsense. I conclude by making the case for reformations and revolutions and for taking the path to faith-science unity through a common quest for truth. The various chapters and subchapters in this book are best read and considered in turn, as steps along this path, as if they form a series of lectures.

The literature in this field is vast. I consulted mainly Western sources and I recognize the deficiencies of this position. All my sources are referenced as footnotes to the pages on which they are cited rather than as endnotes, in order to provide easy access for those who would like to see immediately what I have consulted and to have rapid access to websites.

Readers who are unused to and/or distracted by references as footnotes can choose to ignore them and stick with the text. No one need feel pushed to read my original sources, unless she/he wishes to do so. This book should suffice as an account of the main points that I am making and the necessary evidence.

I hold that every individual is a unique self - a material body-mind integrated with a spiritual soul - existing in and relating to a reality than spans the material realm and the spiritual realm. Many eminent scholars have covered and continue to cover the same ground. I am trying to draw together some existing threads in order to develop what

# INTRODUCTION

I call a theory of Free Thought.

Free Thought is *not* what is commonly called 'freethinking' or 'free thought,' meaning opposition to faith, or to conformity or orthodoxy. Free Thought can have agnosticism, or atheism, or faith as its outcome. Free Thought is the combined mental and spiritual process by which we explore and make our choices about the higher things in life: the development and use of our creativity; faith and unbelief; justice and injustice; morality and immorality.

Free Thought has capital letters to indicate its higher status above our basic thought about the practicalities of life. Free Thought is special. One can postpone a Free Thought decision about faith or about making creative contributions in a given field, but one cannot go for long without basic thought about what to eat, what to wear, how to dodge traffic etc.

By way of comparison, Peter Russell[3] wrote: "*In every moment I have a choice as to how I see a situation. I can see it through eyes caught in the materialist mindset that worries whether or not I am going to get what I think will make me happy. Alternatively, I can choose to see it through eyes free from the dictates of this thought system…The place to go for help is deep within, to that level of consciousness that lies beyond the materialistic mindset - to the God within.*"

Basic thought about our material needs and circumstances occupies most of our waking hours. We spend comparatively little time in Free Thought about the higher things of life, but Free Thought time is time well spent. It is the means by which we make reality checks on the spiritual dimension in our own lives and the human condition in general.

Basic thought and Free Thought are sometimes complementary. We

---

3    Russell, P. 2003. *From Science to God: A Physicist's Journey into the Mystery of Consciousness.* New World Library: Novato CA.129p.p.96.

need money to obtain the necessities of life, which are the targets of our basic thought. We use Free Thought in ethical choices about how we earn money - by a fair day's work for a fair day's pay or by scams. We use Free Thought for decisions about whether to purchase products that damage people and/or nature.

Daniel Kahneman[4] describes us as experiencing selves and remembering selves, with our minds working through *"an uneasy interaction"* between two systems of thinking - the *"automatic"* (fast thinking) *System 1 and the effortful* (slow thinking) *System 2."* My differentiation between basic thought and Free Thought is similar, but I define Free Thought from a much wider perspective, including interactions between the spiritual and material realms across the mind-soul interface. Those interactions are more than *"uneasy."* They are battlefields.

I do not venture here into any discussion of the unconscious. I regard all episodes of Free Thought as conscious experiences. I make no attempts to relate Free Thought to brain structure and states or to R.D. Laing's[5] work on the divided self. I regard the body-mind, including the brain, and the soul as complex works in progress, integrated and working together as a human self.

## Free Will

Each of us is undeniably a self. Each self has a will. Individual free will has been a controversial topic for centuries. The main question is whether the choices made by a human self derive from a sovereign and independent self will or are determined by factors beyond the self's control. These alternative positions are called *free will* and *determinism*.

The position that free will and determinism are incompatible is called

---

4 Kahneman, D. 2012. *Thinking Fast and Slow*. Penguin Books: London.499p.p.415.
5 Laing, R.D. 1960. *The Divided Self: An Existential Study in Sanity and Madness*. Republished in 2010 by Penguin Classics: London.218p.

# INTRODUCTION

*incompatibilism*. Within incompatibilism, *libertarianism* affirms free will and denies determinism, whereas *hard determinism* takes the opposite view. The alternative to incompatibilism is *compatibilism* or *soft determinism*, in which free will is seen as compatible with determinism in various ways. The works of Robert Kane[6] and Bob Doyle's[7] layperson-friendly compilation are excellent guides to the complex debates about free will.

I define free will as every individual's ever-present capacity to *choose* her/his personal goals privately and to decide freely to what extent those goals are divulged to others. The extent to which we have freedom of *action* to profess and/or pursue our freely chosen goals is a separate issue from how we make the *choices* that we keep inside us.

Our family members, peers and others can pressure us into making professions of our goals, but those professions can be either true or false. The choices about what we profess are always ours. We can also be pressured into joining efforts in which we pretend to pursue goals that differ from our chosen ones. On this basis, I hold that we have free will to set our personal goals and that free will is foundational for Free Thought.

I believe that God and the spiritual force for evil fight their battle of wills in our souls and use us to project that battle on to interpersonal and group battlefields, through the outcomes of our Free Thought. Despite all of those human and spiritual pressures, we remain free-willed, sovereign selves. We make free choices in respect of our personal goals and our privately held preferences for actions to pursue those goals.

We share with other living organisms instinctive and natural goals

---

[6] Kane, R. 2005. *A Contemporary Introduction to Free Will*. Oxford University Press: Oxford.196p. Kane, R. Editor. 2008. *Free Will*. Blackwell Publishing: Malden MA.310p.

[7] Doyle, B. 2011. *Free Will: The Scandal in Philosophy*. I-Phi Press: Cambridge MA.458p. Available, with updates, at: www.informationphilosopher.com/scandal

for survival and reproduction. We pursue those goals through basic thought. We also have personal goals concerning the higher things of life. We seek truth about our selves and the world around us. We strive to make creative contributions in our chosen fields and to enjoy the creative contributions of others in many fields. We pursue our higher goals through Free Thought. Their pursuit can help or hinder the attainment of our basic biological goals. Ask any starving artist.

## Free Thought and the Mind-Soul Interface

I define Free Thought as the integrated and iterative process that combines soul processing of and responses to information from the spiritual realm with mental reasoning about and responses to information from the material realm. Free Thought involves three interlinked processes of stimuli and responses: revelations from God and the soul's responses; revelations from the spiritual force for evil and the soul's responses; and revelations from the material realm and the body-mind's responses.

The pivotal element is what I call the *mind-soul interface.* Any dualistic system must have an interface if there is to be any exchange between one subsystem and the other. For mind and soul, John Turl[8] expressed that necessity as follows: *"For the Christian who believes that the Holy Spirit is active in the world, causal closure of the physical domain is not an option.* **There is something in the mind that is designed to interface with the spiritual** (present author's emphasis). *It should not therefore be seen as an inappropriate mechanism if a person's soul, which I contend to be a spiritual entity, can affect her/his own brain."*

That mechanism is much more than appropriate. It *defines* the human self as an integrated whole. During Free Thought, the spiritual

---

8 Turl, E.J. 2010. Substance Dualism or Body-Soul Duality? *Science and Christian Belief* 22(1): 57-86.p.75.

# INTRODUCTION

battlefield in a soul is connected across the mind-soul interface to body-mind experiences of the material realm. Information, in one or more non-material formats, is read, combined and recombined, to make decisions in mind and soul.

## DIVINE INTERVENTIONS?

When there were no humans on Earth, as it was for most of its history, neither God nor the spiritual force for evil would have needed any portals through which to make interventions in the material realm. From the dawn of life, God's 'eye' could have been on the sparrow[9] and on everything else in His creation - the sparrow hawk, the cat, the bacterium, the parasitic worm, the giant squid, the virus, the whale and the rest of biodiversity - but He need not have had His 'hands' on any of it until we came on the scene.

I see no reason to believe that God has ever intervened in the post-Big Bang material realm to effect material change in any of its nonliving components and nonhuman life forms, apart from the period of His Incarnation as Christ. For all other times and places, I believe that God made all the necessary mechanisms for the material realm to work without any physical interventions from Him. It has worked very well in that mode. It has led to the evolution of many species, among which God chose us for His completion, to be spiritual company for Him.

John Polkinghorne[10] used *"the free-process defence"* to explain the *free process* by which the material realm works. This can be readily observed. Neither God nor the spiritual force for evil choreographs accidents, diseases and natural disasters, or indeed bumper harvests and lottery wins. God and the spiritual force for evil intervene only in our souls, which are part of the spiritual realm, after which *we* act on the

---

9 Luke 12:6-7
10 Polkinghorne, J.C. 1996. *The Faith of a Physicist: Reflections of a Bottom-Up Thinker*. Fortress Press: Minneapolis MN.211p.p.83-85.

outcomes of our individual Free Thought, with impacts on ourselves, our fellow humans and the environment.

Whether we perceive them as such or not, we all receive spiritual revelations in our souls, from God and from the spiritual force for evil, as information for our Free Thought. In our Free Thought episodes, we use all available information from the spiritual and material realms and we choose, in private, our goals, our beliefs, our preferred behaviour and how we will actually behave.

Our Free Thought and our mind-soul interfaces are the only portals for information exchange between the spiritual and material realms and the only route for divine interventions. We have evolved and been gifted by God with souls and mind-soul interfaces to be spiritually influenced agents of change in the free process material realm,

**Faith and Organized Religion**

I define faith as personal belief and trust in God, not as an organized religion, or a religious affiliation, or a general sense of spirituality, or the search for something sacred. I define organized religion as an institution or organization, the purpose of which is to further specific beliefs and associated practices in recognition of the spiritual realm, spiritual beings and their alleged influences on the material realm, especially on humans.

Faith is about true freedom and high personal empowerment, living life with God not only as Lord but also as Friend. One cannot be truly free without recognizing the existence of God and the context in which He has given us our freedom. The word religion has origins in Middle English and Old French, where it was used to describe monastic vows. Religion is based on adherence to rules. Faith is a personal and constantly evolving relationship with God.

# INTRODUCTION

A believer is one who has made a decision for faith. A nonbeliever is one who has made a decision against faith or has yet to make a decision for or against faith. Everyone is either a believer or a nonbeliever, but our professions of faith and unbelief can be true or false. No person can ever know for sure what any other person really believes. Memberships of organized religions and participation in religious practices are not proofs of faith. Some nonbelievers comply with religious practices for purely economic and/or social reasons.

Organized religion can support faith, but organized religion allied with political and economic agendas has often perverted faith and led believers away from the love and truth that are the basis of a personal relationship with God. Throughout history powerful religious authorities have established command-and-control over relatively powerless flocks and have served their own selfish interests, rather than serving God. Religious authorities have led and bankrolled many horror shows including the dispossession, killing, maiming and torturing of outsiders, all under religious banners and allegedly in God's name.

Charles Mackay[11] gave chilling accounts of the Christian crusades, inquisitions and persecutions of alleged witches. For a villager in medieval Europe, a trip to Jerusalem or participation in the disposal of an alleged witch would have been an exciting episode in a drab life - especially when paid and assured by political and religious authorities of rewards in heaven. Suicide bombers hear, believe, and respond to similar lies. Present day horror shows include the attacks of 9/11 and the maiming and killing of girls for wanting to choose whom they will marry, or simply for wanting to go to school.

I applaud all who condemn the past and present lies, nonsense and horror shows perpetrated by organized religion. Faith *per se* merits no such condemnation. God does not direct religious horror shows. The

---

11  Mackay, C. 1841. *Extraordinary Popular Delusions and the Madness of Crowds.* Available from Maestro Reprints: Lexington KY.399p.

spiritual force for evil recruits the participants in religious horror shows by dominating their Free Thought.

## SCIENCE

I define science as any intellectual activity, observation or experiment concerning the composition and workings of the material realm and the nature of spirituality in the human condition. The systematic gathering and analysis of subjective evidence from within concerning faith and unbelief qualifies as scientific research. Anyone who seeks an explanation about any aspect of the material realm and/or the true nature of the human condition, including our spirituality, acts as a scientist.

Simon Blackburn[12] noted that science and common sense share: "*a peculiar feature...which is not shared by every 'discourse' such as ethics or aesthetics or pure mathematics. Science and common sense offer **their own*** (original author's emphasis) *explanation of why we do well by using them. It is not the privilege of some second-order, philosophical, subtle and elusive theory called realism. Science and common sense do it all by themselves.*"

In science, we ask questions about the composition and workings of the material realm. We make observations, gather evidence, analyze that evidence, and ask more questions - all in the quest for truth. Early science led to traditional knowledge, mostly about food production and health. Modern science brings us ever closer to truths about the material realm and enables the development of impressive technology for our material wellbeing.

In any given discipline, normal science progresses slowly under a current paradigm. Major advances in science are made through scientific revolutions followed by paradigm shifts, as described by Thomas

---

12  Blackburn, S. 2006. *Truth: A Guide for the Perplexed*. Penguin Books Limited: London. 238p.p.180.

# INTRODUCTION

Kuhn.[13] Under current paradigms *and* during revolutions, all advances in science depend upon the development and application of human creativity, through Free Thought. Human creativity is God-given. God uses the Free Thought of believers and nonbelievers to make truthful disclosures through mathematics and science.

Science is a commonwealth. Science provides the same information to everyone, regardless of age, gender, race, creed and social status. Science requires everyone to provide evidence for whatever findings and conclusions they communicate. No one is considered infallible in science and everyone faces the same rigorous standard of criticism. Science is an activity for all honest seekers of truth, including agnostics, atheists and believers.

John Searle[14] rejected as follows Daniel Dennett's denial that science can investigate subjectivity: "*Science does indeed aim at epistemic objectivity. The aim is to get to a set of truths that are free from our special preferences and prejudices. But epistemic objectivity of* **method** *does not require ontological objectivity of* **subject matter** (original author's emphases).

Searle continued: "*He (Dennett) thinks the objective methods of science make it impossible to study people's subjective feelings and experiences. This is a mistake, as should be clear from any textbook of neurology... There is no reason why an objective science cannot study subjective experiences.*"

Michael Polanyi[15] celebrated the primacy of individual "*commitment*" to "*facts, knowledge, proof, reality etc.,*" as "*the only path for approaching the universally valid*" and emphasized the necessity of objectivity in material realm science. Objectivity in science and the subjective nature

---

13  Kuhn, T.S. 1996. *The Structure of Scientific Revolutions*. University of Chicago Press: Chicago IL.212p.First published in 1962.
14  Searle, J.R. 1997, *The Mystery of Consciousness*. The New York Review of Books: New York.224p.p.114, 123.
15  Polanyi, M.1974. *Personal Knowledge: Towards a Post-Critical Philosophy.* University of Chicago Press: Chicago IL.428p.p.303. First published in 1958; based M. Polanyi's 1951-1952 Gifford Lectures, delivered at the University of Aberdeen, Scotland, UK.

of the internal milieu that drives and sustains us are of course *complementary*. We must safeguard objectivity in science. We must also use science to explore subjective evidence from within concerning faith and unbelief.

## The False Divide Between Faith and Science

I know from my own experiences that the faith-science divide is false. It persists because opinions are polarized between those who hold that the spiritual realm really exists and those who hold that anything alleged to be spiritual in nature is simply imagination, formed entirely by the brain.

Stephen Jay Gould[16] held that science and religion are *"Non-Overlapping Magisteria (NOMA)."* He defined a magisterium as *"a domain where one form of teaching holds the appropriate tools for meaningful discourse and resolution"* and added that: *"we debate and hold dialogue under a magisterium."* Richard Dawkins[17] commented as follows: *"I simply do not believe that Gould could possibly have meant much of what he wrote in Rocks of Ages. As I say, we have all been guilty of bending over backwards to be nice to an unworthy but powerful opponent, and I can only think that this is what Gould was doing".*

I am sure that Gould meant every word. I also reject his NOMA model, but for very different reasons than those of Dawkins. In each magisterium, the teaching, debate and dialogue mentioned by Gould go hand in hand with exploration. If there was no overlap between the material realm magisterium of science and the spiritual realm magisterium of what Gould called religion, but what I call faith, then no human living in the material realm could ever carry out any meaningful exploration of anything in the spiritual realm.

---

16  Gould, S.J. 2002. *Rocks of Ages: Science and Religion in the Fullness of Life.* Vintage: London. 241p. p.5-6.
17  Dawkins, R. 2006. *The God Delusion.* Bantam Press: London. 406p. p.57.

# INTRODUCTION

'No problem!' say the nonbelievers, because nothing exists except the material realm. However, there really is a *huge* problem for humanity if the spiritual realm does exist but has no overlap with the material realm. In that case, no human, made allegedly *"in the image of God,"*[18] would have any hope of contacting Him during her/his earthly life. In truth, the spiritual realm not only exists, it also overlaps with the material realm at the mind-soul interface of every person.

Many believers argue that the evidence for God is best seen when science fails to explain something. This so-called 'God-of-the-gaps' argument always fails. The gaps get filled sooner or later. However, there is a huge *hole* in all attempted explanations of the human condition that deny completely the existence of the spiritual realm. The science of the material realm and the study of faith and unbelief must therefore be brought closer together, with complete honesty about the human condition and the composition and workings of the material realm.

Most believers can acknowledge readily that all truth comes from God, but many are so reliant on their allegedly inerrant sacred texts and/or allegedly infallible religious authorities that they are afraid of making the obvious connection that all truthful disclosures through mathematics and science must also come from God. Many nonbelievers refuse to explore anything of a spiritual nature. Some are afraid of what others might think of them if they did so. No one need ever be afraid to open wide every possible window to truth.

## EXPANDING THE FAITH-SCIENCE QUEST FOR TRUTH

The great historical examples of those who pursued the faith-science quest for truth in the Common Era (CE) include the following, among many: Hildegard of Bingen (1098-1179); Moses Maimonides (1135-1204); Francis Bacon (1561-1676); Galileo Galilei (1568-1642); Sir

---

18  Genesis 1:27

# Free Thought, Faith, and Science

Thomas Browne (1605-1682); Isaac Newton (1643-1727); Michael Faraday (1791-1867); Charles Darwin (1809-1882); Bertrand Russell (1872-1970); and Albert Einstein (1879-1955). I include believers *and* nonbelievers in this distinguished list, because all truthful disclosures through mathematics and science come from God.

Faith-science convergence is seen today in the lives and works of scientists who are believers; for example, Francis Collins,[19] Alister McGrath,[20] John Polkinghorne,[21] and Gerry Schroeder.[22] Ruth Bancewicz[23] and others have compiled further examples in the *"Test of Faith"* project. Faith-science convergence is also seen in organizations including, *inter alia*: the American Scientific Affiliation,[24] Christians in Science;[25] the Society of Ordained Scientists;[26] and the Science and Religion Forum.[27]

Further pointers to faith-science convergence can come from unlikely sources. At a TED (Technology, Entertainment, Design)[28] event in 2010, the prominent atheist author Sam Harris[29] stated: *"The demagogues are right about one thing. We need a universal conception of human values."* Harris proposed science as the source of a standard moral code for everyone. Believers hold that the Universal Moral Code is authored by God and revealed to all of us by Him. Those positions become convergent if one believes that God created everything and that all truthful disclosures through science come from Him.

---

19   Collins, F.S. 2006. *The Language of God: A Scientist Presents Evidence for Belief.* Free Press: New York.294p.
20   McGrath, A.E. 2008.*The Open Secret: A New Vision for Natural Theology.* Blackwell Publishing: Malden MA.372p.
21   Polkinghorne, J.C.2009.*Theology in the Context of Science.* Yale University Press: New Haven CT.166p.
22   Schroeder, G.L. 2009. *God According to God: A Physicist Proves We've Been Wrong About God All Along.* HarperOne: New York.249p.
23   Bancewicz, R. Editor. 2009. *Test of Faith: Spiritual Journeys with Scientists.* Authentic Media Limited: Milton Keynes, UK.119p. Test of Faith: Does Science Threaten God? DVD. Paternoster/Authentic Media: Milton Keynes, UK. Available at: www.testoffaith.com
24   www.asa3.org
25   www.cis.org.uk
26   www.ordainedscientists.org
27   www.srforum.org
28   www.ted.com
29   Available at: www.youtube.com/watch?v=Hj9oB4zpHww

# INTRODUCTION

Richard Feynman[30] wrote the following in the context of whether moral problems could be approached in ways similar to experimental science: "*I believe that it is not at all impossible that there be agreements on consequences, that we agree on the net result, but maybe not on the reason we do what we ought to do...I therefore consider the Encyclical of Pope John XXIII...to be one of the most remarkable occurrences of our time and a great step to the future. I can find no better expression of my beliefs of morality, of the duties and responsibilities of mankind, people to other people...And I recognize this encyclical as the beginning, possibly, of a new future where we forget, perhaps, about the theories of why we believe things as long as we ultimately in the end, as far as action is concerned, believe the same thing.*"

Everyone uses Free Thought to make choices in matters of faith, justice, morality, and creative activities in mathematics, science and the arts. Free Thought includes reasoning about and responses to spiritual revelations. Thomas Henry Huxley worked out his position of agnosticism by his using Free Thought, which included his reasoning about and responses to spiritual revelations, even though he did not recognize them as such. The same applies to Richard Dawkins, Sam Harris and others as they strive to sustain their atheism.

In faith and in science, Free Thought advances our understanding by combining soul processing with mental reasoning. Free Thought provides us with combined access to the spiritual and material realms. Prayers, scientific experiments, and all development and use of human creativity are Free Thought explorations of one unified whole - the spiritual realm and the material realm. In every scientific experiment there is a result, even if it is not a clear-cut result or the desired one. The same applies to prayers. All prayers are heard and answered, even if the

---

30  Feynman, R.P. 2007. *The Meaning of It All*. Penguin Edition: London.133p.p.121-122.First published in 1998. The Encyclical of Pope John XXIII, to which Feynman referred, has the short title *Pacem in Terris* and the full title *On Establishing Universal Peace in Truth, Justice, Charity and Liberty*. It was released on April 11, 1963 and is available at: www.vatican.va/holy_father/john_xxiii/encyclicals

answer is not clear or not the desired one.

I believe that we can glimpse more truths about both the material and spiritual realms by working together and by starting from the premise that those realms form one unified whole, which is our reality. The main bone of contention will always be whether any theory or model like Free Thought can have any real spiritual component. Nonbelievers will say 'No,' on the basis that no spiritual realm exists. I contend that Free Thought *must* have a spiritual component, whether it is about faith, or science, or anything concerning the higher things of life. For now, we will just have to agree to disagree on this.

Most of us dislike being challenged to look beyond our comfort zones. If you think that anything spiritual is nonsensical and/or non-testable, please try to rise above criticizing the ills of organized religion and focus on the meaning of faith *per se*. If you are a believer and you consider a sacred book to be the only source of truth and/or your particular brand of organized religion to be the only path to God, please try to rise above shunning science and focus on what is demonstrably true about the world of nature and the whole of humanity. Despite our various comfort zones, we are all very well equipped to accomplish all of the above.

Expanding the faith-science quest for truth will require science-friendly reformations in organized religion and faith-friendly revolutions across science. The reformations will challenge believers to assess honestly what they claim to know through their faith. The revolutions will challenge nonbelievers, especially scientists, to assess honestly all evidence for the existence of the spiritual realm. The greatest stumbling block will be lack of agreement on the rules of evidence.

There is little point in trying to appraise further the claims from organized religion for supernatural signs and wonders. Christ Himself is reported to have said the following to those who craved for signs and

# INTRODUCTION

wonders during His Incarnation: "*Why doth this generation seek after a sign? verily I say unto you, There shall be no sign given to this generation.*"[31] The same applies today. Bob Dylan[32] sings to those who ask for a sign of the Lordship of Christ that no sign is needed, because all the evidence comes from within. We must learn how to study subjective evidence from within concerning faith and unbelief, as part of science.

Michael Polanyi[33] wrote as follows about the big picture and the personal picture: "*So far as we know, the tiny fragments of the universe embodied in man are the only centres of thought and responsibility in the visible world. If that be so, the appearance of the human mind has been so far the ultimate stage in the awakening of the world; and all that has gone before, the strivings of myriad centres that have taken the risks of living and believing, seem to all have been pursuing, along rival lines, the aim now achieved in us up to this point. They are all akin to us…We may envisage then a cosmic field which called forth these centres by offering them a short-lived, limited, hazardous opportunity for making some progress of their own towards an unthinkable consummation. And that is also, I believe, how a Christian is placed when worshipping God.*" The faith-science quest for truth brings us closer to that "*unthinkable consummation.*"

David Deutsch[34] argued that: "*all progress, both theoretical and practical, has resulted from a single human activity: the quest for…good explanations.*" He added: "*Though this quest is uniquely human, its effectiveness is also a fundamental fact about reality at the most impersonal, cosmic level - namely that it conforms to universal laws of nature that are indeed good explanations. This simple relationship between the cosmic and the human is a hint of a central role of people in the cosmic scheme of things.*"

---

31  Mark 8:12
32  Bob Dylan.1980. *Pressing On.*
33  Polanyi, M.1974. *Personal Knowledge: Towards a Post-Critical Philosophy.* University of Chicago Press: Chicago IL.428p.p.405.First published in 1958; based M. Polanyi's 1951-1952 Gifford Lectures, delivered at the University of Aberdeen, Scotland, UK.
34  Deutsch, D.2011. *The Beginning of Infinity: Explanations That Transform the World.* Viking: New York.487p.p.vii.

# Free Thought, Faith, and Science

I agree with Deutsch, but must add that faith can make those *"good explanations"* better still, because faith explains much more of that *"central role of people,"* in terms of the God-human-cosmic *"scheme of things."* Truthful disclosures through mathematics and science come from God. Faith and science together explain our one whole spiritual and material reality.

E. O. Wilson[35] wrote: *"Science faces in ethics and religion its most interesting and possibly humbling challenge, while religion must somehow find the way to incorporate the discoveries of science in order to retain credibility."* If one substitutes 'faith' for 'religion' in the above statement, the rationale for expanding the faith-science quest for truth is very strong.

---

35  Wilson, E.O. 1999. *Consilience: The Unity of Knowledge*. Vintage Books: New York.367p.p. 262-263,290.

# CHAPTER 2

# THE HUMAN CONDITION

*"And since there is something in us that must still live on, joyn both Lives together; unite them in thy Thoughts and Actions and live in one but for the other."* [1]

### EXISTENCE

Sir Thomas Browne saw the human condition as an opportunity to unite life on Earth with the eternal life to come. In our earthly lives we are as ephemeral as mayflies over a stream. The Apostle James[2] wrote: *"For what is your life? It is even a vapour, that appeareth for a little time, and then vanisheth away."* The Manx poet T.E. Brown[3] concurred as follows: *"And when you look back it's all like a puff. Happy and over and short enough."*

Happy? Some live in appalling deprivation, while others live in luxury. Some experience both in one lifetime. Most of us experience lesser ups and downs. Everyone wonders what might be coming next. Is human

---

1 Browne, Sir Thomas. ca. 1656. *Letter to a Friend*. Closing sentences. Available at: http://penelope.uchicago.edu/letter/letter.html
2 James 4:14
3 Brown, T.E. 1873. From *"Betsy Lee."* The Collected Poems of T.E. Brown. Manx Museum and National Trust: Douglas, Isle of Man. 1976. 736p.p.109. First published by MacMillan and Company, London in 1900. Available at: www.isle-of-man.com/manxnotebook/people/writers/teb/poems.htm

## Free Thought, Faith, and Science

existence then entirely a matter of material circumstances and uncertainty? Many nonbelievers would say 'Yes.' Most believers would say 'No' and point to our spiritual lives - begun on Earth and continued into eternity after death.

Whichever side you are on, I hope that we can all agree that the human condition is not only a state of uncertainty but also a state of unsatisfied curiosity. We ask many questions, including big ones that seem unanswerable. Why do we exist? Why does anything exist? Most believers frame their answers around the prior existence of God - the Original One Who allegedly created everything else to keep Him company. Many nonbelievers seek their answers in mathematics, philosophy and science.

Jim Holt[4] sought answers from interviews with eminent thinkers including, *inter alia*, David Deutsch, Derek Parfit and Roger Penrose. In the course of his exchanges with Parfit, Holt proposed a model in which different types of world - described as *"Nothingness, Axiarchic (Best World), All Worlds* (and) *[Lots of Generic Possibilities]"* - arose from various combinations of so-called *"Meta-Selectors"* and *"Selectors,"* which he named as follows: *"Simplicity, Goodness, Fullness/Non-Arbitrariness, or No Selector."*

One of Holt's models has *"Goodness"* as overarching Meta-Selector and as the Selector at what he called the *"Explanatory Level,"* leading to a so-called Best World. Our particular world might not be the best world, but that particular model fits the concept of a good God and the reality of our existence in a material realm that looks tailor-made for us - the so-called Anthropic Principle.

Holt's preferred model had *"Simplicity"* as the Meta-Selector with *"No Selector"* at the Explanatory Level, yielding worlds with *"Lots of*

---

4   Holt, J. 2012. *Why Does the World Exist?* Liveright Publishing Corporation: New York.307p.p.235, 277, 279.

# THE HUMAN CONDITION

*Generic Possibilities"* and with *"nothingness"* not an option. Theism and material realm science can be fitted to that model. God, as Meta-Selector in the form of Divine Simplicity, creates the material realm at the Big Bang and then becomes God in Self-limiting mode ('No Selector'), letting His creation evolve 'hands-off,' as indicated by the science of the material realm - with the exceptions, for believers, of His spiritual revelations and for Christians His Incarnation. The same model can explain scenarios of 'many worlds,' in other words a material realm multiverse.

Holt described a walk through Paris at night, having just watched a TV chat show in which a Dominican priest, a theoretical physicist and a Buddhist monk discussed the question: *"Why is there Something rather than Nothing?"* Holt mused that the discussants' respective explanations for the beautiful sights of Paris in front of him would probably have been as follows: *"a divine gift;" "an inexplicable quantum fluke;"* and *"an insubstantial dream, an empty illusion."* He concluded that the question was *"really awfully mysterious,"* as indeed it is.

Lawrence Krauss[5] took that very same question as the subtitle to his fine book on *how* the universe could arise from nothing. Using beautiful explanations from cosmology and physics, he described our present state of knowledge and the remaining open questions about the origins and fate of our observable universe - all 'material realm stuff.'

Krauss explained that he preferred not to address the term 'non-existence,' because it: *"takes one off on lots of deep philosophical issues but rather impotent physics ideas."* For Krauss, the term 'nothing' meant 'no-stuff,' which in the context of his arguments meant 'material realm no-stuff.' He stated: *"the remarkable non-miraculous miracle is that combining quantum mechanics with gravity allows stuff to arise from no-stuff."*

---

5   Krauss, L.M.2013. *A Universe From Nothing: Why There Is Something Rather Than Nothing.* Paperback Edition. Atria: New York.210p.p.205, 117-118.

# Free Thought, Faith, and Science

That's fine. We can celebrate the properties of material realm no-stuff and the processes by which it can give rise to material realm stuff. We can applaud the brilliant mathematics and science through which many *how* questions about those properties and processes are being answered. But the *why* questions keep niggling away. Why do we have spiritual needs? Is there any spiritual stuff? Material realm stuff and no-stuff cannot explain everything about the human condition.

There is broad agreement that the universe is expanding at an accelerating rate and that everything will ultimately be unobservable from anywhere. Krauss referred to the following statement from an article written with Bob Scherrer: "*We live at a very special time…the only time when we can verify that we live at a very special time!*" Krauss added: "*We can consider ourselves lucky that we live at the present time.*"

Really? Can it be nothing more than *luck* or coincidence that we just happen to be here at this opportune time, well equipped and motivated to seek explanations for how and *why* things are as they are? I think not. Material realm science alone cannot explain the human condition.

Karl Popper[6] concluded that questions about the human self and brain could be approached by recognizing three worlds: "*World 1*", "*the physical world - the universe of physical entities*"; "*World 2*", "*the world of mental states, including states of consciousness and psychological dispositions and unconscious states*"; and "*World 3*", "*the world of the contents of thought, and, indeed, of the products of the human mind.*"

Roger Penrose[7] described a similar three-worlds model as follows, in which all emphases are his: "*The world that we know most directly is the **world of our conscious perceptions**, yet it is the world that we know least about in any kind of precise scientific terms…There are two other*

---

6 Popper, K. and J.C. Eccles. 1977. *The Self and Its Brain: An Argument for Interactionism.* Routledge: London.597p.p.38.
7 Penrose, R. 2005. *Shadows of the Mind: A Search for the Missing Science of Consciousness.* Vintage Books: London. 457p.p.412-414.

*worlds that we are also cognizant of - less directly than the world of our perceptions - but which we now know quite a lot about...the **physical world** (of) actual chairs and tables...human brains (and)...molecules and atoms, electrons and photons, and space-time... There is also one other world, though many find difficulty in accepting its actual existence...the **Platonic world of mathematical forms**."*

Penrose recognized the following three related *"mysteries:" "why* (do) *precise and profoundly mathematical laws play such an important role in the behaviour of the physical world...how* (can) *perceiving beings arise out of the physical world...*(and) *how* (is) *mentality able seemingly to 'create' mathematical concepts out of some kind of mental model?"*

Scientists like to solve mysteries. Some believers prefer to keep mysteries unsolved and to locate all required truth in sacred texts and/or the pronouncements of religious authorities. Some nonbelievers attack all organized religion as nonsensical and/or harmful. There is intellectual and spiritual laziness on both sides.

I hope that we can all agree that we exist and are all part of *one* big picture, *one* integrated whole. Everyday we predict many material realm events and scenarios with what we can call *certainty*, for all practical purposes. The laws of nature are real and they are fixed. Jump off a high building or hold your hand in a fire and you *will* get hurt. We also live with *chaotic* systems for which we cannot make exact predictions; for example, the paths of weather systems. At the quantum level we encounter uncertainty and turn beables into observeds. We also have spiritual experiences, whether we recognize them as such or not.

Mathematical probabilities are forms of non-materiality, affiliated with something that could have existed, or exists now, or could exist in the future. The big question then is whether *all* such non-materiality is affiliated *only* with materiality, including matter, energy, antimatter, gravity etc. In other words, is all non-materiality 'materially-derived/

derivable,' or is there such a thing as spiritual non-materiality, having no material affiliations? The same questions will arise concerning possible mechanisms for Free Thought.

If all that can ever exist is materiality and 'materially-derived/derivable' non-materiality, then everything is made of or affiliated with material stuff. This material realm-only model has its attractions, including simplicity and elegance. But how can matter, energy and their derived/derivable non-materiality have beliefs, creativity and ethics? More importantly, is the God in Whom I believe and trust just a bit of the material me, made by me somewhere in my brain? 'No,' as I will describe, He is as real to me as anything else, because of His interventions in my life.

The coexistence and interdependence of materiality and 'materially-derived/ derivable' non-materiality in the material realm, as indicated by classical physics and quantum theory, must apply to the whole of a self's body-mind, which is *all* part of the material realm. The self's soul is made of spiritual stuff, which is non-materiality derived entirely from the spiritual realm.

Positing the existence of the spiritual realm and the soul, as spiritual stuff in a self, leads to the question of how the spiritual and material stuffs of self could interact. I suggest that they interact during Free Thought, through the combined processing of information from the spiritual and material realms, in soul and mind, using one or more common non-material formats.

In Free Thought, at the mind-soul interface, the 'materially-derived/derivable' non-materiality of the body-mind overlaps with the spiritual non-materiality of the soul. At the mind-soul interface, the material realm interacts with the spiritual realm in a human self, who is part of both realms and forms a life-long link between them.

# THE HUMAN CONDITION

## HUMANS AND GOD

All humans, past and present, believers and nonbelievers, belong to the same primate species (*Homo sapiens*). We were always vulnerable creatures in a harsh world, yet we spread from small beginnings in Africa to occupy all habitable lands on Earth. Our ancestors interbred with and outcompeted other hominids, including the Neanderthals. Our success depended on having large brains and supplies of nutrients for brain development and function, particularly docosahexaenoic acid (DHA), with fish as a major source.

Katherine Pollard[8] described human and chimpanzee DNA as nearly 99% identical. However, we are very different from chimpanzees and from all other animals. Our differences go far beyond anything that can be accounted for by the unique regions of human DNA that are concerned with the development and activity of the cerebral cortex. Our consciousness makes us highly self-aware and potentially God-aware. Our *souls* confer on us a distinction that dwarfs our biochemical, genetic, morphological, physiological and behavioural similarities with other species.

Christianity and Judaism proclaim that God: *"created man in his own image."*[9] No one really knows what that means. We cannot describe any spiritual beings in anything other than material realm terms. Most depictions of fictitious aliens from space have bodies, limbs and sense organs equivalent to ours. To say that humans are made in God's image makes more sense if it refers not to physical resemblances, but to spiritual resemblances and mutual capacities for communication and relationships. God cannot be lumbered with the limitations of a bipedal, primate body. God is a Free Spirit, just as we are free spirits in our souls. The main attribute we share with God is freedom.

---

8     Pollard, K.S. 2013.What makes us different? Comparisons of the genomes of humans and chimpanzees are revealing those rare stretches of DNA that are ours alone. *Scientific American* 22 (1): 31-25.
9     Genesis 1:27

Gerry Schroeder[10] emphasized that the two names by which God is called in the Hebrew Bible, "*Elokiim*" (Lord) and "*Ja/ko/vah*" (God), indicate that God, allegedly, wished to be called: "*I will be that which I will be.*" That attribute is not captured so well by the translation "*I AM THAT I AM*"[11] in the King James and other English versions of the Bible. The name "*I will be that which I will be*" suggests an *infinite* menu for divine choices, perhaps analogous to an infinite set of quantum probabilities.

God, the spiritual force for evil and humans are all observers. At the quantum level in the material realm, we can turn non-material beables into material observeds. God might be an infinitely variable Beable, coming into reality for any human who experiences Him. In our souls, we might be turning spiritual beables into spiritual observeds, including the revelations that we receive from God and from the spiritual force for evil.

Spiritual revelations from God tell only the truth. Christ is reported to have said the following to his Apostles: "*when he, the Spirit of truth, is come, he will guide you into all truth.*" [12] That makes God the Source of *all* truth, including all truthful disclosures through mathematics and science. Revelations from the spiritual force for evil tell only lies. An episode of Free Thought is typically a battle of good versus evil and truth versus lies. God's truthful disclosures through mathematics and science are good outcomes of the development and use of human creativity, in Free Thought. The spiritual force for evil uses our Free Thought to spread its lies and nonsense.

Some believers claim that *all* events - the mundane, the wonderful and the horrific - are part of God's perfect plan, imposed on everyone in His perfect timing. I disagree. If God considers all believers and potential

---

10    Schroeder, G.L. 2009. *God According to God: A Physicist Proves We've Been Wrong About God All Along.* HarperOne: New York.249p.p.96-97.
11    Exodus 3:14
12    John 16:13

believers to be "*of more value than many sparrows*,"[13] He is unlikely to be bringing about His perfectly timed *good* by causing His beloved to be infected, maimed, raped, robbed and killed, while ignoring their prayers to be spared from such horrors.

The rains really do fall on the just and the unjust.[14] Accidental injury and death, birth defects, diseases, natural disasters, human violence and theft etc. harm the just and the unjust. Those bad outcomes are not acts of God, except insofar as He created the original wherewithal for all of the material and spiritual stuff involved. Those outcomes result from natural free process in the material realm and/or the acts of free-willed humans.

In the 14th century, bubonic plague killed about 25 million people. Was God, or the spiritual force for evil, or karma choreographing the rats, the fleas, the pathogen and its victims? 'No,' a material realm ecosystem containing all of those actors was working by free process. God does not direct the flight path of a mosquito to a divinely chosen human victim.

Each of us lives among family members, friends, peers and authorities, who often recommend or require conformity to their beliefs and behaviour. The information that we receive from material realm sources is always limited and often biased. In some religious schools and homeschooling by fundamentalists, children are at risk of receiving an education deficient in science. Children schooled by atheists are likely to hear a lot about the ills of organized religion but little about faith as a personal choice. Throughout life, however, all information received and all experiences are cumulative, for use in successive episodes of Free Thought. Moreover, God knows all about our unavoidable constraints and the opportunities that come our way.

---

13  Matthew 10:31
14  Matthew 5:45

# Free Thought, Faith, and Science

In our naturally constrained circumstances, I believe that the most important thing that God can ask of any human and that any human can ask of any other human is to be an honest seeker and teller of truth and to act accordingly. The following words of the Old Testament apply to everyone: "*He hath shewed thee, O man, what is good; and what doth the Lord require of thee, but to do justly, and to love mercy, and to walk humbly with thy God.*"[15]

God can be recognized as "*thy God*" in many ways. Christ reportedly said "*I am the way, the truth and the life: no man cometh unto the Father but by me*"[16] and: "*I and my Father are one.*"[17] Christ also reportedly told the Apostle Philip: "*he that hath seen me hath seen the Father.*"[18] Christ was *God* Incarnate, telling the simple men who were His contemporary disciples that the only way to God is *through God*.

I believe that those messages are for *everyone*. Faith as belief and trust in God is not the same as membership of an organized religion. I believe that God makes all souls to seek Him and find the way to Him. I believe that He reaches all souls and accepts the souls of all honest seekers of truth. I believe that God, as the Holy Spirit, pervades *all* humanity.

The loudest voices raised against Paul as he preached the universality of the Gospel were those of the Jewish establishment who could not bear to hear that God's promises were meant for the Gentiles as well as for them.[19] Many Christians deny all non-Christians and even some fellow Christians from other churches the possibility of being accepted by God.

Dorothy Rowe[20] cited the research of Amanda Lohrey, who asked teenage girls from the Hillsong Church in Sydney, Australia whether

---

15  Micah 6:8
16  John 14:6
17  John 10: 30
18  John 14: 9
19  Acts 22: 21-23
20  Rowe, D.2009. *What Should I Believe?* Routledge: Hove, East Sussex U.K.294p.p.12.

anyone who had not accepted Christ must go to hell. One of the girls replied: '*Yeah, much as it sucks, you have to understand that Jesus was the only savior.*" Jesus always was and is God. God is the only Saviour. God can have relationships with whomever He chooses, irrespective of whether they adhere to Christianity, or to any organized religion, or to none. The Apostle Paul wrote: "*But if any man love God, the same is known of Him.*"[21]

If God truly loves all humans past and present, then He *must* have sent and must still be sending spiritual revelations to everyone. The path to God is personal, not institutional. God alone decides the eternal destiny of every human soul. I am not arguing against evangelism, missionary work, or any preaching of the Gospel of Christ. As a Christian, I long for that to be done as sensitively and widely as possible. I am saying only that the potential for humans to have relationships with God and with the spiritual force for evil must be universal and must apply to every human life.

The other enormous problem concerning organized religions is that whereas some of them help to sustain an honest seeker's journey of faith, others insist on doctrines and practices that turn honest seekers away from considering faith and turn some believers to unbelief.

The atheist literature is full of testimonies from those who could no longer accept what their membership in an organized religion or church required of them. They walked away from lies and nonsense and lost their faith, as outcomes of their Free Thought. With or without help or hindrance from organized religion, we all make our sovereign-to-self choices for faith or unbelief and take the consequences.

Michael Shermer[22] wrote that he would make the following case if he met an unexpected God in an unexpected afterlife: "*Lord, I did the best

---

21  I Corinthians 8:3.
22  Shermer, M. 2011.*The Believing Brain*. Times Books/Henry Holt and Company: New York. 385p.p.55.

*I could with the tools you granted me. You gave me a brain to think skeptically and I used it accordingly. You gave me the capacity to reason and I applied it to all claims, including that of your existence. You gave me a moral sense and I felt the pangs of guilt and the joys of pride for the bad and good things I chose to do. I tried to do unto others as I would have them do unto me, and although I fell far short of this ideal too many times, I tried to apply your foundational principle whenever I could. Whatever the nature of your immortal and infinite spiritual essence actually is, as a mortal finite corporeal being I cannot possibly fathom it despite my best efforts, and so do with me what you will."*

I believe that God will deal mercifully with the souls of all honest seekers of truth, but I cannot imagine Him being too impressed with a general case plea such as the following: 'Look here, God, it is entirely your fault that I did not believe in you. You never gave me any credible evidence.'

The evidence for God's existence is *our* existence and the longings in our souls. We are God's *"workmanship."*[23] For believers, the irrefutable evidence for God's existence comes from experience of Him in a two-way relationship. The evidence for the spiritual force for evil is the internal struggle that we feel when making a Free Thought choice to do right or wrong and the pangs of conscience when we choose wrongly. Only God, the spiritual force for evil and the self know the outcomes. Therein lies the necessary and God-given freedom of Free Thought.

## ORIGINAL SIN?

Many believers hold that the first humans, depicted as Adam and Eve, stole forbidden knowledge from God - consciously, willfully and under the influence of the spiritual force for evil; thereby saddling all subsequent humans with so-called original sin. The knowledge that was

---
23  Ephesians 2:10

## THE HUMAN CONDITION

allegedly stolen was called *"knowing good and evil"* and was allegedly forbidden because its acquisition would have elevated humans to the status of *"gods."*[24]

Assuming for the sake of argument that the Garden of Eden story is a tenable, though figurative, description of the earliest state of the human condition and assuming there really was no human knowledge of good and evil prior to the theft of the symbolic apple, then one could say that the theft of the apple could not be construed as either moral or immoral.

But God had already laid out clearly for humans what was right and what was wrong. Adam and Eve had a *choice* between continuing in obedience to God, taking sustenance from the trees and other food around them, including specifically *"the tree of life,"* and alienation from God through taking fruit from the distinct and very tempting *"tree of the knowledge of good and evil."*[25]

They were *de facto* moral choice makers from the very start. They thought about the material realm of the garden and chose how they would act. They must have been using more than basic thought. They had body-minds and souls and lived in close communion with God. They must have reasoned about and responded to spiritual revelations from God and the spiritual force for evil. God wanted them to choose a life of obedience to Him. The spiritual force for evil urged the opposite.

From the very start then, Adam and Eve must have engaged in Free Thought. Their Free Thought led to wrong outcomes, but it was a God-given *process*. We are similarly equipped. I conclude that neither Adam nor Eve, nor any early humans, nor any human at any time, ever needed to steal from God the capacity for what I call Free Thought, which is the God-given human mechanism for choosing between good and evil.

---

24  Genesis 3:5
25  Genesis 2:9

# Free Thought, Faith, and Science

I believe that the spiritual force for evil was the *only* original sinner. It allegedly fell from grace and was booted out of heaven, thereafter to wage unceasing spiritual warfare with God.[26] From the dawn of humanity, humans with Free Thought were coopted into the battle between good (God) and evil (the spiritual force for evil) just as we are today - all with God-given Free Thought, but *not* stained with original sin from conception or babyhood.

If indeed God made us in His image, then He must have made us as free choice makers. We live in the material realm. Our souls come from God and are part of the spiritual realm, to which they will go for an eternal life after death on Earth. God and our individual souls can communicate, reason and be friends together during our earthly lives. That was the scene in the Garden of Eden story and it continues through us. God has *given* us our souls, our higher consciousness, our Free Thought and the Universal Moral Code, which really *is* the knowledge of good and evil.

In any case, was the symbolic theft of the apple really such a bad happening? The old English poem "*Adam lay ybounden*"[27] describes Adam in a close contractual relationship with God, which was broken *only* because of: "*an apple that he took.*" The writer noted that if the apple had not been taken, there would have been no Incarnation, no Christianity and no Blessed Virgin Mary elevated to become the "*heaven'e queen.*"

The poem concludes: "*Bless&eacute;d be the day that apple taken was. Therefore we moun singen* (must sing) *Deo gracias!*" That makes a fine argument for dropping the focus on original sin and focusing instead on the need for all of us to establish better relationships with God, one soul at a time.

Whatever interpretation you might choose to put on any or all of the above, including its dismissal as utter nonsense or heresy, I hope that

---

26   For example, Revelation 12:7-9
27   Anon. ca.1400. *Adam lay ybounden*. Sloane Manuscript No. 2593. The British Library: London.

# THE HUMAN CONDITION

we can all agree that what I call Free Thought, or what you might prefer to call critical reasoning, cannot be considered as wrong or sinful. It is our hallmark as humans to be honest seekers of truth and critical, curious, sceptical inquirers. We are made that way.

The atoms and molecules of an unfertilized egg, a sperm, a zygote, an embryo, a foetus and a baby cannot possibly be bearing any guilt. I suppose that God could choose to gift every new person with a stained and guilty soul, but that hardly seems likely for a just and loving God, having made the ultimate sacrifice to take away the guilt of all humans. Babies do not choose to be born and I cannot believe that any are steeped in original sin. Until their Free Thought begins, humans make no choices that can be called sinful.

## The Self

Apart from identical twins and the possibility of cloned humans, everyone has a unique genome. Everyone has a set of tissues and organs that can be termed 'self,' as opposed to 'non-self.' Every self has an immune system shaped by a unique history of raising antibodies. However, when we get beyond our basic biology and into self-awareness and psychological criteria for self-identity, there is a wide diversity of opinion about whether the self has low or high importance in the big picture of human existence. I am with those who give the human self very high importance. I believe that God gives very high importance to every self and to the destiny of every soul.

I define the self as the totality of every individual human being, comprising a unique spiritual soul, which is integrated during life on Earth with a unique material body-mind. The body-mind, memory, and all detectable mental maps and images are parts of the material realm. In common with everything else in the material realm, our body-minds are made of fundamental particles.

The soul is part of the spiritual realm and must be made of spiritual stuff. Whatever the spiritual stuff of self might be, it is *non-material*. Information from the spiritual realm is acquired only by individual souls. Our experiences of God and the spiritual force for evil and our denials of spiritual things are entirely personal. We can keep those experiences and denials private or we can share them as subjective evidence from within.

Information from the material realm is acquired through the physical senses of a self and sent for reasoning in the mind. The self experiences material realm reality as states of matter and energy, atoms and molecules, solids, fluids, flesh and blood etc. We share and reach consensus on many human experiences and observations of the material realm, by applying common sense, mathematics and science.

Patrick McNamara[28] wrote about the brain, the self and "*religious experiences*" as follows: "*there is considerable anatomical overlap between the brain sites implicated in religious experience and the brain sites implicated in the sense of Self and self-consciousness…religious practices often operate to support transformation of the Self such that the Self becomes more like an 'ideal Self' that the individual hopes to become. This hoped-for Self is a more centralized and unified sense of Self. Religious practices also help one to avoid becoming a 'feared Self.' This combination of a positive 'approach' motivational element toward a hoped-for Self and a negative 'avoidance' motivational element away from a feared Self makes religion a powerful tool for processes of self-regulation more generally… religious practices contribute to the creation of a unified self-consciousness and an ideal 'executive Self.'*"

McNamara concluded the following: "*religion was one of the forces (indeed a primary force) that created the executive Self… The executive Self is a social Self and is a master of social cooperation, but looks to its own goals*

---

28   McNamara, P. 2009. *The Neuroscience of Religious Experience*. Cambridge University Press: Cambridge.301p.p.xi-xii, 258.

*to guide its behavior. Religion uses the decentering process to help transform the Self and to resolve internal conflict...It is unlikely that technical or computational intelligences of human beings could have evolved as far as they have if the executive Self, the 'agent intellect,' had never been developed by our ancestors. Our ancestors used religion to do so."*

McNamara wrote about religion, which is not the same as faith. Nevertheless, his descriptions of the self fit with my distinction between basic thought and Free Thought. The *"ideal Self"* can choose her/his goals in private, using free will and Free Thought. The *"executive Self"* can use basic thought *and* Free Thought to act in pursuit of those goals, while cooperating prudently with fellow humans and securing the necessities of life.

Thomas Merton[29] distinguished between the *"inner self"* and *"the exterior 'I,' the 'I' of projects, of temporal finalities, the 'I' that manipulates objects in order to take possession of them."* He stated further: *"The inner self is as secret as God and, like Him, it evades every concept that tries to seize hold of it with full possession. It is a life that cannot be reached and studied as object, because it is not a 'thing'."*

Merton characterized *"Christian, Jewish and Islamic mysticisms"* as recognizing: *"an infinite gulf between the being of God and the being of the soul, between the 'I' of the Almighty and our own inner 'I'."* He added: *"Yet paradoxically our inmost 'I' exists in God and God dwells in it."* I believe that we bridge that gulf in our Free Thought.

Merton compared Christianity, Zen and Eastern mysticisms and concluded the following: *"In Zen, there seems to be no effort to get **beyond*** (original author's emphasis) *the inner self. In Christianity, the inner self is simply a stepping stone to an awareness of God...the fact that the Eastern mystic, not conditioned by centuries of theological debate, may not*

---

29  Merton, T. 2004. *The Inner Experience: Notes on Contemplation.* HarperOne: New York.176p.p.5, 7, 5, 11, 12-13.

*be inclined to reflect on the fine points of metaphysical distinction does not necessarily mean that he has not experienced the presence of God when he speaks of knowing the Inmost Self.*" I agree. I believe that God can reach every self and that every self can reach God.

Antonio Damasio[30] wrote the following, in which all emphases are his, unless stated otherwise: "*I believe conscious minds arise when a self process is added onto a basic mind process...We can consider the self process from two vantage points. One is the vantage point of an observer appreciating a dynamic object - the dynamic **object** constituted by certain workings of minds, certain traits of behaviour, and a certain history of life. The other is the vantage point of the self as **knower**, the process that gives a focus to our experiences and eventually lets us reflect on those experiences.*" He defined the "*self-as-object*" or "*the material me*" as: "*a dynamic collection of integrated neural processes, centred on the representation of the living body, that finds expression in a dynamic collection of integrated mental processes.*"

Damasio continued: "*The self-as-subject, as knower, as the 'I', is a more elusive presence, far less collected in mental or biological terms than the **me**, more dispersed, often dissolved in the stream of consciousness, at times so annoyingly subtle that it is there but almost not there...The self-as-subject-and knower is not only a very real presence but **a turning point in biological evolution*** (present author's emphasis). *We can imagine that the self-as-subject-and-knower is stacked, so to speak, on top of the self-as-object, as a new layer of neural processes giving rise to yet another layer of mental processing. There is no dichotomy between the self-as-object and self-as-knower; there is, rather, a continuity and progression. The self-as-knower is grounded on the self-as-object.*"

I agree with Damasio that there was a "*turning point in biological evolution*" when humans (*Homo sapiens*) arrived on the scene. I regard that

---

30   Damasio.A. 2010. *Self Comes to Mind: Constructing the Conscious Brain*. William Heinemann: London. 367p.p.8-10.

turning point as God's gift of the human soul, which distinguishes a human self from any kind of animal self. Damasio saw it as some kind of a higher complexity in neural processes. I also agree with Damasio on the difference between processing information about objects and reflecting on experiences. I would say that his 'self-as-object' engages in basic thought and his 'self-as-knower' in Free Thought. Damasio's 'self as object/self as knower' and Merton's 'exterior I/inner self' are models from very different worldviews. If one recognizes only the material realm, as Damasio does, it is indeed hard to define the 'self as knower' adequately and to capture the whole of the self.

In Free Thought as I define it, the self processes information from the material and spiritual realms and is part of both those realms. The outcomes of Free Thought are private to the self, to God and to the spiritual force for evil. They cannot be deduced from the extents to which an individual self cooperates socially and/or participates in religious practices.

## Happiness and Morality

Judith Rich Harris[31] identified three systems and goals, as an individual human self interacts with others: the *"relationship system* (to) *establish and maintain favorable relationships,"* the *"socialization system* (to) *be a member of a group,"* and the *"status system…to be better than one's rivals"*. The extents to which we accomplish those three goals contribute to our levels of material wellbeing, peace of mind and happiness.

Terry Eagleton[32] explained as follows the difficulties of getting a handle on happiness: *"What counts as happiness? What if you find it in terrorizing old ladies? Someone who is determined to become an actor may spend*

---

31  Harris, J.R. 2006. *No Two Alike: Human Nature and Human Individuality.* W. W. Norton & Company: New York, 329p.p. 242-243.
32  Eagleton, T.2008. *The Meaning of Life: A Very Short Introduction.* Oxford University Press: Oxford. 109p.p.81.

*fruitless hours auditioning while living on a pittance. For much of the time she is anxious, dispirited, and mildly hungry. She is not what we would usually call happy. Her life is not pleasant and enjoyable. Yet she is, so to speak, prepared to sacrifice her happiness to her happiness."*

John Stuart Mill[33] described as follows the promotion of happiness as the foundation of human morality: "*The creed which accepts as the foundation of morals, Utility, or the Greatest Happiness Principle, holds that actions are right in proportion as they tend to promote happiness, wrong as they tend to produce the reverse of happiness… To give a clear view of the moral standard set up by the theory, much more requires to be said; in particular what things it includes in the ideas of pain and pleasure; and to what extent this is left as an open question. But these supplementary explanations do not affect the theory of life on which this theory of morality is grounded - namely,* **that pleasure, and freedom from pain, are the only things desirable as ends** *…According to the Greatest Happiness Principle…the ultimate end, with reference to and for the sake of which all other things are desirable (whether we are considering our own good or that of other people), is an existence exempt as far as possible from pain, and as rich as possible in enjoyments…* **This being, according to the utilitarian opinion, the end of human action, is necessarily also the standard of morality** (present author's emphases)."

Mill's utilitarian recipe for morality cannot define adequately the things that cause pleasure and pain. Some things cause both. The 'no pain/no gain' adage for achieving success in athletics is a good example. Moreover, one person's chosen pleasure can require another's pain. Individual and collective experiences of pleasure and pain are not sufficiently quantifiable to be adopted as an objective basis for morality.

Some physiological and biochemical indicators of happiness can be monitored; for example, the fluxes of the 'feel good' chemical serotonin

---

33   Mill, J.S.1861. *Utilitarianism*. Published in 1964, as p.1-60. In *Utilitarianism, Liberty, Representative Government*. J.M. Dent & Sons Ltd., London.398p.p.6, 11.

in the brain. Our individual capacities for biochemical happiness appear to have some genetic basis.[34] But happiness has a spiritual basis as well as a material one. It cannot be measured through biochemistry and genetics alone.

On the same trail as Mill, Sam Harris[35] made the case that human morality is simply maximizing science-based wellbeing across our *"moral landscape."* In a related discussion with Richard Dawkins,[36] Harris stated: *"We must appeal to the value of understanding - the value of evidence"*. He also stated that subjective evidence is the mainstay of much of psychology and consciousness research and that despite the limitations of having to *"take somebody's word for their experience"* and the difficulty that *"people can be wrong about their subjectivity - we can be bad witnesses about what it's like to be us"*, we must try to: *"seek objectivity about subjective facts."* I agree strongly and emphasize the same point throughout this book.

Nevertheless, I take issue with Harris' main position, which he stated as follows: *"a universal morality can be defined with reference to the negative end of the spectrum of conscious experience: I refer to this as 'the worst possible misery for everyone'."* In other words, morality = maximized science-based wellbeing = minimized misery = maximized happiness. Harris referred to members of Taliban throwing acid on the faces of women. The problem for Harris' position is that some of the Taliban increase their self-assessed wellbeing and happiness by throwing acid and become more miserable when they see girls becoming happier and less miserable by going to school.

On a lighter note, the fans of Manchester United get miserable if their team is beaten by Manchester City and the fans of Manchester City

---

34   De Neve, J-E. 2011. Functional polymorphism (5-HTTLPR) in the serotonin transporter gene is associated with subjective well-being: evidence from a US nationally representative sample. *Journal of Human Genetics* 56: 456-459.
35   Harris, S. 2010. *The Moral Landscape: How Science Can Determine Human Values.* Free Press; New York.291p.p.39.
36   Available at: http://richarddawkins.net

get miserable if the result goes the opposite way. The English fans of both clubs get miserable if France beats England and the French fans get miserable if England wins. Each of us has a unique, sovereign-to-self set of criteria for deciding how happy or miserable we are. We all make our own subjective assessments about the impacts of our chosen behavior on our own happiness and on the happiness of others.

Morality cannot be defined by expediency or by majority vote. Morality is not defined by maximized happiness/wellbeing or by minimized misery, regardless of whether any of the above is science-based or not. Morality comes from a Source above all of those scenarios. Morality has come to us from God, through the evolution of His creation and His spiritual gifts to us as the species chosen by Him to have the capacity to make moral choices rather than merely pursuing expediency.

Morality is built on foundations from the evolution of sharing, trust and aversion to injustice in animals. However, real morality and genuine altruism are found only in humans. The quasi-morality that we see in our primate relatives and other animals is driven only by basic thought, in pursuit of what seems expedient. The gap between the animal condition and the human condition is huge. It cannot be explained as a result of evolution through natural selection and descent with modification. God completed us and made us unique, by giving us souls, higher consciousness, Free Thought, and the Universal Moral Code - the so-called Golden Rule, 'do as you would be done by', which is expressed in the Bible as: *"love thy neighbour as thyself."*[37]

Nicholas Wade[38] summarized the global picture on morality as follows: "...*the commonalities in morality are generally more striking than the variations. The fundamental moral principle of 'do as you would be done by' is found in all societies, as are prohibitions against murder, theft and incest."* Moral codes are very similar in Buddhism,

---
37   Leviticus 19:18 and Matthew 19:19
38   Wade, N. 2009. *The Faith Instinct: How Religion Evolved and Why It Endures.* Penguin Press: New York.310p.p.24.

# THE HUMAN CONDITION

Christianity, Hinduism, Islam, Judaism and Taoism and among many agnostics and atheists.

The Universal Moral Code is the standard for moral conduct, given by God for recognition and compliance by all humans. During Free Thought, God urges compliance with the Universal Moral Code, while the spiritual force for evil urges non-compliance. Non-compliance is common in adverse circumstances such as ecological collapse, natural disasters, famines, resource wars and the coming to power of authorities endorsing or demanding immoral acts. Nevertheless, the Universal Moral Code remains, in every human soul. Our best evidence is the human conscience, which the Oxford Dictionary[39] defines as follows: *"an inner feeling or voice viewed as acting as a guide to the rightness or wrongness of one's behaviour."*

The outcomes of Free Thought in an actor's conscience and those in a person affected adversely by her/his action can be hard to reconcile. For example: John - *"My conscience is clear."* Jane - *"Either you have no conscience, or you cannot see the awful impact of what you did."* John - *"What did I do? What awful impact?"* Jane - *"You did that, and caused this awful impact!"* John - *"Yes, I did that, but was the impact so awful? Anyway, I never intended it."* Jane - *"So you apologize?"* John - *"No, my conscience is clear."*

Prior to and during that exchange, two soul battlefields and two assessments of the material realm led to a series of opposing outcomes of Free Thought. Note also that the profession of having a clear conscience and/or the appeal to someone's conscience can be false statements, disguising different opinions held within.

Killing, stealing, cheating, betraying, bullying and bankrupting can bring great material gains to their perpetrators, but such behaviour

---

39   Pearsall, J., and P. Hanks, Editors. 1998. *The New Oxford Dictionary of English: Thumb Index Edition.* Oxford University Press; Oxford. 2152p.

not only looks bad to its victims and to most onlookers, it *feels* bad to most of its perpetrators. Immoral behaviour rarely brings peace of mind. Moral behaviour looks good to almost everyone, feels good to its perpetrators, and usually brings peace of mind, though sometimes at the cost of reduced material wellbeing.

Moral persons strive to dispense justice and expect justice from their fellows, but justice is often delayed beyond the lifetime of a person who has been treated unjustly and can be denied indefinitely. In His parable about an unjust judge and a widow seeking justice, Christ reportedly asked the following question and made the following promise: *"And shall not God avenge his own elect, which cry day and night to him, though he bear long with them? I tell you that he will avenge them speedily."*[40] No one can identify with any certainty *"his own elect,"* but it is obvious that some victims of injustice are not avenged speedily or even at all during their earthly lives.

I believe that God sees and records everything. As was written in the Epistle to the Hebrews: *"Neither is there any creature that is not manifest in his sight: but all things are naked and opened unto the eyes of him with whom we have to do."*[41] I trust God to correct *in the spiritual realm* all of the uncorrected injustice that pervades the material realm. Without that ultimate resolution of injustice by God, much of human existence would be a horror show. Beyond my own experiences of God, I take that image of Him, as the Ultimate Corrector of all injustice, as a strong argument for His existence.

## Consciousness and the Brain

I define human consciousness as an individual's awareness of self and surroundings. Within the human awareness of self we either

---
40 Luke 18:7-8
41 Hebrews 4:13

miss, or find and develop our awareness of God. Human consciousness is clearly on a much higher plane than the consciousness of any animal, including our nearest primate relatives. Nevertheless, animal consciousness is still an important part of the story of human consciousness and future research on animal consciousness might yet hold big surprises.

Human consciousness must have arisen either through evolution, by natural selection and descent with modification, or through divine intervention or, as I believe, through *both*. Daniel Dennett[42] was right when he stated: "*Recognizing our uniqueness as reflective, communicating animals does not require any human 'exceptionalism' that must shake a defiant fist at Darwin and shun the insights to be harvested from that beautifully articulated and empirically anchored system of thought.*"

I thank God for the truthful disclosures about life on Earth that He made available to us through Charles Darwin. Any believer who shakes a "*defiant fist*" at Darwin, whether she/he is a Christian or a Muslim or a member of any organized religion, is insulting God, Who made the wonderful processes that Darwin did so much to elucidate.

As discussed above, Antonio Damasio[43] defined consciousness as: "*a state of mind in which there is knowledge of one's own existence and of the existence of surroundings.*" He concluded that: "*Consciousness is a state of mind with a self process added to it.*" I must add awareness of one's soul to what Damasio called: "*knowledge of one's own existence.*" The human self is complete not only because, as Damasio put it, "*Self Comes to Mind,*" but also because *soul* interfaces with mind. With the soul as the spiritual core of self, what Damasio called "*self process*" becomes what I call Free Thought.

---

42  Dennett, D.C. 2003. *Freedom Evolves*. Penguin Books: New York.347p.p.308.
43  Damasio, A. 2010. *Self Comes to Mind. Constructing the Conscious Brain*. William Heinemann: London.367p.p.157.

## Free Thought, Faith, and Science

In his review of the success and scope of mechanistic biology and its religious implications, Joseph Needham[44] stated: "*The name which must command our chief respect…is that of Descartes, who first saw clearly that the body was really a machine governed not by any vital force but by the soul or mind or whatever the non-material part of man may be called…we are not tied to any of the details of his philosophy or his physiology; indeed, both of them have for two centuries been merely of historical interest. Nor are we in the least compelled to accept his absolute dualism of matter and spirit.*"

Needham continued: "*The earlier mechanistic biology before the era of experimental science was indeed somehow incompatible with religion… But now that the assumptions on which the triumph of mechanistic biology in the last century is based have been well examined, it is seen that for its own sphere it is a real triumph but, at the same time, its jurisdiction over other fields cannot be admitted…The biochemist and the biophysicist, therefore, can and must be thorough-going mechanists,* **but they need not on that account hesitate to say with Sir Thomas Browne, 'Thus there is something in us that cannot be without us and will be after us though indeed it hath no history what it was before us and cannot tell how it entered into us'** *(present author's emphasis).*" Browne's "*something in us that cannot be without us and will be after us*" is the soul.

René Descartes mage huge contributions to science, but his dualistic body-mind model was wrong. Antonio Damasio[45] described Descartes' error as: "*the abyssal separation between body and mind…the suggestion that reasoning, and moral judgement, and the suffering that comes from physical pain or emotional upheaval might exist separately*

---

44 Needham, J. 1926. *Mechanistic Biology and Religious Consciousness*, p.219-257.In (J. Needham, ed.) *Science Religion and Reality*. The Sheldon Press: London. 396p.p.250 and 256-257. Available at: www.kessinger.net Needham quoted from Sir Thomas Browne's *Religio Medici*. 1643. First Edition. First Part; Section 35. My source has slightly different wording but identical meaning, available at: http://penelope.uchicago.edu/relmed/relmed.html
45 Damasio, A. 2005. *Descartes' Error: Emotion, Reason and the Human Brain*. Penguin Books: London.312p.p.249-250.

*from the body.*" That 'error' is corrected when we take the body-mind as material realm stuff and the soul as spiritual realm stuff, with no "*abyssal separation*" between them, but rather a mind-soul interface across which they interact.

Russell Shorto[46] noted Descartes' own problems with body-mind dualism: "*Descartes…realized that there was a difficulty with his division of reality into mind and body – the difficulty being to figure out how the two substances interacted…His conclusion was that there is a connective tissue between the two… The seventeenth-century terminology for this…was 'passion'…Love, joy, anguish, remorse: we experience these in both body and mind, and somehow, Descartes became convinced, these passions link our two selves.*"

Shorto referred to the following statements in Descartes'[47] final published work: "*I consider that we observe not any thing which more immediately agitates our soul, than the body joined to it, and consequently we ought to conceive that what in that is a Passion, is commonly in this an Action: so that there is no better way to attain to the understanding of our Passions, than by examining the difference between the soul and the body, that we may know to which of them each function in us ought to be attributed.*" Descartes took the soul to be the seat and generator of our emotions, moving our "*(animal) spirits*" to effect bodily changes in muscle movements etc.

The human body-mind, including the entire brain, is *all* material realm stuff. The human soul is the *spiritual* core of self and is made of spiritual realm stuff. I propose a fully integrated and holistic state for every human self, as body-mind plus soul. I propose that the body-mind and soul work in concert during Free Thought. I recognize the co-existence

---

46 Shorto, R. 2008. *Descartes' Bones: A Skeletal History of the Conflict Between Faith and Reason.* Doubleday: New York.299p.p.252-253.
47 Descartes, R. 1649. *Les Passions de L'âme.* Second Article.p.2-3.Translated in London, May 24, 1650. Printed for A.C. (sic), to be sold by J. Martin, and J. Ridley, at the Castle in Fleet Street near Ram Alley. In the quotation, spelling is modernized for clarity. Available from: http://net.cgu.edu/philosophy/descartes/Passions_Letters.html

of two realms and two kinds of stuff, the spiritual and the material.

My model for Free Thought is *not* Descartes' body-mind dualism revisited. It can be described as interactive dualism or dual-aspect monism and I accept either of those labels. However, I prefer to call it an all-embracing monism that positions the entire human self in the totality of material and spiritual reality.

John Searle[48] described human consciousness research as being pursued typically with the following two interrelated presuppositions: that scientific reductionism (i.e., materialism or naturalism) will eventually provide all the answers and that any form of dualism (i.e., separation of the mental from the physical) must be avoided. He criticized this approach as follows: *"the urge to reductionism and materialism derives from the underlying mistake of supposing that if we accept consciousness as having its own real existence, we will somehow be accepting dualism and rejecting the scientific worldview."*

I agree. Rejecting on principle anything that resembles dualism is *de facto* a constraint and can lead to a mistaken, material realm-only view of reality. Nevertheless, most current research on consciousness is indeed being conducted from the perspective that the 'brain is all' and without recognizing the possibility of a spiritual soul. It is assumed that everyone has similar 'hardware' (the physical brain and neural networks) and suites of changing 'software' and sources of information that together provide for all of our memory, skill centres, stored knowledge and beliefs, through mapping, images and multiple interactions.

Roger Penrose[49] summarized as follows four viewpoints about the extents to which human consciousness and awareness are computational:

---

48  Searle, J.R. 1997, *The Mystery of Consciousness*. The New York Review of Books: New York.224p.p.xiii.
49  Penrose, R. 2005. *Shadows of the Mind: A Search for the Missing Science of Consciousness*. Vintage Books: London.457p.p.12, 16, 36. First published in 1995.

# THE HUMAN CONDITION

*"A. All thinking is computation; in particular, feelings of conscious awareness are evoked merely by the carrying out of appropriate computations.*

*B. Awareness is a feature of the brain's physical action; and whereas any physical action can be simulated computationally, computational simulation cannot by itself evoke awareness.*

*C. Appropriate physical action of the brain evokes awareness, but this physical action cannot even be simulated computationally.*

*D. Awareness cannot be explained by physical, computational, or any other scientific terms."*

Penrose continued: *"According to C, the problem of conscious awareness is indeed a scientific one, even if the appropriate science may not yet be at hand. I believe that it must indeed be by the methods of science - albeit extended in ways that we can perhaps only barely glimpse at present - that we must seek our answers."*

I agree, but I believe that the methods of science will eventually take us to D, which is the only option that would allow inclusion of the spiritual realm. Penrose stated his preference for a 'strong' version of viewpoint C, but admitted that it has a gap and called for: *"some fundamentally new physics to be involved."*

In the following passage, where the emphases are from Penrose unless stated otherwise, I think he sailed close to soul territory: *"The legal issue of 'responsibility'* **seems** *to imply that there is indeed, within each one of us, some kind of independent 'self' with its* **own** *responsibilities - and, by implication, rights - whose actions are* **not** *attributable to inheritance, environment, or chance. If it is other than a mere convenience of language that we speak as though there were such an independent 'self', then there must be an ingredient missing from our present-day physical understanding.* **The discovery of such an ingredient would surely profoundly**

***alter our scientific outlook*** (present author's emphasis)."

Penrose and others applied quantum theory to research on brain microtubule structure and functions. Penrose added the possibility of wavefunction self-collapse from quantum coherence, which he called "*objective reduction (OR)*." Penrose and Stuart Hameroff[50] proposed "*orchestrated*" OR in microtubules and the "*Orch-OR*" model of consciousness. Consistent with the Orch-OR model, Hameroff[51] proposed the so-called "*Conscious Pilot*" model, in which dendritic synchrony is said to move through the brain's neural networks, bringing conscious experience and choice to otherwise non-conscious cognitive modes - like a conscious human pilot moving around an aircraft on autopilot. Using the Orch-OR model, Hameroff and Deepak Chopra[52] proposed a "*quantum soul,*" having a form that I would call 'materially-derived non-materiality.' Laura McKemmish and others[53] found the Orch-OR model "*not biologically feasible.*" In all such work, one finds inspiring attempts to relate human consciousness and the brain to the composition and workings of everything else in the universe.

We are conscious beings studying our own consciousness. We are part of the total fabric of the universe, trying to describe and understand that total fabric. These tasks call for the involvement of all relevant disciplines. Research on consciousness and the brain is arguably the most exciting and most multidisciplinary field in science. Its greatest challenge is to include faith as a contributory discipline.

---

50  Hameroff, S.R. and R. Penrose. 1996. Orchestrated objective reduction of quantum coherence in brain microtubules: the "*Orch-OR*" model for consciousness, p. 507-540. In S.R. Hameroff, A.W. Kaszniak and A.C. Scott (eds.) *Toward a Science of Consciousness - The First Tucson Discussions and Debates*. MIT Press: Cambridge MA.
51  See: www.quantumconsciousness.org/pilot.htm.
52  Hameroff, S. and D. Chopra. 2012. The 'quantum soul': a scientific hypothesis, p.79-93. In A.M. Morena-Almeida and F.S.S. Santos (eds.) *Exploring Frontiers of the Mind-Brain Relationship*. Springer Science + Business Media LLC: New York. 243p.
53  McKemmish, L.K., Reimers, J.R., McKenzie, R.H., Mark, A.E. and N.S. Hush. 2009. Penrose-Hameroff orchestrated objective reduction proposal for human consciousness is not biologically feasible. *Physical Review E* 80 (2): 021912.6p.

# THE HUMAN CONDITION

Matthew Alper[54] proposed experiments to monitor the prayers and spiritual contemplations of believers through brain scans and blood chemistry and to make comparisons with nonbelievers, paying particular attention to alleged religious conversions. Such approaches might indicate the size of the hole in explanations of the human condition that omit the spiritual realm and the soul. We cannot detect or contact the spiritual realm through our material realm senses. We can only look within and try to 'feel' our souls, as our spiritual receivers and transmitters. I assume that even some of the hardest-nosed of hard-nosed scientists feel like more than the sum of their material parts.

## THE SOUL

I define the soul as the unique spiritual and immortal essence of self that is given by God to each individual human, integrated during life in the material realm with that individual's unique body-mind and taken after death in the material realm to an eternal life in the spiritual realm. The soul is the spiritual core of self, created by God in the spiritual realm. The Book of Proverbs puts that beautifully: "*The spirit* (i.e., soul) *of man is the candle of the Lord.*"[55]

There are popular myths that X-ray images can reveal the soul as some kind of aura and that the soul has actually been weighed. In 1907, Duncan MacDougall[56] measured the body weights of six humans and 15 dogs just before and after what he took to be the point of death. Ascertaining an exact point of death is always difficult and measuring accurately the weight of a large object that is either hydrating or dehydrating is well nigh impossible. However, MacDougall claimed to have measured a tiny reduction in weight of 21 grams for just *one* of his human subjects immediately after death and took that to be the weight

---

54 Alper. M. 2006. *The God Part of the Brain: A Scientific Interpretation of Human Spirituality and God.* Sourcebooks Incorporated: Naperville IL.273p.p.233.
55 Proverbs 20:27
56 See: www.snopes.com/religion/soulweight.asp

of the soul. The other five subjects had various weight losses and gains, but MacDougall found reasons for discarding all his other data except the one loss of 21 grams. The dogs died without any apparent loss of weight, which was taken as proof that dogs lack souls.

MacDougall's work was published in *"American Medicine"* and the *"New York Times."* His claim to have weighed a soul is of course nonsense. You are the only person who can detect your soul and I am the only person who can detect mine, as we engage in our private Free Thought. Each of us is a one-time, *de novo* experiment, performed by God - a unique spiritual and material self is given a brief life of freedom on Earth, after which a soul lives on forever somewhere in the spiritual realm.

The soul's state of faith or unbelief is always self-chosen and reflects majority occupation by God or by the spiritual force for evil. At any point in the development of the spiritual *curriculum vitae* of a soul, the most important entry is a 'Yes,' 'No,' or 'Don't know' answer to the question: 'Do I believe and trust in God?' The outcome is always faith or unbelief. The other entries are records of the spiritual consequences of all things done or left undone, with everything repented having been forgiven by God and deleted.

Among nonbelievers, especially scientists, anything called the soul is likely to be equated with the psyche, meaning a human mind and its properties and processes. Eric Fromm[57] used the word soul to mean the psyche or mind and called the psychoanalyst a *"physician of the soul."* Psychoanalysts and other mental health professionals are physicians of the *mind*. God is the only physician of the soul. The spiritual force for evil is the source of the soul's acute and chronic disorders. I cannot equate what is now called the psyche, meaning the mind and mental states, with the soul.

---

57   Fromm, E.1967.*Psychoanalysis and Religion*. Yale University Press: New Haven CT.119p.

# THE HUMAN CONDITION

John Polkinghorne[58] described the soul as: *"the almost infinitely complex, dynamic, information-bearing pattern, carried at any instant by the matter of my animated body and continuously developing throughout all the constituent changes of my bodily make-up during the course of my earthly life. That psychosomatic unity is dissolved at death by the decay of my body, but I believe it is a perfectly coherent hope that the pattern that is me will be remembered by God and its instantiation will be recreated by him when he reconstitutes me in a new environment of his choosing."*

I would take *"psychosomatic unity"* to mean integration of mind and body, not body-mind and soul. Moreover, I cannot imagine why after the death of a body-mind an ongoing soul would need for her/his eternal spiritual life any re-embodiment to become again something like a renovated primate in a renovated Earth-type ecosystem. If our souls really do need something like that for their eternal lives, I suppose that the required information could be passed across the mind-soul interface from earthly mental and somatic records, as the last event in an earthly life, but I consider that unlikely.

The material realm body-mind changes markedly through life and is indeed *"dissolved at death"* - typically after a period of senility, unless one has died younger through illness, accident, murder or suicide. When we die, the chemicals that had come together to make us go back into the cosmic melting pot. Some might appear in new human selves. The chemical and physical composition of a material realm body-mind is always in flux.

When you are next in a crowded place, such as an airport, concert hall, football ground, market or school, take a look at the multitude around you and remember that all of them, like you, are constantly exchanging molecules with the surrounding universe. The same applies to your cat, dog, potted plant and the rest of biodiversity on Earth, but only you

---

58  Polkinghorne, J.C. 1996. *The Faith of a Physicist: Reflections of a Bottom-Up Thinker.* Fortress Press: Minneapolis MN. 211p.p.163.

and your fellow humans have souls. This might give some nonbelievers pause for thought as to whether there is something more to a human being than material stuff.

The transitory composition of the material realm body-mind is precisely what *precludes* it from being the entire *"information-bearing pattern"* that is the human self. I can envisage how the brain keeps swapping new atoms, molecules and electrical circuitry for old, but I cannot see how that material realm stuff could ever form a *spiritual* self.

I conclude that the soul is entirely spiritual and therefore entirely non-materially affiliated, from its implantation by God into a new human life in the material realm to its eternal life in the spiritual realm.

Keith Ward[59] discussed whether the soul could be the brain and concluded the following: *"That would probably be too crude a view, for it might mislead us into thinking that all conscious states were determined by predictable laws of physics alone, or by brain states as thus predictable. All of our evidence shows that this is not the case; for our conscious processes often seem to be causally primary…The most important characteristic of the soul is its capacity for transcendence. It has the capacity to 'exist',* **to stand outside the physical processes that generate it, and of which it is part** (present author's emphasis).*"*

Ward continued: *"…the doctrine of the soul reminds us that it is not only in the supreme creativity of Einstein or Beethoven that we touch the realm of the spirit through the veils of sense…each of us can distinguish, in our own way, between truth and falsehood, beauty and squalor, the worthwhile and the trivial, love and selfishness…*(the soul) *may be able to disentangle itself from the public spatial properties of the brain and exist either alone or in some different form of materialization."*

I am with Ward all the way in his defence of the soul and his perspectives

---

59   Ward, K. 1998. *In Defence of the Soul.* Oneworld Publications: Oxford.175p.p.142, 144-145.

on the integrated nature of the self and all the higher things that I call the territory of Free Thought, but I cannot accept that *"physical processes"* generate the soul and that the soul is some *"form of materialization."* The simplicity and seamlessness of that model are attractive, but I cannot see how it could lead to the reconstitution of a spiritual soul after the death and dispersal of an entire material self. I perceive the soul as made entirely of spiritual stuff. The soul develops its spiritual *curriculum vitae* while integrated with a body-mind on Earth and lives forever after death as a unique spiritual form.

John Eccles[60] expressed a similar view: *"Since materialist solutions fail to account for our experienced uniqueness, I am constrained to attribute the uniqueness of the Self or Soul to a supernatural spiritual creation… each Soul is a Divine creation which is implanted into the growing foetus at some time between conception and birth. It is the certainty of the inner core of unique individuality that necessitates the 'Divine creation'. I submit that no other explanation is tenable; neither the genetic uniqueness with its fantastically impossible lottery, nor the environmental differences which do not* **determine** (original author's emphasis) *one's uniqueness, but merely modify it."*

In a dialogue with Karl Popper,[61] Eccles argued for the soul as follows: *"The self-conscious mind is to my way of thinking in a position of superiority over the brain in World 1 (i.e., the world of physical entities). It is intimately associated with it and of course is dependent on the brain for all detailed memories, but in its essential being it may rise superior to the brain as we have proposed in creative imagination. Thus there may be some central core, the inmost self, that survives the death of the brain to achieve some other existence which is quite beyond anything we can imagine."*

Popper held that the uniqueness of the human self evolves with the brain, in an emergent evolutionary process. Eccles responded thus: *"I*

---

60   Eccles, J.C. 1991. *Evolution of the Brain: Creation of the Self.* Routledge: London. 282p.p.237.
61   Popper, K. and J.C. Eccles. 1977. *The Self and Its Brain: An Argument for Interactionism.* Routledge: London.597p.p.557, 559-561.

*believe that my personal uniqueness, that is my own experienced self-consciousness, is not accounted for by this emergent explanation of the coming-to-be of my own self. It is the **experienced** (present author's emphasis) uniqueness that is not so explained. Genetic uniqueness will not do...So, I am constrained to believe that there is what we might call a supernatural origin of my unique self-conscious mind or my unique selfhood or soul...It is the uniqueness of the experienced self that requires this hypothesis of an independent origin of a self or soul, which is then associated with a brain, that so becomes my brain."*

Popper disagreed and gave his view of the most likely soul-less evolutionary sequence as follows: "*we have to assume that animal consciousness has developed out of non-consciousness...At some stage this incredible invention was made...Now the self-conscious mind (as opposed to animal consciousness, which possibly may go back to pre-brain forms) seems to me very clearly a product of the human brain. But in saying this I know very well that I am stating very little...it must not be taken as an explanation. We have the same situation with the emergence of life from something non-living.*"

William James[62] took the soul trail as follows: "*The world of our experience consists at all times of two parts, an objective and a subjective part...The objective part is the sum total of whatsoever at any given time we may be thinking of, the subjective part is the inner 'state' in which the thinking comes to pass...That unsharable* (sic) *feeling which each one of us has of the pinch of his individual destiny as he privately feels it rolling out on fortune's wheel may be disparaged for its egotism, may be sneered at as unscientific, but it is the one thing that fills up the measure of our concrete actuality, and **any would-be existent that should lack such a feeling, or its analogue, would be a piece of reality only half made up*** (present author's emphasis)." The whole self is a body-mind integrated with a soul.

---

62  James, W. 2002. *The Varieties of Religious Experience: A Study in Human Nature.* The Modern Library: New York. 602p.p.542-543.First Published in 1902 as the Gifford Lectures on Natural Religion, delivered at Edinburgh in 1901-1902.

# THE HUMAN CONDITION

Those who question the existence of the soul cite cases of large personality and behavioural changes brought about by brain injury, brain surgery and psychotropic drugs and ask what happened to the victims' souls, if any? For example, the 1848 American railroad worker Phineas Gage lost most of his brain's left pre-frontal lobe when a large iron bar was blasted through his head. He made an extraordinary recovery, thanks to the antiseptic properties of traces of explosive left on the iron, but his personality changed drastically from uprightness to moral laxity. The Gage case is still being debated.[63]

Matthew Alper[64] argued against the existence of the soul as follows, based on his experience with LSD: *"I had a bad trip that led to a severe clinical depression compounded by a dissociative, (sic) depersonalization, and anxiety disorder...until, finally, with the aid of pharmacological drugs, I was restored to my previous, relatively healthy self...According to the various belief systems (religions) I had thus far encountered, the human soul was supposed to be spiritual in nature, a fixed and permanent agent, unalterable and everlasting. Again and again, I was told that when I died, though my physical body would perish, 'I' – the sum of my conscious experience, the essence of my thoughts and feelings, what was perceived as my soul or spirit – would persist for all eternity. The fact, however, that my conscious self had been so drastically altered convinced me that there was no fixed or eternal essence in me."*

It is not surprising that serious physical and chemical insults to the brain and any form of brain deterioration can bring large changes in personality and behaviour. Major injury, chemical disruption or sickness of the rest of the body can do the same - for example, loss of sight, loss of limbs, and chronic pain from trauma or disease. Damage to a body-mind can affect a mind-soul interface and Free Thought. However, the soul is made of *spiritual* stuff and is impervious to damage from any material event or state.

---

63 See http://en.wikipedia.org/wiki/Phineas_Gage
64 Alper. M. 2006. *The God Part of the Brain: A Scientific Interpretation of Human Spirituality and God.* Sourcebooks Incorporated: Naperville IL.273p.p.10-11.

John Turl[65] put that as follows: "*What has become necessary for the dualist with modern insight is a revision of understanding of where the boundary lies between essential soul and contingent personality. In this respect, far from undermining traditional belief, science may well have done Christianity a service. An area of concern for some Christians is what happens to a person's standing before God if the balance of mind is disturbed. Great distress can be caused to relatives when a person of faith is so affected by dementia that all evidence of the original commitment appears to be lost. However, if the psychological phenomena are the result of effects on the brain beyond the control and outside the responsibility of the person affected, there is no reason to believe that the soul has been compromised.*"

Stewart Goetz and Charles Taliaferro[66] discussed arguments against the soul and substance dualism and quoted Malcolm Jeeves as follows: "*The nature of the [causal] interdependence increasingly uncovered by scientific research makes a substance dualism harder to maintain without tortuous and convoluted reasoning.*" Goetz and Taliaferro countered that soul functions and events can have their own identity, correlated with brain functions and mind events as long as the brain is functioning well, and can retain their identity when the brain has degenerated.

I agree. I can envisage deleterious spiritual occupation of a soul as the spiritual force for evil gains soul territory, but I cannot envisage how *spiritual* damage to a soul could cause a mind and brain to lose any of their *material* identity or integrity. Neither can I envisage how *material* damage to a body-mind could cause a soul to lose any of its *spiritual* identity or integrity, as known to God.

---

65  Turl, J. 2010. Substance Dualism or Body-Soul Duality? *Science and Christian Belief* 22: 57- 86.p.77.
66  Goetz, S. and C. Taliaferro. 2011. *A Brief History of the Soul*. Wiley-Blackwell: Chichester UK.228p.p.153-155. These authors cited, respectively: p.240, in Jeeves, M.2004.Towards a Composite Portrait of Human Nature, p.233-249. In M. Jeeves (ed.) *From Cells to Souls - and Beyond: Changing Portraits of Human Nature*. Eerdman's Publishing Company: Grand Rapids MI; and p.333-334 in Evans, C.S. 2005. Separable Souls: Dualism, Selfhood and the Possibility of Life After Death. *Christian Scholar's Review* 34:327-340.

# THE HUMAN CONDITION

Goetz and Taliaferro quoted C. Stephen Evans, as follows: "*We did not need neurophysiology to come to know that a person whose head is bashed in with a club quickly loses his or her ability to think or to have any conscious processes. Why should we not think of neurophysiological findings as giving us detailed, precise knowledge of something that human beings have always known, or at least could have known, which is that the mind (at least in this mortal life) requires and depends on a functioning brain?*"

Goetz and Taliaferro also made the following very important point: "*Without question, those in the Plato-Augustine-Descartes line...believe in the soul's existence on the basis of what they are aware of from the first-person perspective.* **There is not the least bit of evidence for the idea that they arrive at their belief in the soul's existence after failing to explain various experiences in terms of what goes on in the physical world** (present author's emphasis)."

In other words, those great thinkers and believers in the existence of the human soul and others who followed the same line did *not* make their cases for the soul's existence from a 'soul of the gaps' perspective. They made their cases based on subjective evidence from within; in other words, their own spiritual experiences and spiritual experiences shared by others.

Marcus Aurelius[67] wrote the following beautiful words concerning the soul: "*The properties of the rational soul are these: it sees itself, forms itself, renders itself what it will, and enjoys itself the fruit it bears itself...It attains its own goal, let the boundaries of life be fixed where you will.*" That comes close to what I call Free Thought - the body-mind and soul receive and process information from the spiritual and material realms, respond accordingly and make free choices about what to believe, what to trust and what to do.

---

67   *The Thoughts of Marcus Aurelius Antoninus.* Translated by John Jackson. 1906. The World's Classics LX. Humphrey Milford; Oxford University Press: London.135p.p.103-104.

The vast ancient and modern literature on the soul, from which I have cited only a few sources here, indicates clearly the human need to ask the following questions. Are there such things as souls? If so, do I have one? If so, how did I get it, what might it made of and what might be the implications for me, now and in the future? If not, what might be the basis for my spiritual nature and what might it be worth my while to explore further concerning my spiritual self? A soul composed of spiritual stuff is a vital element for my attempt to develop a theory of Free Thought, as described in the next two chapters.

## CHAPTER 3

# FREE THOUGHT: SCOPE AND FOUNDATIONS

*"The finer things"* [1]

### FREE THOUGHT TERRITORY

Steve Winwood sings it well - the finer things shine through, within us. The finer things of life include love and truth, the creation and enjoyment of beautiful and uplifting art, justice, morality, and truthful disclosures about the composition and workings of our world, through mathematics and science. All of the above are inputs to and outcomes of Free Thought. Free Thought also draws from and leads to more of the evil things in life including devious crimes, fraudulent claims, Nazi art and pornography.

The process and outcomes of any episode of Free Thought in any individual are known completely and with certainty only to that individual, to God and to the spiritual force for evil. As the Apostle Paul put it: *"For what man knoweth the things of a man, save the spirit of man which is in him?"* [2]

---

1  Steve Winwood.1986. *The Finer Things.*
2  I Corinthians 2: 11

# Free Thought, Faith, and Science

Free Thought probably commences at about seven years - the so-called age of discretion or age of reason in the Roman Catholic Church. In Free Thought, an individual pursues her/his personal goals as a free-willed, highly self-aware and potentially God-aware, material and spiritual being. The baselines for Free Thought shift from episode to the next. Free Thought can lead to faith or to unbelief, which includes agnosticism and atheism.

Free Thought has an infinite diversity of inputs and outcomes, as information from multiple sources in the material and spiritual realms is combined, interactively and iteratively, with responses to both. Free Thought operates in an open system - the totality of the spiritual and material realms. Those two realms are linked only across individual mind-soul interfaces, during Free Thought.

Our privately held beliefs, goals and preferences for action are always freely chosen. The same applies to whatever we choose to profess and how we choose to behave in the prevailing external circumstances. We use Free Thought to assess how free we are to profess our true goals and beliefs. We then speak and behave as we decide, but appropriate to the prevailing external circumstances.

Terry Eagleton[3] made essentially the same point as follows: "*There is no absolute distinction between being influenced and being free. A good many of the influences we undergo have to be interpreted in order to affect our behaviour; and interpretation is a* **creative** (present author's emphasis) *affair… We can act as free agents only because we are shaped by a world in which this concept has meaning, and which allows us to act upon it.*" Free Thought is very definitely "*a creative affair*" - a mix of human, divine and evil creativity.

---

3    Eagleton,T. 2010. *On Evil.* Yale University Press: New Haven CT.176p.p.11-12.

# FREE THOUGHT: SCOPE AND FOUNDATIONS

## Free Thought and free thought/freethinking

The terms 'free thought' and 'freethinking' have long been used to describe the worldviews and life styles of those who reject faith and/or some pattern of conformity or orthodoxy. Many authorities and members of organized religions have long considered what they call freethinking as sinful and freethinkers as undesirable characters - as mentioned in an old American gospel song: "*This train is bound for glory... This train don't carry no drinkers...or freethinkers.*"

John Robertson[4] wrote the following on the history of the term free thought: "*In England, as in the rest of Europe...the phenomenon of freethought had existed, in specific form, long before it could express itself in propagandist writings, or find any generic name save those of atheism and infidelity; and the process of naming was fortuitous as it generally is in matters of intellectual evolution...In 1667, we find Sprat, the historian of the Royal Society, describing the activity of that body as having arisen or taken its special direction through the conviction that in science, as in warfare, better results had been obtained by a 'free way.'*"

Robertson concluded that the words "*freethinking*" and "*freethinker*" first came into use in English literature in the last quarter of the seventeenth century. The members of the newly formed Royal Society, who were a mixture of believers and nonbelievers, sought to take science forward in a "*free way.*" One of the Royal Society's first rules prohibited the debating of any theological question. Similar tensions over what can and cannot be debated are seen throughout the history of faith-science and organized religion-science relationships, even when parties on both sides subscribe to the same organized religion.

Robertson described as follows the fate of the philosopher, all-round scientist and faith-science bridge builder Averroës (Ibn Rushd) (1126-1198 CE) under repressive Islam: "*...he* (Averroës) *lived in an*

---

4   Robertson, J.M. 1914. *A Short History of Freethought: Ancient and Modern.* Volume I. Watts and Co., London.484p.p.2-4, 271, 275-276. Available from: www.kessinger.net

*age of declining culture and reviving fanaticism; and all his conformities could not save him from proscription, at the hands of a Khalif who had long favoured him, for the offence of cultivating Greek antiquity to the prejudice of Islam. All study of Greek philosophy was proscribed at the same time and all books on the subject were destroyed."*

Robertson also assessed freethinking in late nineteenth and early twentieth century Islamic society. The following passage is interesting for comparison with present day scenarios and views: *"the comparative prosperity or progressiveness of Islam, as a proselytizing and civilizing force in Africa...is not strictly or purely a religious phenomenon...Even in Africa...a systematic observer* (citing here Harry Johnston's 1899 'History of the Colonization of Africa by Alien Races,' p. 283) *notes and predicts the extension of 'a strong tendency on the part of the Mohammedans towards an easy going rationalism'... Thus at every culture level* (i.e., in Islam) *we see the persistence of that force of intellectual variation* (i.e., freethinking) *which is the subject of our enquiry."*

In the history of Christianity, we find that same *"force of intellectual variation,"* for example, in the following statements by the Unitarian Reverend Frank Walters:[5]

> I. *"The soul, with its divine consciousness, its spiritual experience, its interior laws, must become the final arbiter of Church, Bible, Creed, and Sacrament."*

> II. *(Quoting from Bishop Butler's Analogy, ii. 3)* *"Reason is indeed the only faculty we have wherewith to judge anything, even revelation itself...Reason can, and it ought to, judge, not only of the meaning, but also of the morality and the evidence of revelation."*

---

5   Walters, F.1890. *Rationalism: What It Is and What It Is Not*, p.2-28.In R. B. Drummond (ed.) *Free Thought and Christian Faith, Four Lectures on Unitarian Principles*. Williams and Norgate: Edinburgh.123p.p.9, 20. Available at: www.kessinger.net

# FREE THOUGHT: SCOPE AND FOUNDATIONS

> III. *(Quoting from Dr. Channing's Discourse on Self Denial)* *"I am surer that my rational nature is from God than that any book is the expression of his will."*

Walters and his fellow Unitarians contested established church doctrines and affirmed that faith is a decision made freely by the self, based on reasoning *and* responses to spiritual revelations. In the Bishop Butler quotation above, *"Reason"* means substantially the same as Free Thought. Walters held that freedom of thought and freedom from indoctrination are essential and obtainable and that sacred text literalism must yield to individual soul searching, in a relationship with the Living God. I agree.

Adam Storey Farrar's[6] 1862 Bampton Lectures gave the more mainstream Christian perspective, as shown by the following excerpts, in which all emphases are his:

> I. *"The word 'free thought' is now commonly used…to express the result of the revolt of the mind against the pressure of external authority in any department of life or speculation…It will be sufficient now to state, that the cognate term,* **free thinking***, was appropriated…in the last century to express Deism. It differs from the modern term* **free thought***, both in being restricted to religion, and in conveying the idea rather of the method than of its result, the freedom of the mode of inquiry rather than the character of the conclusions attained; but the same fundamental idea of independence and freedom from authority is implied in the modern term."*

> II. *"The history of recent doubt has brought before us some whose minds doubt wholly of the supernatural…their convictions have become so fixed that they manifest a fierce spirit of proselytism, and*

---

6   Storey. Farrar, A.1895. *A Critical History of Free Thought in Reference to the Christian Religion: Eight Lectures Preached Before the University of Oxford, in the Year M.DCCC.LXII, on the Foundation of the Late Rev. John Bampton, M.A. Canon of Salisbury.* D. Appleton and Company: New York.487p.p.v, 358, 376-378. Available at: www.kessinger.net

> can dare to point the finger of scorn at those who still believe in the unseen and supernatural relations of God to the human soul…We can have no sympathy with them: we can rejoice that they retain a moral standard…but must tremble lest their unbelief end in thorough animalism; lest Epicureanism be their final philosophy."

III. *"Let us strive to use the two methods of finding truth, - study and prayer. Let us gain more knowledge, and consecrate it to the investigation of the highest problems of life and of religion; especially applying ourselves, by the help of the ripest aid which miscellaneous literature or church history can afford us, to the study of the sacred scriptures. But above all these intellectual instruments, let us add the further one of prayer…The infinite God condescends to enter into communion with our spirits, as really as a man talketh with a friend."*

Seeking truth through prayer, Bible study and daily guidance from God and seeking truth through science are indeed mutually reinforcing. They all use the same process - Free Thought. However, some present day religious authorities continue to preach against what they call freethinking and hold it to be contrary to the will of God. They miss the point that God, Who gave us free will and Free Thought, surely did so in order that we could use those gifts and so strive to align our wills and the outcomes of our Free Thought with His will, as far as is possible in the human condition.

When one has made a decision for faith and become a believer, one's free will and Free Thought are not switched off. A believer asks God for guidance, which comes not as a constraint to freedom but as truthful inputs to Free Thought. Personal freedom is widened through faith. Faith connects the believer with the Creator of everything, in friendship.

Most so-called free thought organizations exclude believers. The

# FREE THOUGHT: SCOPE AND FOUNDATIONS

North Texas Church of Free Thought[7] equates faith with superstition and claims to offer: *"atheists and unbelievers all the social, emotional and inspirational benefits of traditional faith-based religions, but without the superstition."*

The Freedom from Religion Foundation[8] defines *"freethinkers"* as: *"skeptics - atheists, agnostics, rationalists and secular humanists - who form their opinions about religion based on reason, rather than on tradition, authority or established belief."* The false implication there is that believers do not reason about religion. This Foundation works: *"to ensure that the voice of freethought is not censored and that reason and free enquiry are not stifled through religious intimidation."*

The Filipino Freethinkers[9] publish mostly atheist material, but profess a more inclusive perspective: *"We practice and promote reason, science, secularism, and, of course, freethinking...Freethought is not the same as atheism. Freethinking is a way of thinking that can lead to different conclusions. Although most freethinkers are nontheistic - atheists, agnostics, deists - many freethinkers still reach religious conclusions. However, theistic freethinkers do tend to have more liberal or progressive religious views than other believers."*

Most so-called free thought publications exclude contributions from believers. The *"Freethinker Journal"*[10] describes itself as: *"The voice of atheism since 1881."* Most websites that claim to be serving so-called freethinkers are maintained by and for nonbelievers, though some provide for debates with believers.[11] Jeffrey Mark[12] listed websites for *"atheists and agnostics and freethinkers,"* beginning with his own site[13]

---

7   www.churchoffreethought.org
8   www.ffrf.org
9   www.filipinofreethinkers.org
10  http://freethinker.co.uk
11  See, for example: www.theinfidelguy.com
12  Mark, J. 2008. *Christian No More: A Personal Journey of Leaving Christianity and How You Can Leave It Too*. Cogspage Media LLC: Cincinnati OH.284p. See: www.ReasonablePress.com
13  www.escapingchristianity.com

and continuing with about 20 others, *all* of which preclude faith as an outcome of so-called free thought.

G.K. Chesterton[14] criticized the conventional definition of a freethinker as meaning only: "*a man who, having thought for himself, has come to one particular set of conclusions, the material origin of phenomena, the impossibility of miracles, the improbability of personal immortality and so on.* He noted that: *"almost all of these ideas are definitely **illiberal*** (i.e., formulated to limit freedom; present author's emphasis and comment)."

Chesterton stressed that no thinking can be properly called *free* while insisting that it must lead to a particular outcome and/or take place in a particular type of person. I agree. Everyone uses Free Thought. So-called freethinkers, pillars of the establishment, believers, nonbelievers, conformists, nonconformists, the orthodox and the unorthodox, the subservient and the rebellious, *all* make their choices about the higher things of life through Free Thought.

## Information

Information can be defined using rigorous philosophical and/or operational criteria such as having an intelligent source and/or being structured and not chaotic. I define information broadly and simply as comprising *everything* that can be sensed materially or transmitted and received spiritually.

This broad definition of information encompasses the spoken and written word in any language, as well as all music, and all measurements and relationships that can be disclosed through mathematics and science; for example, the genetic code. In the material ream, all detectable matter and energy can be considered as information.

---

14  Chesterton, G.K. 2009. *Orthodoxy*. Moody Publishers: Chicago IL239p.p.188.First published in 1908.

# FREE THOUGHT: SCOPE AND FOUNDATIONS

Information is the resource for all basic thought and all Free Thought. Information for basic thought comes material realm sources. Information for Free Thought comes from material realm and spiritual realm sources. In basic thought and Free Thought, all of our decisions reflect the information available and our diligence in accessing and using as much of it as possible.

Vlatko Vedral[15] considered the universe as quantum information and stated: "*Quantum mechanics opens the door to genuine randomness (i.e. events which, at their most fundamental level, have no underlying cause).*" He continued: "*If indeed the randomness in the Universe, as demonstrated by quantum mechanics, is a consequence of our generation of reality then it is as if we create our own destiny. It is as if we exist in a simulation, where there is a program that is generating us and everything that we see around us.*"

Vedral concluded: "*Within our reality everything exists through an interconnected web of relationships and the building blocks of this web are bits of information. We process, synthesize and observe this information in order to construct the reality around us. As information spontaneously emerges from the emptiness we take this into account to update our view of reality. The laws of Nature are information about information and outside of it there is just darkness.*"

I am with Vedral in turning to quantum theory for explaining the big picture and the personal picture. I agree also that all of our basic thought and Free Thought are the processing of information. However, I cannot agree that all the information available to us and assembled by us comes from a random mix, which originated in emptiness and is surrounded by darkness.

We receive and process information from a stunningly beautiful material

---

15  Vedral, V. 2010. *Decoding Reality: The Universe As Quantum Information*. Oxford University Press: Oxford.229p.p.170, 217-218.

realm and a spiritual realm in which good ultimately defeats evil and offers an eternal life of love. Randomness? Emptiness? Darkness? One does not have to be a highly educated scientist to see that the human condition demands better explanations and that there is a wealth of information, means and motivation for seeking them.

The revelations sent to our souls by God and the spiritual force for evil are all forms of information. Revelations from God comprise guidance and assurance in response to prayer, or as He supplies those needs even without being asked. Revelations from God are often abundantly clear, as with the so-called: *"fruit of the Spirit...love, joy, peace, longsuffering, gentleness, goodness and faith."*[16]

Revelations from the spiritual force for evil come in complex disguises. Some of the best campaigners *against* faith are misguided believers. C.S. Lewis[17] depicted the spiritual force for evil as Uncle Screwtape, who wrote to his nephew Wormwood, a junior fieldworker for evil, as follows: *"We want the church to be small not only that fewer men may know the Enemy but also that those who do may acquire the uneasy intensity and defensive self-righteousness of a secret society or a clique."* Religious authorities and rank and file believers who deny the truth of evolution and other truthful disclosures from God through science risk turning their churches into such cliques, unaware that this serves the wrong side.

Ignorance results from a lack of information and/or a laziness to acquire all available information and use it diligently. James Carse[18] discussed three kinds of ignorance: *"ordinary ignorance"*, meaning simply lack of knowledge; *"willful ignorance"*, meaning awareness that certain knowledge (for example, evidence for evolution) exists, but avoiding it on purpose; and *"higher ignorance"*, also called *"learned ignorance"* and

---

16　Galatians 5:22
17　Lewis, C.S. 2007. *The Screwtape Letters: with Screwtape Proposes a Toast*, p.178-296. In *The Complete C.S. Lewis Signature Classics*. HarperOne: New York.746p.p.204.First published in 1942.
18　Carse, J.P. 2008. *The Religious Case Against Belief*. Penguin Press: New York.227p.p.12-20.

# FREE THOUGHT: SCOPE AND FOUNDATIONS

considered as an *"awakening"*, obtained by thorough personal reflection on what is known, unknown and possibly unknowable.

Ordinary ignorance is widespread. Truthful disclosures through science are increasing every day, but many people are unable to hear about and/or to understand some of them, because of their isolation and/or inadequate education. Willful ignorance is also common. Many people waste opportunities to explore and learn about faith and science. Higher or learned ignorance/awakening is less common. Carse cited Nicholas of Cusa (1401-1464 CE) as its author and used Galileo as an example. I hold that this form of ignorance is not limited to academics. It can develop in any honest seeker who comes to realize how little is known about our reality, compared to what is still unknown and/or unknowable.

## THE FREE WILL DEBATE

Roderick Chisholm[19] wrote: *"Human beings are responsible agents; but this fact appears to conflict with a deterministic view of human action (the view that every event that is involved in an act is caused by some other event); and it **also** (original author's emphasis) appears to conflict with an indeterministic view of human action (the view that the act, or some event that is essential to the act, is not caused at all). To solve the problem, we must make somewhat far-reaching assumptions about the self or the agent - about the man who performs the act. Perhaps it is needless to remark that, in all likelihood, it is impossible to say anything about this ancient problem that has not been said before."*

Despite its messiness and unresolved nature, I cannot avoid some discussion of the free will debate. Free will is foundational for what I call Free Thought. If free will is an illusion, then my model for Free

---

19 Chisholm, R.2008. *Human Freedom and the Self*, p.47-58.In (R. Kane, ed.) *Free Will*. Blackwell Publishing: Malden MA.310p.p.47.

# Free Thought, Faith, and Science

Thought is based on false assumptions.

I define free will as every individual's ever-present potential to choose personal goals independently and to hold them internally. In Free Thought we process all of the available information that is pertinent to the pursuit of our free-willed goals and then make appropriate decisions for action or inaction.

Jonathan Bricklin[20] reviewed William James' position on free will and concluded the following, in which citations other than those from James are omitted: "*Having found no evidence for free will, James declared that the free will controversy was 'insoluble on strictly psychologic[al] grounds.'...James believed that there was no psychological proof of determinism either...Our actions, before they are made, are 'ambiguous or unpredestinate', in a word, 'indeterminate'. But while James claimed that in 'common parlance' this was the equivalent of saying 'our wills are free', he himself knew better. That no external or internal* **stimulus** *commands a predictable sway over attention does not substantiate free will. To accept attention (active or passive) as 'an independent variable' is to acknowledge only a radical ignorance concerning it. On strictly psychological grounds... the 'question of fact in the free will controversy' is neither free will nor determinism. It is indeterminism.* **What we believe to be acts of will are automatic reactions to stimuli of unascertainable origin** (present author's emphases)."

I disagree with those conclusions and offer Free Thought as an alternative model. Free Thought is an integrated and iterative process in which three cycles of *stimuli* and responses are interlinked. The origin of one set of stimuli is clear. Our senses give us stimuli as information from the material realm and we process that information in our minds. Some of our "*acts of will*" in response to those stimuli can be called

---

20   Bricklin, J. 2004. A variety of religious experience: William James and the non-reality of free will, p.77-98. In B. Libet, A. Freeman and K Sutherland (eds.) *The Volitional Brain: Towards a Neuroscience of Free Will.* Imprint Academic: Exeter, U.K. 298p.p.97. First published in 1999 in the *Journal of Consciousness Studies* 6 (8-9): 77-98.

"*automatic*" for all intents and purposes. We have a will to survive and to reproduce. We choose to eat and drink and to live as far as possible in 'Goldilocks' surroundings and then we act accordingly. This is the territory of basic thought, not Free Thought.

The other two sets of stimuli originate from God and the spiritual force for evil. It can be difficult distinguish between them in the mix of spiritual revelations that comes into our soul battlefields, but they are not completely "*unascertainable.*" We can distinguish some truths from some lies. In any case, our responses to those stimuli are definitely not "*automatic.*" They are part of our Free Thought.

Again, there is a fundamental difference between arriving at privately held *choices* and performing *acts* chosen to suit external circumstances. Everyone has a sovereign-to-self will, through which she/he makes free choices about goals. Everyone uses basic thought and Free Thought to make her/his choices about preferred behaviour and actual behaviour.

Eric Fromm[21] emphasized human herd instincts, as follows: "*Rationalization is a compromise between our sheep nature and our human capacity to think…inasmuch as we are sheep, reason is not our real guide; we are guided by an entirely different principle, that of herd allegiance…The unfolding and full emergence of reason is dependent on the attainment of full freedom and independence. Until this is accomplished man will tend to accept for truth that which the majority of his group want to be true; his judgement is determined by need for contact with the herd and by fear of being isolated from it.*"

With all due respect to Fromm and while recognizing that humans sometimes exhibit herd behaviour, I reject his dismal analysis, as quoted above. As Charles Mackay[22] documented well, many of us *behave* as if we are being ruled by herd instincts, but our individual Free Thought

---

21  Fromm, E.S. 1967. *Psychoanalysis and Religion.* Yale University Press: New Haven CT.126p.p.59.
22  Mackay, C. 1841. *Extraordinary Popular Delusions and the Madness of Crowds. Available from Maestro Reprints*: Lexington KY.399p.

## Free Thought, Faith, and Science

continues *internally*. Through our Free Thought we choose the extents to which our actions or inaction will match our privately held choices and preferences. The baselines for new episodes of Free Thought evolve through life. Anyone can think herself/himself out of the apparent 'herd thought' and can choose whether or when to reveal the non-herd outcomes of her/his Free Thought by non-herd behaviour.

After a strong defence of "*The Right to Not Believe,*" the atheist André Comte-Sponville[23] concluded the following: "*Freedom of thought is the only good that is perhaps more precious than peace, for the simple reason that, without it, peace would merely be another name for servitude.*" Nevertheless, some hold that free will and free thought in the conventional sense do not exist and that each of us is governed completely by combinations of the following: our genes; our accumulating experiences and external influences through what we hear, read and see; our doings unto others and their doings unto us.

Godless denials of free will are similar to the Christian doctrine of God's predestination as negations of personal freedom. From his atheist perspective, Sam Harris[24] argued that free will is an illusion because: "*You will do whatever it is you do, and it is meaningless to assert that you could have done otherwise.*" In other words, a murderer must murder, a cheater must cheat and a Mother Teresa must help the poor. Harris' examples included choosing whether or not to have a second cup of coffee and whether to participate in criminal behaviour.

According to Harris, one's belief in or denial of free will is an inevitable consequence of one's identity and no one has any freedom to choose otherwise. Harris made his case as follows: "*The brain is a physical system, entirely beholden to the laws of nature - and there is every reason to believe that changes in its functional state and material structure entirely dictate our thoughts and actions. But even if the*

---

23  Comte-Sponville, A. 2007. *The Little Book of Atheist Spirituality.* Translated by Nancy Huston. Penguin Books: London.212p.p.133.
24  Harris, S. 2012. *Free Will.* Free Press: New York. 83p.p.44,11-12.

# FREE THOUGHT: SCOPE AND FOUNDATIONS

*human mind were made of soul-stuff, nothing about my argument would change. The unconscious operation of a soul would grant you no more freedom than the unconscious physiology of your brain does. If you don't know what your soul is going to do next, you are not in control… The soul that allows you to stay on your diet is just as mysterious as the one that tempts you to eat cherry pie for breakfast."*

For Harris, anything that might be called the soul or spiritual stuff must be a material realm extension of mind and brain. His denial of the reality of free will is actually a denial of the existence of the spiritual realm, including the soul as the spiritual core of self. Even if the soul was nothing but an extension of the material mind and brain - an explanation that I reject, but must recognize that even some believers accept - the operations of our souls and our soul states could not be described accurately as entirely *"unconscious."* The human conscience is not an unconscious experience. Our soul states and events are integral to our Free Thought, all of which takes place during states of consciousness.

Roy Varghese[25] considered the human *"phenomenon of thought, of understanding* (and) *seeing meaning"* to be *"Beyond consciousness."* He held that: *"thinking in concepts is by its very nature something that transcends matter."* However, if one accepts the possibility of a mind-soul interface as the link between the spiritual and material realms, there is neither need nor basis to position anything that I call basic thought or Free Thought as something that is *"beyond consciousness."* Varghese continued: *"a denial of the self cannot even be claimed without contradiction. To the question, 'How do I know I exist?' a professor famously replied, 'And who's asking?'"*

Our attempts to deny personal responsibility for our decisions and actions are bound to fail. Try this appeal by an apprehended criminal:

---

25  Varghese, R.A. 2007. *The 'New Atheism': A Critical Appraisal of Dawkins, Dennett, Wolpert, Harris and Stenger*, Appendix A, p.161-183. In Flew, A. 2007. *There Is a God. How the World's Most Notorious Atheist Changed His Mind.* HarperOne: New York. 222p. p.176-177, 181.

# Free Thought, Faith, and Science

"*Sorry, dear Officer of the Law, please put the blame for my crime entirely on my parents for supplying my genes and for my disastrous upbringing and on my buddies for leading me astray. You cannot blame me, because all that I am today was made by them, not by me.*" Or try a lawyer's defence, based on predestination: "*Your Honour, my client holds that he should not be punished because it was God who made his criminal actions inevitable.*"

To accept such arguments would be to deny that the conscious human self has any true freedom of thought or of action. Now please ask yourself: 'Am I *free* to think about free will?' If you say 'Yes,' then to what do you attribute that freedom? If you say 'No,' then how and/or why are you being constrained? All answers will be outcomes of your Free Thought.

### *The Debate in Organized Religion*

Matthew Gordon[26] described, as follows, some of the historical debate about free will in Islam: "*Opponents of the Umayyad Caliphate (661-750 CE) argued strongly for the existence of human choice and the requirement that all humans accept responsibility for their actions. However, Umayyad partisans espoused the idea of predestination, arguing that God had ordained all things beforehand...The emphasis on human free will was taken up in the ninth century CE by a group of scholars known as the Mutazilites. Their concern was to promote the idea that evil in the world could never be of divine origin but was the outcome of human behaviour alone. Lengthy debate led to a certain compromise by later Sunni thinkers, among them al-Ashari (died 935 CE), who argued that God, while retaining his omnipotence, also provides humankind with a 'modicum' (or, in other readings, a 'brief moment') of freedom and thus responsibility.*"

Christianity past and present has a similar mix of opinions about free

---

26  Gordon, M.S. 2002. *Understanding Islam. Origins. Beliefs. Practices. Holy Texts. Sacred Places.* Duncan Baird Publishers: London.112p.p.94.

will. Many Christians believe that God predetermines everything and that each day is therefore a divinely precooked dollop of life, offering no option other than to live it out exactly as He planned it. On that basis, God can sit back and enjoy the show, in which He planned every scene and knows every ending. Some Christians believe that God has predestined not only everything that will happen to us on Earth, but also whether our souls will go to heaven or hell, with no possibilities for us to alter any of His historical choices.

The main arguments for this all-embracing precooking of temporal and eternal life are biblical quotations, for example: *"For whom he did foreknow, he also did predestinate to be conformed to the image of his Son...Moreover whom he did predestinate, them he also called: and whom he called, them he also justified; and whom he justified, them he also glorified"*[27] (and) *"For the children being not yet born, neither having done any good or evil, that the purpose of God according to election might stand, not of works, but of him that calleth...As it is written, Jacob have I loved, but Esau have I hated."*[28]

Some Christians who believe in the predestination of every soul as bound for heaven or hell recommend nonetheless that everyone should still look to Christ and ask Him whether the Good News is for them or not: 'Dear Lord, am I one of the elect or damned?' That awful doctrine has brought untold misery to those who find reasons to classify themselves as damned irrevocably to hell. Those who preach that doctrine should remember that Apostle Paul also wrote the following: *"For God hath concluded them **all** in unbelief, that he might have mercy **upon all** (present author's emphases)."*[29]

If everyone got a daily, unchangeable dollop of life, then no one would have any real freedom or any scope to change anything beyond playing a role in the enactment of God's unchanging will. Why then would any

---
27   Romans 8:29-30
28   Romans 8:11,13
29   Romans 11:32

believer need to say: 'Not my will, but Thy will be done'? If everything was predetermined, there would be only one will that could determine anything - the will of God. Predestination would make us robots and deny even God the freedom to change His mind.

I cannot believe that God and the world that He made are working like that. I cannot believe that a God of Love would create a system in which all love and all hate were preprogrammed and choice-less. We do not get precooked daily dollops of life. We live free-willed lives. We engage in basic thought and Free Thought in a free process material realm and on battlefields of good versus evil. The dramas that unfold within us and around us are unscripted.

Our free will, God's will, His spiritual revelations and our responses determine whether we establish or miss a relationship with Him, through our Free Thought. The outcomes cannot be known in advance. For some believers, that detracts too much from God's omnipotence, but I cannot see how it could be otherwise.

God's chosen workings for His creation do not lessen His omnipotence. John Polkinghorne[30] explained beautifully how God has chosen to limit Himself, on our behalf: *"My argument is not that God's not knowing the future is essential to guarantee free will, but that a world that can contain freely choosing beings must be open to the future so that it is a world of true becoming. The argument then is that God will know that world truly, that is, according to its actual nature in its actual becoming-ness. The consequence is a divine choice* **to engage with time and not know the details of the future** (present author's emphasis)."

God is beyond time, but He has chosen to engage with time in His relationships with humans. Free-willed human goals and Free Thought then shape the course of each human life in the midst of free process

---

30  Polkinghorne, J.C. and N. Beale. 2009. *Questions of Truth: Fifty-One Responses to Questions About God*. Westminster John Knox Press: Louisville KY.186p.p.33.

## FREE THOUGHT: SCOPE AND FOUNDATIONS

in the material realm and the good or evil behaviour of one's fellows. After death, the fate of every human soul is entirely God's decision. If He had not endowed all humans with free will and Free Thought, He would have nothing to judge.

Free will is not surrendered at conversion from unbelief to faith or vice versa. At conversion to faith, the individual's free will is put it into closer harmony with God's will. Both can then be pursued in concert. We receive spiritual revelations from God and from the spiritual force for evil throughout our lives, as we continue in Free Thought. Faith is not an abnegation of freedom. Faith is a higher form of freedom - freedom in relationship with the Giver of freedom. In faith, the believer is a new self but remains a self, with a self will that remains free.

The following statement[31] summarizes well my position on how our free will and God's will are related: "*God, in giving us free will, said to us: 'your will be done.' Some of us turn back to him and say: 'My will is that your will be done.' That is obedience to the first and greatest commandment. Then, when we do that, he turns to us and says: 'And now, your will be done.' And then he writes the story of our lives with the pen strokes of our own free choices.*"

The only precondition that I recognize as applying to all souls and overarching all free will and Free Thought is the *grace* of God - the free gift of forgiveness and salvation, just for the asking and never merited. God's grace is His pre-emptive gift as we engage in battles in the spiritual and material realms. Grace has huge impacts on the outcomes of our battles, but it does not determine them and they are not predetermined. They are shaped by us, living under God's grace but living in a free process material realm, among free willed humans, with the whole show literally bedevilled by the spiritual force for evil.

---

31   www.spirithome.com. For the first and greatest commandment, see Matthew 22:37.

# Free Thought, Faith, and Science

M. Scott Peck[32] wrote: "*the issue of free will, like so many great truths, is a paradox. On the one hand, free will is a reality. We can be free to choose without 'shibboleths' or conditioning or many other factors. On the other hand, we cannot choose freedom...We must ultimately belong either to God or the devil.*" He continued: "*Each of us is ultimately free to choose how we are going to behave. We are free to reject what we have been taught and what is normal for our society. We may even reject the few instincts that we have, as do those who rationally choose celibacy or submit themselves to death by martyrdom. Free will is the ultimate human reality.*" I agree. We cannot choose or deny freedom. We all *have* our freedom. Our freedom is God-given.

## *The Debate in Science*

Russell Stannard[33] wrote: "*As regards the workings of the brain, these will be governed by the laws of physics...This means that for a given physical state of the brain at a given point in time, and from a knowledge of all external influences upon it, one ought to be able to predict what the succeeding physical state will be from a simple application of those laws. We say that the future state is determined.*" In the same vein, Stephen Hawking and Leonard Mlodinow[34] wrote against free will as follows: "*It is hard to imagine how free will can operate if our behaviour is determined by physical law, so it seems that we are no more than biological machines and that free will is just an illusion.*"

Benjamin Libet and others[35] showed that a type of brain electrical activity, which they called the "*Readiness Potential,*" begins several hundred milliseconds (ms) *before* the earliest time at which an actor can report

---

32  Scott Peck, M.1985. *People of the Lie: The Hope for Healing Human Evil*. Touchstone: New York.269p.p.83, 244.
33  Stannard, R. 2010. *The End of Discovery*. Oxford University Press: Oxford.228p.p.12-13.
34  Hawking, S. and L. Mlodinow. 2010. *The Grand Design: New Answers to the Ultimate Questions of Life*. Bantam Press: London.199p.p.32.
35  Libet, B., Gleason, C.A., Wright, E.W. and D.K. Paul. 1983. Time of conscious intention to act in relation to onset of cerebral activity (Readiness Potential); the unconscious initiation of a freely voluntary act. *Brain* 196 (3): 623-642.

# FREE THOUGHT: SCOPE AND FOUNDATIONS

a *conscious intention* to act. Libet and others concluded that: *"cerebral initiation of a spontaneous, freely voluntary act can begin unconsciously, that is, before there is any (at least recallable) subjective awareness that a 'decision' to act has already been initiated cerebrally. This introduces certain constraints on the potentiality for conscious initiation and control of voluntary acts."*

Libet[36] wrote the following in a subsequent paper: *"Human subjects became aware of intention to act 350-400 ms **after** (original author's emphasis) RP (Readiness Potential) starts, but 200 ms before the motor act. The volitional process is therefore **initiated** (original author's emphasis) unconsciously. But the conscious function could still control the outcome; it can veto the act. **Free will is therefore not excluded** (present author's emphasis)... These findings put constraints on views of how free will may operate; it would not initiate a voluntary act but it could control performance of the act. The findings also affect views of guilt and responsibility. But the deeper question remains: Are voluntary acts subject to macro-deterministic laws or can they appear without such constraints, non-determined by natural laws and 'truly free'?"*

My answer is that an actor's self-willed and privately held preferences for action are always truly free, but her/his overt acts are constrained by further internal choice making, relative to external circumstances and natural laws. At risk of sounding naïve or even stupid in commenting on a field that is not my own, I suggest that the lag times observed by Libet and others between a detectable Readiness Potential for an act and the actor's self-reported decision to act are not surprising. The immediacy of a single electrical measurement is being compared with the time needed to complete an internal sequence - choosing a specific Readiness Potential to act and then packaging it from a state of less than fully conscious registration into a self-aware and reportable message.

---

36   Libet. B. 1999. Do we have free will? *Journal of Consciousness Studies* 6 (8-9): 47-57.

Peter Clarke[37] discussed whether quantum (Heisenbergian) indeterminism might provide a basis for free will and termed that position "*quantum libertarianism,*" which he explained as follows: "*Quantum libertarians propose that mind-directed changes occur 'hidden' within the cloud cover of Heisenbergian uncertainty. According to standard quantum physics, such hidden effects are assumed to be random, but the unconventional proposal of quantum libertarianism is that they are **non-random*** (original author's emphasis), *directed by the mind (or soul etc.).*" Clarke found problems of scale for using Heisenbergian uncertainty in attempts "*to free the brain from the shackles of deterministic law*" and raised doubts as to whether chaos-induced amplification of Heisenbergian indeterminism could be specific enough to provide free will.

Daniel Wegner[38] concluded as follows that our conscious will and feeling of selfhood is all an illusion: "*The fact is, it seems to each of us that we have conscious will. It seems we have selves. It seems we have minds. It seems we are agents. It seems we cause what we do. Although it is sobering and ultimately accurate to call all this an illusion, it is a mistake to conclude that the illusory is trivial. On the contrary, the illusions piled atop apparent mental causation are the building blocks of human psychology and social life. It is only with the feeling of conscious will that we can begin to solve the problems of knowing who we are as individuals, of discerning what we can and cannot do, and of judging ourselves morally right or wrong for what we have done…**Our sense of being a conscious agent comes at a cost of being technically wrong all the time. The feeling of doing is how it seems, not what it is - but that is as it should be. All is well because the illusion makes us human*** (present author's emphasis)."

The notions that we do not really cause what we do and that we are essentially living lives of illusion are not much of a basis for expanding the faith-science quest for truth or even the science-only quest for

---

37  Clarke, P.G.H. 2010. Determinism, brain function and free will. *Science and Christian Belief* 22:133-149.p.141, 142, 145. I acknowledge with thanks Peter Clarke's help in furthering my appreciation of this debate and pointing me to important literature.
38  Wegner, D. 2002. *The Illusion of Conscious Will.* The MIT Press: Cambridge MA. 405p.p.341-342.

# FREE THOUGHT: SCOPE AND FOUNDATIONS

truth. If it is *"technically wrong"* but feels humanly right to ask the question whether a human being has free will, what would the answer be worth, even if we could get one? This whole perspective seems to me like a road to nowhere, in the contexts of science and faith. My experiences cause me to conclude that humans have free will, which is exercised through what I call Free Thought. On that basis, I can move to suggest a possible model and mechanisms.

# CHAPTER 4

# FREE THOUGHT: MODEL AND MECHANISMS

*"I will praise thee; for I am fearfully and wonderfully made: marvellous are thy works; and that my soul knoweth right well."*[1]

### Combining Mental Reasoning and Soul Processing

Free Thought would have to be *"wonderfully made"* in order to be an integrated and iterative process, combining soul processing of and responses to information from the spiritual realm with mental reasoning about and responses to information from the material realm.

Reasoning is an individual's mental processing of information so as to acquire understanding and/or form judgments. She/he obtains information for reasoning from whatever she/he senses in the material realm and whatever is passed to the mind from the soul. The information processed in the soul comprises spiritual revelations from God and/or the spiritual force for evil and further information passed to the soul from the mind.

---

[1] Psalm 139: 14

# FREE THOUGHT: MODEL AND MECHANISMS

A threatened animal makes decisions through basic thought about whether it will snarl or keep silent, fight or flee. An expression of human anger can be an animalistic outcome of basic thought or an outcome of Free Thought, brewed in the soul as well as the mind. A rapid decision to snap back at a threatening stranger is an outcome of basic thought. Premeditated abuse between partners and insults hurled by racists are outcomes of Free Thought.

Antonio Damasio[2] distinguished between *"conscious deliberation"* and *"lightning-speed choices* (based on) *common knowledge"* and stated the following: *"The emotional action program we call fear can get most human beings out of danger, in short order, with little or no help from reason."* He concluded that: *"the reasoning system evolved as an extension of the automatic emotional system, with emotion playing diverse roles in the reasoning process...* (and that) *Emotion also assists with the process of holding in mind the multiple facts that must be considered in order to reach a decision."*

I agree that some of our basic thought is an *"emotional action program"* and that emotion, which can also be called feeling, provides information that enters the mix for an individual's Free Thought. However, emotion and feelings are generic terms. They encompass not only objectively measurable, material realm biochemical and electrical indicators but also subjectively felt states, derived from spiritual revelations received and spiritual responses made.

Most believers know that they receive and respond to spiritual revelations, though there are major differences of opinion about how soul processing of spiritual revelations might affect mental reasoning and vice-versa. For example, Douglas Groothuis[3] disagreed with William Willimon's message that Christian truth is found not through clear

---

2 Damasio, A.2005. *Descartes' Error: Emotion, Reason and the Human Brain.* Penguin Books: London.312p.p.xi-xii.
3 Groothuis, D. 2000.*Truth Decay: Defending Christianity Against the Challenges of Postmodernism.* Intervarsity Press: Downers Grove IL.303p.p.145.

thinking, but as a gift from the Holy Spirit. Groothuis stated: "*This is a false dichotomy. The Spirit can lead us to truth through clear thinking… God himself says through Isaiah to his covenanted people, 'Come let us reason together'.*" I agree with Groothuis.

During Free Thought, an individual has dialogues with God and the spiritual force for evil, through soul processing. An episode of Free Thought starts with pre-existing states of mind and soul and ends with outcomes in which those states have been changed to varying extents. Free Thought episodes can sustain or change a soul's state of faith or unbelief.

A state of mind is a combination of past and ongoing basic thought and Free Thought. A state of mind is contained within the brain and is a material realm state. As such, it is subject to the laws of classical physics, chaos theory and quantum theory. A state of soul is contained within the soul and is a spiritual realm state. The soul is the spiritual core of self - the receiver, processor and transmitter of spiritual information. Each state of soul and each state of mind about the higher things of life reflect the outcome of the previous episode of Free Thought and serve as baselines for the next episode.

Antonio Damasio[4] defined two kinds of human consciousness as follows, in which the emphases are his: "*The minimal scope kind I call* **core** *consciousness, the sense of the here and now, unencumbered by much past and by little or no future. It revolves around a core self and is about personhood but not necessarily identity. The big scope kind I call* **extended** *or* **autobiographical** *consciousness* (which) *manifests itself most powerfully when a substantial part of one's life comes into play and both the lived past and the anticipated future dominate the proceedings. It is about both personhood and identity. It is presided over by an autobiographical self.*" What Damasio called "*big scope,*" "*extended,*" or "*autobiographical*"

---

4   Damasio, A. 2010. *Self Comes to Mind: Constructing the Conscious Brain.* William Heinemann: London.367p.p.168-169, 270-271.

# FREE THOUGHT: MODEL AND MECHANISMS

consciousness would include what I call a state of soul.

Damasio described as follows what he saw as the commonalities between learning practical skills and moral behaviour: *"When we walk home thinking about the solution to a problem rather than about the route we take, but still do get home safe and sound, we have accepted the benefits of a nonconscious skill that was acquired in many previous conscious exercises, following a learning curve...the conscious-unconscious interplay also applies in full to moral behaviors. Moral behaviors are a skill set, acquired over repeated practice sessions and over a long time, informed by consciously articulated principles and reasons but otherwise 'second-natured' into the cognitive unconscious."*

Damasio's distinction between the skills of getting home safely and the skills relating to higher things, such as making moral choices, resembles my distinction between basic thought and Free Thought. I agree that some of our moral behaviour is reinforced through practice. However, I must put our internal moral *choice making* on a different level from making choices about our biological needs. The former are made through Free Thought and the latter through basic thought. What Damasio called *"second natured"* is what I would call: 'Checked against the Universal Moral Code, implanted by God in the soul.'

## Turning to Quantum Theory

Quantum theory has been invoked in all manner of pseudoscientific attempts to explain the allegedly paranormal. I am aware of the pitfalls in turning to quantum theory to attempt explanations for anything of an allegedly spiritual nature. For example, some believers invoke quantum theory to explain how God intervenes in the material realm in order to accomplish His will and/or to respond to believers' requests for specific material outcomes and even miracles.

# FREE THOUGHT, FAITH, AND SCIENCE

Ian Barbour[5] summarized as follows William Pollard's statement of that mistaken position: *"God…determines which actual value is realized within the range of probability distribution…Since an electron in a superposition of states does not have a definite position, no force is required for God to actualize one among the set of alternative potentialities. By a coordinated guidance of many atoms, God providentially governs all events. God, not the human mind, collapses the wave function to a single value."* Barbour[6] rejected that explanation and concluded that: *"the ideas of divine self-limitation and process theology are more consistent with both scientific evidence and central Christian beliefs."* I agree with Barbour.

I believe that any quantum observer - a human self, or God, or the spiritual force for evil - can collapse a quantum wavefunction and bring an observed state or event into existence from a pre-existing menu of probabilities (beables). I do not believe that God collapses quantum wavefunctions to highly improbable states in order to do miracles that contravene the laws of nature.

In quantum theory, Heisenberg's Uncertainty Principle states that we can *either* know where an electron is *or* we can know its momentum (i.e., what it is doing), but we cannot know *both* at the same time. For present purposes, it will suffice to recognize the wide acceptance and importance of quantum theory for explaining how the material realm works at the scale of fundamental particles. I make no attempt to explore the main schools of thought in quantum theory, beyond naming them as follows, omitting their variations, and I apologize if my inexpert summaries give any offence to experts:

    I. *Niels Bohr and his colleagues' and followers' Copenhagen interpretation, which emphasizes states of becoming and links measurements to outcomes that could be just about anything;*

---

5  Barbour, I.G. 1997. *Religion and Science: Historical and Contemporary Issues.* HarperOne: New York.368p.p.187-188. First published in 1990 as the 1990-1991 Gifford Lectures. Barbour cited Pollard, W. 1958.*Chance and Providence.* Charles Scribner's Sons: New York.

6  Barbour, I.G. 2000. *When Science Meets Religion: Enemies, Strangers, Or Partners?* HarperOne: New York. 205p.p.89.

# FREE THOUGHT: MODEL AND MECHANISMS

II. *David Bohm's deterministic interpretation, which proposes further hidden variables within quantum states; and*

III. *Hugh Everett's 'many worlds' interpretation, in which reality is said to comprise many parallel but disconnected 'worlds,' with every possibility happening in its own particular 'somewhere.'*

Henry Margenau[7] stated: "*some fields, such as the probability field of quantum mechanics, carry neither energy nor matter... The probability field is compatible with the existence of paths and therefore also real, but only in a derivative, secondary sense. In quantum theory, where there is no path, probabilities take on the character of primary, of ultimate concepts or observables... Probability fields therefore take the place of paths, and if the latter were regarded as real in classical mechanics the former deserve the attribute of reality in the domain ruled by quantum theory.*" He concluded the following: "*In very complicated physical systems such as the brain, the neurons and sense organs, whose constituents are small enough to be governed by probabilistic quantum laws... The mind may be regarded as a field in the accepted physical sense of the term. But it is a non-material field, its closest analogue is perhaps a probability field.*"

John Eccles[8] followed Margenau's work and tried to show that synaptic events in the brain take place at the quantum scale. Eccles then worked with Friedrich Beck[9] to propose a quantum mechanical model for exocytosis - the all-or-nothing discharge of the neurotransmitter substance. Peter Clarke[10] discussed and extended the subsequent criticisms of the Beck-Eccles model, including the huge mismatches between the scale for Heisenbergian uncertainty and the scale, time and

---
7   Margenau, H. 1984. *The Miracle of Existence*. Oxbow Press: Woodbridge CT.143p.p.22, 91, 96-97.
8   Eccles, J.C. 1986. Do mental events cause neural events analogously to the probability fields of quantum mechanics? Abstract. *Proceedings of the Royal Society of London B*, Volume 227, no. 1249: 411-428.
9   For example: Beck, F. and J.C. Eccles. 1992. Quantum aspects of brain activity and the role of consciousness. *Proceedings of the National Academy of Sciences of the USA*.89: 11357-11361.
10  Clarke, P.G.H. 2010. Determinism, brain function and free will. *Science and Christian Belief* 22:133-149.

energy needed for anything to be effected in neural processes.

In response, Michael Brownnutt[11] suggested that superposition might remove the problems of scale - as in the quantum coherence proposed as explanations for the high efficiency of bacterial photosynthesis and the magnetic compasses of birds. Clarke[12] urged caution in extending such proposals to brain function, but added: *"if quantum phenomena could be shown in brain neurons, this could have far-reaching implications for our understanding of brain determinism and conscious thought more generally."*

Modelling Free Thought requires inclusion of the spiritual realm. We can consider the possibility that the quantum world of the material realm extends into or has a counterpart in the spiritual realm. In quantum theory concerning the material realm, non-material probabilities for material 'beables' become observed material events and states. Spiritual probabilities for spiritual beables, which are not only non-material but also non-materially-derived, might become experienced spiritual revelations and responses by similar mechanisms.

## Model and Mechanisms

I envisage Free Thought as the iterative sorting and recombination of information from the spiritual and material realms, through the mind-soul interface. All the information inputs, mixes and outputs would have to be available in one or more common non-material formats. Using a common non-material language, mind probability fields affiliated with neural events and states could then 'talk to' soul probability fields affiliated with soul events and states, and vice-versa. The mind and soul could then dine together from shared menus of probabilities for neural and soul events and states, with mind and soul changing

---

11  Brownnutt, M. 2012, Response to Peter Clarke on 'Determinism, Brain Function and Free Will.' *Science and Christian Belief* 24: 81-86.
12  Clarke, P.G.H. 2010. Indeterminism beyond Heisenberg. *Science and Christian Belief* 24:85-86.

# FREE THOUGHT: MODEL AND MECHANISMS

dynamically, until a Free Thought outcome eventuated. From shared quantum probability fields, the self would have chosen paths from probabilities, in mind and soul and across the mind-soul interface.

The superposition of quantum wavefunctions affiliated with the material realm (mind) and the spiritual realm (soul) might account for part or all of that process. From superpositioned soul, neural and shared quantum probability fields, new beables could be brought into existence as observeds, by the individual and/or God and/or the spiritual force for evil. All of that activity could relate to its predecessors - locally and across the material realm universe and the spiritual realm, through quantum entanglement. Free Thought would then be an integrated process in an integrated human self, receiving information from the material and spiritual realms and making responses in both.

A Free Thought episode can leave the individual with changed states of mind and/or soul. Thereafter, the material realm-influenced, God-influenced and spiritual force for evil-influenced, but still sovereign, self has impacts in the material realm - on fellow humans and the environment, by moral or immoral behaviour and the use or misuse of creativity. By suggesting that God acts through quantum probability fields in our individual Free Thought, I am not limiting His capacity to influence a human self. God stands at the door of every soul and knocks.[13] The sovereign self, which is also influenced by the spiritual force for evil, decides whether to open or not, by the process of Free Thought.

The same model explains divine action in human affairs, as God and the spiritual force for evil use *us* as their material realm agents. It also explains how we respond to God and to the spiritual force for evil as we develop and use our creativities. It provides for God, the spiritual force for evil and the individual soul to act as co-observers - calling spiritual beables into existence sequentially, as revelations and responses, in an

---
13   Revelation 3: 20

iterative manner and linked to a similar system in the material realm, with dynamic sorting and mixing at the mind-soul interface.

Figure 1 A-F builds the model and mechanisms for Free Thought. Figure 2 shows the three cycles of stimuli and responses through which Free Thought obtains information and responds, in the spiritual and material realms.

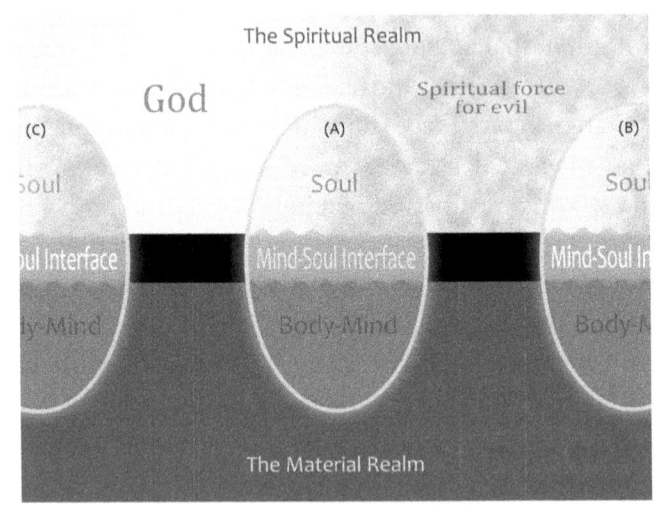

Figure 1A. The barrier between the spiritual and material realms is crossed only within individual humans. God and the spiritual force for evil inhabit the spiritual realm. Humans, depicted as persons A, B and C, have material realm body-minds and spiritual realm souls, which overlap at their mind-soul interfaces.

# FREE THOUGHT: MODEL AND MECHANISMS

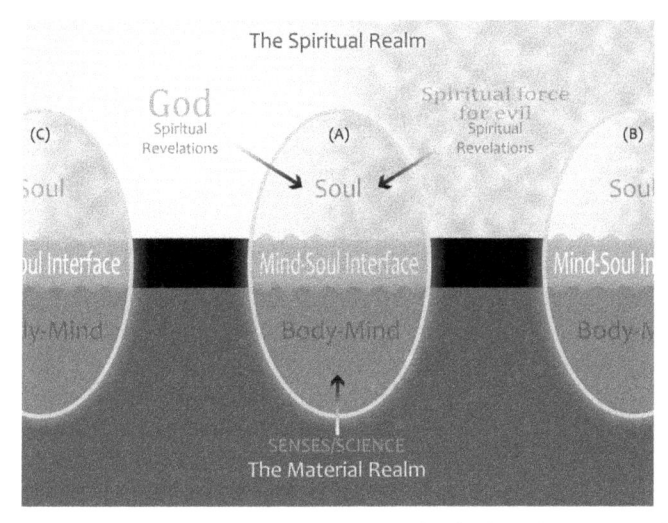

Figure 1B. As shown only for person A, each individual soul receives information from the spiritual realm, as spiritual revelations from God and from the spiritual force for evil, and each individual body-mind receives information from the material realm, including disclosures through science, via the senses.

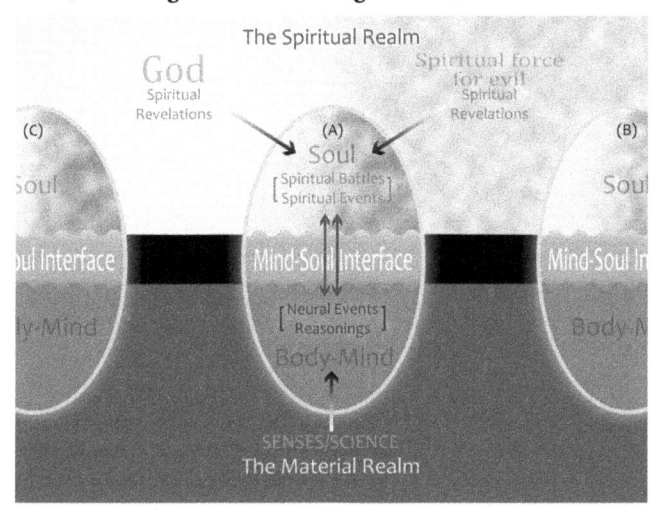

Figure 1C. As shown only for person A, information from God and the spiritual force for evil comes for processing in the soul, from which information about spiritual events and states passes through the mind-soul interface to the mind. Information about neural events and states in the mind passes through the mind-soul interface to the soul. Information exchange across the mind-soul interface is iterative, enabling the mixing of spiritual realm (soul) and material realm (body-mind) information.

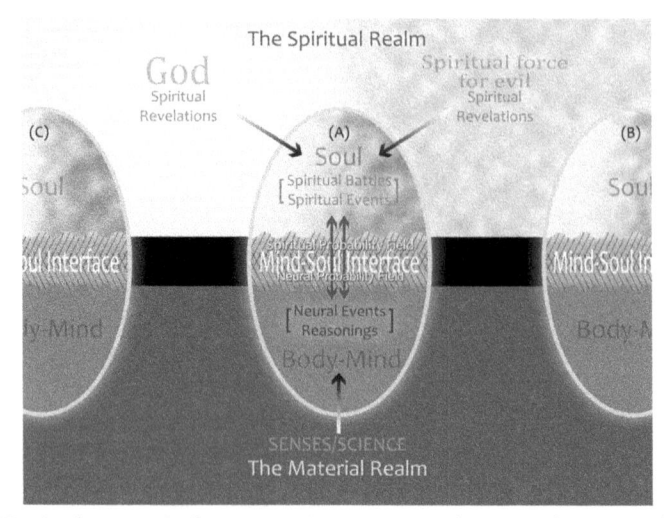

Figure 1D. As shown only for person A, iterative exchange and mixing of spiritual realm information (soul events and states) and material realm information (neural events and states) at the mind-soul interface is accomplished in one or more common non-material formats, here termed as spiritual and neural probability fields.

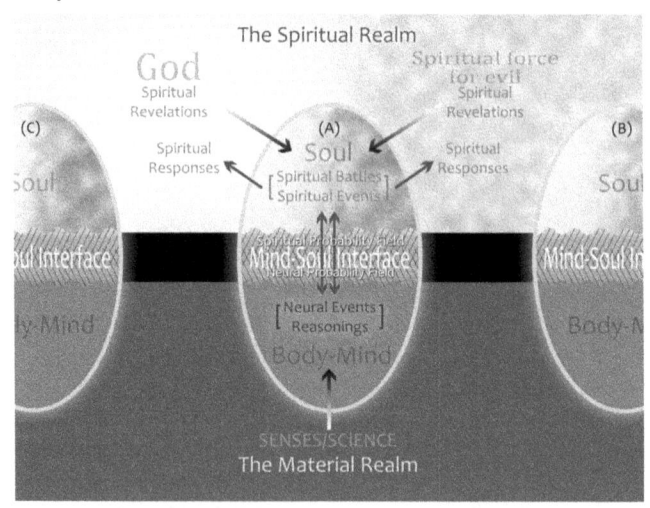

Figure 1E. As shown only for person A, the iterative exchange and mixing of information at the mind-soul interface leads to responses from the soul back to God and/or the spiritual force for evil. The responses bring more spiritual revelations from either or both sides.

# FREE THOUGHT: MODEL AND MECHANISMS

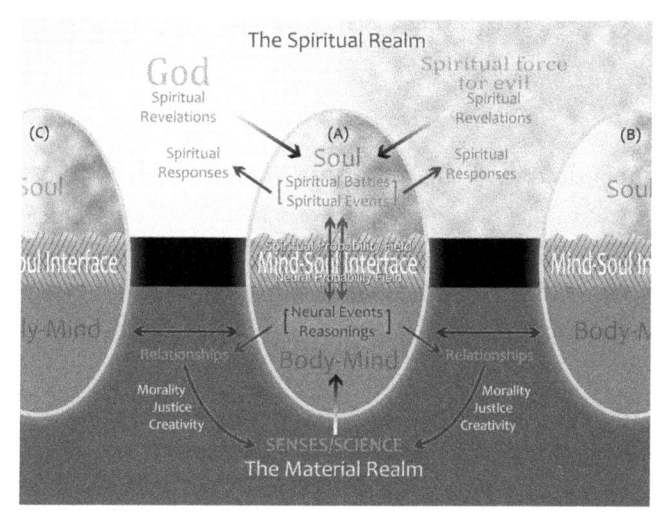

Figure 1F. The whole process of Free Thought is shown only for person A. Information from the material realm, including disclosures through science, is received by the senses and passes to the mind for reasoning. In the soul, revelations are received from God and the spiritual force for evil. The results of mental reasoning (neural events and states) and soul processing (spiritual events and states) are exchanged and mixed iteratively at the mind-soul interface. The body-mind's actions reflect the outcomes of Free Thought. There are three interlinked stimulus and response processes in this open system: spiritual revelations from God and the spiritual force for evil and the soul's responses; and the individual/interpersonal processes of experiencing and acting in the material realm.

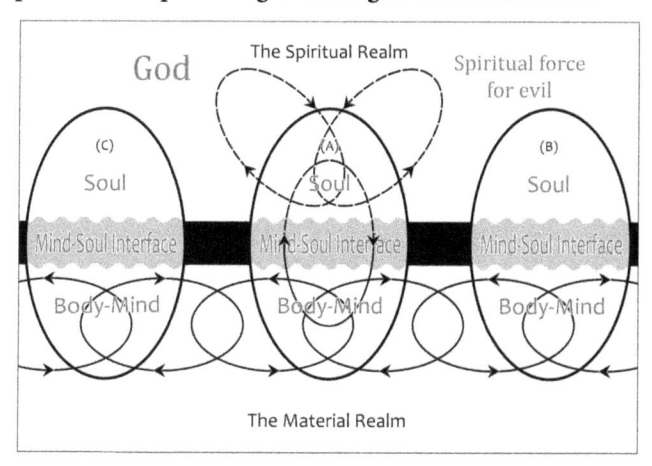

Figure 2. The whole process of Free Thought is shown only for person A. Internal and external relationships are summarized as three interlinked stimulus and response processes in an open system, spanning the material and spiritual realms.

# Free Thought, Faith, and Science

## Comparisons with Other Perspectives and Findings

### *The Quantum Observer*

Henry Stapp[14] described the role of the quantum observer as follows: *"The observer in quantum theory...chooses which question will be put to Nature: which aspect of nature his inquiry will probe.* Stapp called this: *"'The Heisenberg Choice', to contrast it with the 'Dirac Choice', which is the random choice on the part of Nature that* (Paul) *Dirac emphasized.*

Stapp continued: *"According to quantum theory, the Dirac choice is a choice between alternatives that are specified by the Heisenberg choice: the observer must first specify what aspect of the system he intends to measure or probe...In quantum theory, it is the observer who both poses the question and specifies the answer. Without some way of specifying what the question is...the quantum process grinds to a halt."*

Stapp noted further that: *"the patterns of brain activity actualized by an event unfold not only into instructions to the motor cortex to institute intended motor events. They unfold also into instructions for the creation of conditions for the next experiential event."* He held that Heisenbergian uncertainties at nerve terminals and throughout the brain would: *"necessarily engender a quantum diffusion in the evolving state of the brain...* (so that) *the dynamically generated state that is the pre-condition for the next question will not correspond to a well-defined unique question: some 'scatter' will inevitably creep in."*

Despite that apparent problem, the individual still poses her/his next question. The required Heisenberg choice is made somehow. Stapp proposed that each experience has an *"intentional aspect,"* which is its experiential goal or aim, and an *"attentional aspect,"* which is an

---

14  Stapp, H.P. 2004. *Attention, Intention, and Will in Quantum Physics*, p.143-164.In B. Libet, A. Freeman and K. Sutherland (eds.) *The Volitional Brain*. Imprint Academic: Exeter, UK.p.153-154, 157-158. First published in 1999 in the Journal of Consciousness Studies 6 (8-9): 143-164. On p.155, Stapp quoted from p.1062 in William James' 1910 *Some Problems in Psychology*; Chapter X in William James' *Writings 1902-1910*; published in 1987 by the Library of America: New York.

# FREE THOUGHT: MODEL AND MECHANISMS

experiential focussing on an updating of the current status of: *"the person's idea of his body, mind and environment."* An *"appropriate question"* can then be asked.

For Stapp, that allowed what he called a way of: *"closing the causal gap associated with the Heisenberg Choice* (by introducing) *two parallel lines of causal connection in the mind/brain/body system…the physical line that unfolds - under the control of the local deterministic Schrödinger equation - from a prior event…that generates the physical* **potentialities** (original author's emphasis) *for succeeding possible events…*(and) *a mental line of causation that transfers the experiential intention of an earlier event into an experiential attention of a later event. These two causal strands, one physical and one mental, join to form the physical and mental poles of a succeeding quantum event. In this model, there are three intertwined factors in the causal structure: (1), the local causal structure generated by the Schrödinger equation; (2), the Heisenberg Choice, which is based on the experiential aspects of the body/brain/mind subsystem that constitutes a person; and (3), the Dirac choice on the part of nature."*

Stapp's model was for a soul-less 'subsystem' of a person, but I can see in its recipe for iterative sharing of physical and mental information a resemblance to the iterative sharing of spiritual realm (soul event and state) and material realm (neural event and state) information that I am proposing for Free Thought. In Free Thought, the Dirac choice from what Stapp called *"nature"* must be based on whatever the material realm has on offer. If nature also includes whatever the spiritual realm has on offer, then its Dirac choices impact the mind and the soul. The Heisenberg Choices of a free-willed self then determine material and spiritual events and states, culminating in an outcome of Free Thought.

Lee Rozema and others[15] showed that whereas Heisenbergian uncertainty prevails throughout the quantum world, *weak* measurements

---

15  Rozema, L.A., Darabi, A., Mahler, D.H., Hayat, A., Soudagar, Y. and A.M. Steinberg. 2012. Violation of Heisenberg's measurement-disturbance relationship by weak measurements. *Physical Review Letters* 109, 100404.5p.

allow results that violate Heisenberg's original measurement-disturbance relationship. Since the 2012 Nobel Prize in Physics was awarded to Serge Haroche and David Wineland for their progress in studying simple quantum systems of single photons or ions and for paving the way to quantum computing, the door has opened wider for further studies on quantum superposition states and weakly 'nudging' them to new probabilities and new outcomes.

All of the above is of course material realm stuff, but I can still speculate that some measurements in a human mind and/or soul might be necessarily weak. God and/or the spiritual force for evil might be giving to our souls 'spiritual nudges' that are very weak by any material realm comparisons, but that are still able to predispose for and continue to influence our soul states, prior to and during our Free Thought episodes. A weakly nudged soul could nudge weakly the state of a mind-soul interface, which was getting weak nudges from a mind.

A small change from a weak spiritual nudge in a soul might impact a mind-soul interface and therefore change a mind. A small change in a mind might impact a mind-soul interface and cause a weak nudge to be passed to a soul. There could be roles for weak measurements/nudges in the processing of non-material forms of information on either side of and within the mind-soul interface.

The weak measurements/nudges could be allochthonous (originating from God and/or the spiritual force for evil, in the spiritual realm external to the self) and autochthonous (originating from the mind and/or soul within the free-willed self). Free Thought might therefore involve weak observations of and/or weak nudges to quantum states in the soul, plus stronger self-made choices that determine outcomes across mind and soul. All of this is of course pure speculation.

The notion that God and the spiritual force for evil might be making weak measurements and giving weak nudges to our souls is not likely to

# FREE THOUGHT: MODEL AND MECHANISMS

go down well with nonbelievers, or indeed with believers who see God as omnipotent and the spiritual force for evil as occasionally powerful, though ultimately defeated. Nevertheless, what might be considered as a weak measurement or nudge in material realm terms could very well be a powerful one in spiritual terms. Divine and diabolical self-limitations might be needed to make connections between the spiritual and material realms connections in us - and nowhere else.

## *Dual-Aspect Monism*

John Polkinghorne[16] wrote in support of a: *"more ambitious metaphysical programme represented by **dual-aspect monism** (original author's emphasis)...(in which) the duality of energy and information that science is beginning to embrace might prove to be part of a movement that takes with equal seriousness our basic human experiences of physical embodiment and of personal agency...it could also refer to a physical world within whose open grain it would be fully conceivable that the God who is that that world's Creator is providentially at work through the input of active information."*

The model and mechanisms that I propose for Free Thought have much in common with Polkinghorne's positions, especially his recognition that God has Self-limited His interventions in the material realm. Polkinghorne conceived the physical world (material realm) as having an *"open grain,"* which allows our explorations and God's interactions. I suggest that our mind-soul interfaces are the *only* places where that grain lies truly open - the only points of spiritual realm and material realm interaction.

In response to Ignacio Silva's[17] review of his positions on how God might act in the world, Polkinghorne[18] wrote: *"Various forms of 'causal*

---

16  Polkinghorne, J.C. 2005. *Exploring Reality: The Intertwining of Science and Religion.* Yale University Press: New Haven CT.181p.p.35-36.
17  Silva, I. 2012. John Polkinghorne on divine action: a coherent theological evolution. *Science & Christian Belief* 24(1): 19-30..
18  Polkinghorne, J.C. 2012. Divine action - some comments. *Science & Christian Belief* 24(1): 31-32.

*joint' were proposed. God was seen as either the extraphysical determinator of quantum outcomes or as the selector of the pattern in which a chaotic system traversed its strange attractor, determined by a divine input of pure information."* Polkinghorne explained as follows why he saw chaos as a likely mechanism for the causal joint: *"because of its manifest macroscopic relevance, while knowing, of course, that the* **mathematical** (original author's emphasis) *theory of chaos is derived from deterministic equations."* He continued: *"However, we know that these equations are themselves only an approximation to what we believe to be a more subtle and supple physical reality…we are not in a position to give a fully accurate account of the details of agency, either human or divine. What we can say is that appeal to the models of causal joint explanation show that these possibilities are not excluded by what an honest science can actually tell us about the physical process."*

Roger Penrose[19] wrote the following on chaos: *"Although ordinary chaotic systems are completely deterministic and computational they can,* **in practice** (original author's emphasis), *behave as though they are not deterministic at all. This is because the accuracy according to which the initial state needs to be known, for a deterministic prediction of its future behaviour, can be totally beyond anything that is conceivably measurable."* He also stated: *" 'pure randomness' indeed does nothing useful for us… (and that) if anything it would be better to stay with the pseudo-randomness of chaotic behaviour."*

Chaos has attractions for explanations of the material realm and divine interventions, but does the writ of chaos really run so comprehensively? I could be wrong, but I find it easier to envisage quantum probabilities and outcomes pervading the entire material and spiritual realms.

Whatever might be the mechanism or mix of mechanisms in Free Thought, the spiritual revelations received by a soul, from God and

---

19  Penrose, R. 2005. *Shadows of the Mind: A Search for the Missing Science of Consciousness.* Vintage Books: London.457p.p.21, 26.

# FREE THOUGHT: MODEL AND MECHANISMS

the spiritual force for evil, cannot be random. They reflect the wills of their senders and they are likely to be tailored to an existing state of soul. When we add to that scenario the sovereign will of the self and her/his state of mind, we have all the ingredients for an episode of Free Thought in which nothing can really be called random.

### CONSCIOUSNESS AND THE BIG PICTURE

Stuart Kauffman[20] described his theory of consciousness as follows, in which my clarifications and comments are added in brackets: "*I will base my theory* (of consciousness) *on the view of decoherence as due to interaction of a quantum system* (i.e., an infinite menu of probabilities/probability fields/wavefunctions) *with a quantum environment* (i.e., mind and, if nonbelievers will allow, soul) - *or a quantum plus classical environment — perhaps something like a quantum oscillator bath* (i.e., quantum level neural states and brain/nervous tissue), *the loss of phase information* (i.e., the selection of particular beables from the infinite menu), *and the emergence of classical behaviour* (i.e., brain and nervous tissue following the laws of classical physics above the quantum level)."

Kauffman restated his theory as follows, with his emphasis throughout: "*...the conscious mind is a persistently poised quantum coherent-decoherent system, forever propagating quantum coherent behaviour, yet forever decohering to classical behaviour...mind - consciousness, res cogitans - is identical with quantum coherent immaterial possibilities, or with partially coherent quantum behaviour, yet via decoherence, the quantum coherent mind has consequences that approach classical behaviour so very closely that the mind can have consequences that create actual physical events by the emergence of classicity. Thus, res cogitans has consequences for res extensa! Immaterial mind has consequences for matter.*"

---

20  Kauffman, S.A. 2008. *Reinventing the Sacred: A New View of Science, Reason, and Religion.* Basic Books: New York.320p.p.177, 209.

# Free Thought, Faith, and Science

I prefer the descriptor non-material to immaterial. If the soul is added to the above scenario as another *"immaterial"*/non-material/spiritual component of self and as part of consciousness, then Kauffman's *"poised quantum coherent-decoherent system"* and *"quantum oscillator bath"* look similar to my model and mechanisms for Free Thought. Both are open systems, working through integrated and iterative mechanisms. Both could work through changes in quantum probability fields and related outcomes. Kauffman's model links quantum physics to classical physics. My model indicates the only link and portal for interactions between the spiritual and material realms.

David Bohm[21] proposed an *"unbroken wholeness,"* expressed most deeply by: *"implicate or enfolded order…(where) space and time are no longer the dominant factors determining the relationships of dependence or independence of different elements. Rather, an entirely different sort of basic connection of elements is possible, from which our ordinary notions of space and time, along with those of separately existent particles, are abstracted as forms from the deeper order."* Bohm regarded those *"ordinary notions"* as: *"the explicate or unfolded order, which is a special and distinguished form contained within the totality of all the implicate orders."*

Those *"ordinary notions"* and that *"special and distinguished form"* are what we experience as conscious beings; in other words, our accessible reality. If everything spiritual is deleted from my model and mechanisms for Free Thought, Bohm's big picture fits well with the remaining soul-less, mind-only, material realm-only human condition, in which quantum wavefunctions are collapsed and other forms of material realm information are processed in various ways, with or without a full understanding of their nature.

Putting back the spiritual realm, the soul, God, the spiritual force for evil, and a strong barrier between the spiritual and material realms,

---

21  Bohm, D. 2002. *Wholeness and the Implicate Order.* Routledge Classics: London.284p.p.xviii. First published in 1980.

# FREE THOUGHT: MODEL AND MECHANISMS

crossed only at mind-soul interfaces, gives a much bigger dualistic picture. Nevertheless, Bohm's *"unbroken wholeness,"* comprising explicate/unfolded and implicate/enfolded order, need not conflict with my Free Thought model if the largely hidden nature of his implicate/enfolded order is seen as extending throughout the material and spiritual realms, with our soul states providing (subjective) evidence for the latter.

The sensed material realm and soul-inhabited spiritual realm are *one* self-experienced reality. The spiritual and material realms interact at every mind-soul interface. In my Free Thought model, God, the spiritual force for evil and the soul make spiritual beables into spiritual observeds, while the body-mind makes material beables into material observeds.

## Postscript

Please try this Free Thought experiment. Have in mind a basic moral question such as the following: *Do I support capital punishment as the morally justified penalty for a premeditated murder committed during the course of a robbery?* Before you try in earnest to answer that question, please ask yourself this preliminary question: *In seeking an answer to this moral question, will I be reasoning about any information from the material realm?*

I assume that you have answered 'Yes' and that you perceive yourself as truly existing in the material realm and as fully conscious. Now please ask yourself this further preliminary question: *In seeking an answer to this moral question, will I also be reasoning about any past and present revelations concerning this issue that could come from a spiritual source; such as guidance from a deity?* Your answer to this can be a 'Yes' or a 'No' or a 'Maybe.' Any of those answers will make you think further about how you make your moral choices and your status as a believer or a nonbeliever.

## Free Thought, Faith, and Science

Please now ask yourself the original moral question about support for capital punishment, as the penalty for a premeditated murder committed during a robbery. Your answer will again be a 'Yes' or a 'No' or a 'Maybe.' When you have decided on your answer, and irrespective of whatever answers you gave to any questions so far, please ask yourself now the following final question: *What did I feel during the process that I used to reach my final answer?* Try to describe your experience in writing. Answers to any and all of these types of questions can be sought from individuals in a large and diverse population and then analyzed for possibilities of significant clustering.

## CHAPTER 5

# FAITH

*"faith is the substance of things hoped for, the evidence of things not seen"* [1]

### What Is Faith?

The writer of the Epistle to the Hebrews described faith as providing evidence for the existence of things beyond the material realm. That evidence is the believer's subjective evidence for her/his personal experiences of God. One cannot have a two-way relationship with something that does not exist.

Some define faith as a feeling that all will be well or as well as can be expected, as long as one strives to do the right thing. Others define faith as religion in general or as one particular organized religion, such as Christianity or Islam.

Faith is defined here as personal belief and trust in God. Believing in something means being certain that it is true. Trusting in something means having complete confidence in its reliability. Belief in God is the foundation of faith. Trust in God is the believer's way of life.

---

[1] Hebrews 11:1

# Free Thought, Faith, and Science

A popular Christian hymn states: *"Only believe and thou shalt see that Christ is all in all to thee."*[2] *"Only believe"* and a life walked with God will follow. It will be a changed life, rich in spiritual benefits that will last forever. If believers had not experienced that reality, faith would have died out long ago.

Faith is usually accompanied by theism, but a deist who trusts God only to welcome her/his soul after death and expects no divine interventions of any kind during life on Earth also has faith. Belief in God as a mathematical probability or as a prudent bet in game theory is not a state of faith, because trust is lacking.

In faith and unbelief, Free Thought enables each of us to seek for truth and to decide what we believe and trust. Choosing faith does not require belief in all of the angels, demons, ghosts, intercessory saints, miracles and other supernatural happenings described in sacred texts and religious dogma. Choosing unbelief does not mean choosing to ridicule everything concerning spirituality.

Terry Eagleton[3] wrote about faith and choice as follows: *"Faith…is not in the first place a matter of choice. It is more common to find oneself believing something than to make a conscious decision to do so - or at least to make such a conscious decision because you find yourself leaning that way already. This is not…a matter of determinism. It is rather a question of being gripped by a commitment from which one finds oneself unable to walk away…The Christian way of indicating that faith is not in the end a question of choice is the notion of grace. Like the world itself from a Christian viewpoint, faith is a gift."*

With all due respect to Eagleton, I disagree. I cannot accept that faith and unbelief are states in which a self has become enmeshed and has no option but to accept. In particular, I do not believe that

---
2   Monsell, J.S.B. 1863. Words from the hymn, *Fight the Good Fight*.
3   Eagleton, T.2009. *Reason, Faith, and Revolution: Reflections on the God Debate*. Yale University Press: New Haven CT.185p.p.137-138.

# FAITH

one makes a choice for faith when covered by God's grace and left with no other option.

If choosing faith was like that, it would indeed be *"a matter of determinism"* - God's determinism, as in the doctrine of predestination. I hold that faith and unbelief are always *choices*, made by a human self, through Free Thought. Many believers and nonbelievers can explain well why they chose faith or unbelief.

According to Michael Shermer:[4] *"We form our beliefs for a variety of subjective, personal, emotional, and psychological reasons in the context of environments created by family, friends, colleagues, culture and society at large; after forming our beliefs we then defend, justify, and rationalize them with a host of intellectual reasons, cogent arguments, and rational explanations. Beliefs come first, explanations for beliefs follow. I call this process* **belief-dependent realism** (original author's emphasis), *where our perceptions about reality are dependent on the beliefs that we hold about it."*

I agree that we form our beliefs on a *"subjective"* basis and in the midst of complex, diverse and dynamic *"environments."* I agree also that we like to defend our self-chosen beliefs to others. But we are always sovereign selves. Moreover, what we say to others about our beliefs need not match what we truly believe and will always be partial explanations, compared with our internal explanations to ourselves. No one knows what anyone else truly believes. The outcomes of our episodes of Free Thought are complete in themselves until we embark on further episodes. They are our private repositories of our choices.

Testimonies from believers and nonbelievers defeat the argument that something akin to brainwashing - by family, friends and authorities at church and school - determines our lifetime choices of faith and

---

[4] Shermer, M.2011. *The Believing Brain: From Ghosts and Gods to Politics and Conspiracies - How We Construct Beliefs and Reinforce Them as Truths.* Time Books: New York.383p.p.5.

unbelief. Mother Teresa of Calcutta[5] admitted to a crisis of faith lasting about 50 years. Just before his resignation, Pope Benedict XVI said: "*It seemed like the Lord was sleeping.*" Daniel Dennett and Linda LaScola[6] studied five working Pastors who had become atheists. The ongoing "*Clergy Project - Moving Beyond Faith*" maintains a website[7] for "*clergy who do not hold supernatural beliefs.*"

## BELIEF WITHOUT DOUBT

Paul Tillich[8] described faith as follows: "*Faith has a cognitive element and is an act of the will. It is the unity of every element in the centered self.*" He also made the following case that faith and doubt must coexist: "*The doubt which is implicit in every act of faith...is the doubt which accompanies every risk. It is not the permanent doubt of the scientist, and it is not the transitory doubt of the skeptic, but it is the doubt of him who is ultimately concerned about a concrete content. It does not reject every concrete truth, but it is aware of the element of insecurity in every existential truth. At the same time, the doubt which is implied in faith accepts this insecurity and takes it into itself in an act of courage.*"

Alvin Plantinga[9] included the following in his description of the Calvinist position on faith: "*It is part of Calvinism to hold that Christians are not complete; they are in process. John Calvin himself...points out that believers are constantly beset by doubts, disquietude, spiritual difficulty, and turmoil. 'It never goes so well with us,*' he (Calvin) *says, 'that we are wholly cured of the disease of unbelief and entirely filled and possessed by faith'... There is an unbeliever within the breast of every Christian; in the believing mind, says Calvin, 'certainty is mixed with doubt'.*"

---

5    Van Biema. 2007. The Secret Life of Mother Teresa: Her Agony. *Time*. September 3, 2007:26-33.
6    Dennett, D. and L. LaScola. 2010. Preachers who are not believers. *Evolutionary Psychology* 8 (1): 122-150.
7    www.clergyproject.org
8    Tillich, P.2009.*Dynamics of Faith*. HarperOne: New York.147p.p.8-9.First published in 1957.
9    Plantinga, A.1998. Christian Philosophy at the End of the 20th Century, p.328-352. In J. Sennett (ed.) *The Analytic Theist: An Alvin Plantinga Reader*. William B. Eerdmans Publishing Company: Grand Rapids MI.369p.p.336: citing Calvin's *Institutes*, III, ii, paragraph 18.

# FAITH

With all due respect to all of the above and to anyone who believes that faith and doubt can coexist, I disagree. 'Belief,' 'trust,' and 'doubt' can of course be defined in different ways, but faith defined as belief and trust in God has no gradations and cannot coexist with doubt. One who trusts in the complete reliability of something has no doubts. Anyone can make any number of changes from unbelief to faith and vice versa during her/his lifetime, but her/his status at any given time will always be either believer or nonbeliever.

James Carse[10] argued that: "*there is a religious case to be made against belief. Belief has been defined as the place where we stop our thinking.*" John Patrick Shanley[11] wrote: "*There is the culture of doubt and the culture of dogma...Doubt keeps the doors and window open. Belief is one room with no way out...Doubt is not paralysis. Certainty is.*" Richard Dawkins[12] considered faith to be a: "*process of non-thinking...(which) demands a positive suspension of critical faculties.*"

I disagree with all of the above. Faith and unbelief are outcomes of Free Thought. Neither is a state of paralysis. Either can be changed by further Free Thought. I accept that those who criticize organized religion, rather than faith *per se*, can find grounds for accusing some of the so-called faithful of appearing to have stopped thinking, but the human race cannot be bisected artificially and neatly into fortunate doubters who are thinking critically and unfortunate believers who have allegedly stopped thinking.

In his excellent TV series on Darwin, Dawkins[13] stated: "*The more we discover...how petty our little private beliefs seem.*" Like Dawkins, I am amazed at the beauty and grandeur of the natural world and the authority and elegance of the natural laws that govern it. But an individual's

---
10   Carse, J.P. 2008. *The Religious Case Against Belief.* The Penguin Press: New York. 227p.p.146.
11   Shanley, J.P. 2007. I am, therefore I doubt. *International Herald Tribune.* Friday, February 9, 2007: p.7.
12   Dawkins, R. 2008. *The Root of All Evil.* Channel 4 TV. Available at: www.richarddawkins.net
13   Dawkins, R. 2008. *The Genius of Charles Darwin: God Strikes Back.* Channel 4 TV. Available at: www.richarddawkins.net

private beliefs are never "*petty.*" They define the state of a precious and immortal soul, in a unique life of free choice.

Science gives us wonderful experiences and explanations, but God has much more to say to us besides His truthful disclosures through science. Through science we learn more about the composition and workings of the material realm, but science says nothing about *why* the material realm exists and *why* we are motivated and equipped to ask.

Nonbelievers, especially scientists, regard those why questions as meaningless. They see no basis for hypothesizing that the existence of the material realm and the human condition might have some higher purpose. They hate the prospect of any unnecessary complexity and like to wield Occam's razor to prune away any talk and thought about the possible existence of the spiritual realm.

Occam's razor has helped the quest for truth about the composition and workings of the material realm, but it is not always applicable. Even with no hypotheses about the spiritual realm, our material reality is often highly and necessarily complex. For a mind-boggling illustration, take a look at the charts published by Boerhinger Mannheim and others[14] to depict our metabolic pathways.

Nicholas Everitt[15] noted the following possibility, which he distilled from William James'[16] famous essay on belief: *"in at least some cases, if you believe in an open proposition* (such as the existence of God), *then that can put you in a position to discover some hard evidence in favour of it, evidence that you could not have discovered if you had not initially believed it."*

In other words, a Free Thought trial balloon exploring the prospects

---

14  Available at: www.expasy.org
15  Everitt, N. 2004. *The Non-Existence of God*. Routledge: London.326p.p.201.
16  James, W. 1896.*The Will to Believe: An Address to the Philosophical Clubs of Yale and Brown Universities.* Transcribed by W. O'Meara in 1997.

# FAITH

for faith - 'God, are you there?' - might lead to making a choice for faith. I agree, but exploring faith is not yet faith. Faith and unbelief are always completed choices. Faith and unbelief are outcomes of Free Thought episodes. They become baselines in mind and soul for re-examination in subsequent episodes.

Bertrand Russell[17] criticized James and other pragmatists on the grounds that a so-called 'will to believe' or a 'right to adopt a believing attitude,' maintained against logic and intellect, is unlikely to be a path for finding truth. Russell stated: "*The precept of veracity, it seems to me, is not such as James thinks. It is, I should say: 'Give to any hypothesis just that degree of credence which the evidence warrants.' And if the hypothesis is **sufficiently important*** (present author's emphasis) *there is the additional duty of seeking further evidence.*"

I agree with Russell that an idea or hypothesis such as the existence of God cannot be deemed to be true simply by observing how well it seems to benefit human lives. However, subjective evidence from within, as believers and nonbelievers recount their personal experiences of faith and unbelief, is clearly a valid form of evidence for scientific research. The hypothesis that God exists is obviously "*sufficiently important.*"

## Faith and Organized Religion

Bhikku Payuto[18] held that there are two kinds of faith: the religious faith, which "*obstructs wisdom*" by requiring unquestioning belief in dogma, and the faith that is "*a channel for wisdom,*" because it "*stimulates curiosity.*" Some religious authorities obstruct wisdom from God, by preaching and teaching from willful ignorance, by denials of God's truthful disclosures through science, and by discouraging curiosity.

---

17  Russell, B. 2004. *History of Western Philosophy*. Routledge Classics: London.778p.p.727.
18  Payuto, B.P.A. 1993. *Toward Sustainable Science: A Buddhist Look at Trends in Scientific Development*. Translated by B.G. Evans. Buddhadhamma Foundation: Bangkok, Thailand. 175p.p.75-76.

Some nonbelievers like to ask believers why any alleged God or gods would have chosen tiny planet Earth as the place for humans to be brought into being and to face making a choice between faith and unbelief? No one can *know* with any certainty why God did that, but we can see *how* it became possible. After the Big Bang, conditions turned out just right for evolution of the cosmos and life on Earth. God created the physics and chemistry that led to life and a Goldilocks zone for us.

In the same vein, why would God have chosen, allegedly, to make His most important initial revelations to tribes in the Middle East? Why did He also choose, allegedly, to become Incarnate, as His Son Jesus Christ, in Roman-occupied Palestine? Again, no one can *know* with certainty why God did any of that, but His timing and choice of locations were excellent. Settled agriculture and associated community development began in the Middle East. Christ became God Incarnate at a major crossroads of commerce, cultural exchange and organized religion.

A key question concerning faith and organized religion is whether it is possible for a person to become a believer and to sustain faith without signing up to membership in an organized religion or church? I must answer 'Yes,' but most of my fellow Christians, together with most believers who are members of other organized religions would say 'No,' because they cannot envisage faith being sustained in isolation from a religious community.

I am sure that God can sustain the faith of any believer, from a prisoner in solitary confinement or a lone shipwrecked mariner to a member of a vibrant church family. It can be a small step from insisting that faith requires church membership to regarding all nonmembers as nonbelievers. Religious exclusivity breeds bigotry, discrimination and conflict.

# FAITH

The following paragraphs from the Catechism of the Catholic Church[19] express well some of the main aspects of a personal choice for faith and fit well with the concept of Free Thought:

I. *"(150) Faith is first of all a personal adherence of man to God.*

II. *(154) Believing is possible only by grace and the interior helps of the Holy Spirit. But it is no less true that believing is an authentically human act.*

III. *(155) In faith, the human intellect and will cooperate with divine grace: Believing is an act of the intellect assenting to the divine truth by command of the will moved by God through grace.*

IV. *(157) Faith is **certain** (original authors' emphasis).*

V. *(166) Faith is a personal act - the free response of the human person to the initiative of God who reveals himself."*

However, the Catholic Catechism also insists that no one can find and sustain faith without being a Church member, for example: "(166) *No one can believe alone*; (181) *Believing is an ecclesial act… The Church is the Mother of all believers. No one can have God as Father who does not have the Church as Mother.*" I disagree. No church can claim a monopoly on providing the conditions under which God can reach an individual and vice versa. Faith is not church membership. Faith is first, foremost and forever a *personal* relationship with God.

Christ is reported to have said: "*Not every one that saith unto me, Lord, Lord, shall enter into the kingdom of heaven, but he that doeth the will of my Father which is in heaven.*"[20] In one of his most powerful gospel

---

19  *Catechism of the Catholic Church.1994.Definitive Version: Based on the Latin Editio Typica.* Episcopal Commission on Catechesis and Catholic Education: Catholic Bishops Conference of the Philippines. Word & Life Publications: Makati City, Philippines.828p.
20  Matthew 7: 20

songs, Bob Dylan[21] asks whether Jesus will know us and welcome us when we meet him, or will say: "*I never knew you: depart from me.*" Faith is the way to get ready.

## Faith in the Quest for Truth

N.T. Wright[22] described faith as follows: "*To believe, to love, to obey (and to repent of our failure to do those things)…Christian faith isn't a general religious awareness. Nor is it the ability to believe several unlikely propositions.* **It is certainly not a kind of gullibility which would put us out of touch with any genuine reality** (present author's emphasis)." Faith *must* find something real. If I had not experienced God through faith, I would not have remained a believer.

Nonbelievers often argue that an allegedly happy state of faith is delusional and/or due to placebo. Believers can argue that the same applies to states of agnosticism and atheism. Fabrizio Benedetti and others[23] suggested that placebo works through a general circuit of the human brain and is one of the: "*human self-regulating faculties with which evolution has equipped us for effective social, emotional and physical health.*" Through placebo, we *think* ourselves into more comfortable states. If a badly wounded soldier is told that she/he has received an injection of morphine, but has actually been given nothing more than a saline solution, there can still be significant reduction of pain. Conversely, if a real painkiller is given but the one suffering is not told, pain reduction can be limited.

From my experiences in faith and the experiences that believers have shared with me, I know that the shift from unbelief to faith changes life in ways far beyond anything that could be considered as placebo.

---

21 Bob Dylan.1980. *Are You Ready?* Matthew 20:23
22 Wright, N.T. 2006. *Simply Christian: Why Christianity Makes Sense.* HarperOne: New York. 240p.p.209.
23 Benedetti, F., Mayberg, H.S., Wager, T.D., Stohler, C.S. and J-K. Zubieta. 2005. Neurobiological mechanisms for the placebo effect. *The Journal of NeuroScience* 25 (45): 10390-10402.

# FAITH

Believers are not simply imagining a non-existent God and imagining escaping harm from a non-existent spiritual force for evil. The shift from unbelief to faith changes one's entire worldview into a personal relationship with God, taking His side in the war between good and evil. Everyone participates in that war, knowingly or not.

From the Reformed Epistemology (RE) movement, Alvin Plantinga[24] stated the following, in which the emphases are his: "*On the A/C (Aquinas/Calvin) model...theistic belief as produced by the sensu divinitatis (our natural knowledge of God) is basic. It is also **properly** basic... for a person in the sense that it is indeed basic for him (he doesn't accept it on the evidential basis of other propositions) and, furthermore, he is **justified** in holding it in the basic way: he is within his epistemic rights, is not irresponsible, is violating no epistemic or other duties in holding that belief in that way.*"

Put more simply, it really is 'OK' for rational people, including scientists, to choose faith and theism. Our *sensu divinitatis* is part of our reality. A human self can recognize the existence of her/his soul and receipt of spiritual revelations. I see abundant evidence that God is a Living God and an ever-present Actor in the lives of believers and nonbelievers.

Nicholas Everitt[25] criticized RE at length and argued that even if Plantinga was right about theism being a properly basic and warranted belief: "*the consequences for the role of reason in religion will be relatively small.*" That might well apply in much of organized religion, but the role of reason will always be large in Free Thought towards a decision for or against faith *per se* and also during Free Thought episodes in subsequent lives of faith or unbelief.

Everitt accepted that a stamp of warranty on Christian, theistic belief:

---

24  Plantinga, A.1999. *Warranted Christian Belief.* Oxford University Press: Oxford.508p.p.178.
25  Everitt, N. 2004. *The Non-Existence of God. Routledge*: London.326p.p.26, 28.

# Free Thought, Faith, and Science

*"can indeed be a comfort to the believer, in assuring her that she is free from reproach in holding on to her theism even though she cannot offer any supporting evidence."* However, he argued that the believer will still want to ask this very important question: *"'Given that I do not violate any epistemic duties in holding this belief, what are the reasons for thinking that this belief is **true** (present author's emphasis)?'* I agree that the believer must ask this question. Experiencing God in a life of faith is how the believer knows that her/his belief is true.

Many nonbelievers, especially scientists, profess a dislike for discussing anything about faith, though this might not reflect fully how they feel inside. In truth, no one needs any warrant or justification to explore faith. *Everyone* has a God-given and sovereign right to do so. The most important and *proper* questions are whether faith is based on truth or on lies and delusions, whether God truly exists and, if so, whether any human can have any communication with Him?

How could it be anything other than basic, proper, justified and warranted to search for truthful answers to these questions, which are the biggest questions in human existence? We are all alive, curious, creative and seeking truthful answers. We all have the capacity and the need to seek for truth, through faith *and* through science. We all have Free Thought. The sovereignty of selfhood, exercised through Free Thought, gives everyone the basis, justification, permission, warrant, or whatever else you might wish to call it, to believe in God or to deny His existence.

The information upon which we base our Free Thought about faith is not chosen entirely by us. God and the spiritual force for evil send us spiritual revelations. We can ignore them to various extents, but we cannot blot them out completely. Honest quests for truth take place in the presence of lies. All material realm consequences of the outcomes of human Free Thought down the ages to the present day, including the evolution of organized religions and the institutions of science, reflect

# FAITH

spiritual revelations *and* spiritual warfare. All humans, including all philosophers, scientists and theologians, must have received spiritual revelations from both sides.

Spiritual revelations can be hard to recognize. Nonbelievers deny their existence. Some believers, especially deists, hold that God is remote from us. I disagree. God is in the spiritual realm. Our souls are also in the spiritual realm. God is not remote. We fail to recognize spiritual revelations mainly because we are so preoccupied with basic thought about material things. Anyone who tries can tune in to God.

## Soul States: Baselines, Shifts and Leaps

Free Thought does not cease upon having made a decision for either faith or unbelief. After any episode of Free Thought, the self, body-mind and soul, continues to receive information from the material and spiritual realms. Further episodes of Free Thought follow, producing a dynamic succession of soul states - from the onset of Free Thought in childhood to its cessation at death.

The soul is always in a state of faith or unbelief, regardless of whether that state is recognized or not. The soul's state can change from unbelief to faith or vice versa at any time up to the point of death. Christ likened God to the good householder who paid the same wages to those who began work in his vineyard only *"at the eleventh hour"* as to those who had started work early in the day.[26]

My friend and colleague Daniel Pauly[27] coined the brilliant term *"shifting baselines"* as he described the changes that occur in fisheries ecosystems. In fishing families, parents and grandparents tell their children and grandchildren how much bigger the catches were and how much

---
26  Matthew 20:1-16
27  Pauly, D.1995. Anecdotes and the shifting baseline syndrome of fisheries. *Trends in Ecology and Evolution (TREE)* 10 (10): p.430.

better the environment used to look in the old days. There is abundant evidence for shifting baselines in ecosystems. Successive generations of humans have abused nature and suffered the consequences.

Shifting baselines are part of evolution. They go hand-in-hand with selection processes - not only in ecosystems and natural resource systems, but also in the arts, commerce, politics, and organized religion. Organized religions have waxed and waned, but the human need for a spiritual life seems to have been relatively constant. Even in seemingly secular societies, people search for something spiritual. When people lose their faith and/or defect from organized religion, they continue look for spirituality elsewhere.

In material terms, our personal circumstances can shift substantially - from rags to riches, sickness to health and vice versa. In the different stages and material circumstances of life, we seek to establish interpersonal relationships and group identities, but our personal decisions about faith *per se* are private and are known with certainty only by ourselves, God and the spiritual force for evil.

In strict religious societies, the consequences of professing beliefs that are different from doctrinal norms can range from mild ostracism to execution. In secular societies, overt professions of faith can lead to alienation from family, friends and employers. However, our Free Thought decisions about faith *per se* are not constrained by societal norms and/or deference to human authority and institutions, including organized religions. We choose what we believe and we choose the extents to which we will share our true beliefs or keep them secret.

Before any episode of Free Thought, such as making a moral choice, the soul will have a particular baseline within its overall state of faith or unbelief. During an episode of Free Thought and subsequent to its outcome, the soul baseline shifts. Part of the soul battlefield has been gained by one side and ceded by the other. Free Thought about faith

# FAITH

begins with a pre-existing soul baseline and ends with a lesser or greater soul shift to a new baseline. If the soul shift goes far enough, its state can change from unbelief to faith or vice-versa.

Conversion happens when the soul state shifts from unbelief to faith. Some believers argue that a conversion, such as the one described by the Apostle Paul,[28] occurs entirely through revelations, with reasoning having played no part. I disagree. Paul's conversion experience on the road to Damascus involved spiritual revelations about his past misdeeds and the divinity of Christ *and* a call to further mental reasoning about the world around him. He made a Free Thought choice for faith in Christ, based on soul processing *and* mental reasoning.

Decisions for faith cannot be based entirely on soul processing with no mental reasoning involved. Everybody reasons about the realities of the material realm, about how to cope with problems and achieve personal goals. Everybody also processes spiritual revelations in the soul. Free Thought combines mental reasoning and soul processing. The result can be a leap to or from faith, or no decision. Whatever the outcome, mental reasoning and soul processing continue in further episodes of Free Thought.

Some believers hold that one's free will is surrendered to God just prior to and during conversion, because one is then covered by God's grace, which overwhelms all other influences. I disagree. Faith is a personal choice.

There are divergent views about what happens to one's free will after conversion. Some say that the new believer continues to act out what was always predestined for her/him and therefore continues to fulfill only God's will, there being no other option. Others say that the new believer, who allegedly surrendered her/his free will to God during conversion, is then allegedly told by God something like: 'OK, now have

---

[28] Acts 9:1-20

## Free Thought, Faith, and Science

your own free will back, but with My help please try to use it from now on to set all your goals in accordance with My will.' The latter explanation is far better than the catch-all explanation of predestination, but my preferred explanation is that all converts who have leaped to faith and all who leap away from faith to unbelief retain their free will throughout those self-chosen major changes and thereafter.

A soul shift to a state of faith or unbelief is a *choice* - an outcome of Free Thought in a free-willed self. Thereafter, the soul's baseline continues to shift through further episodes of Free Thought. Some Christians like to sing: "*Take my will and make it Thine; it shall be no longer mine*"[29] and "*I surrender all.*"[30] That sounds nice, but it is clearly impossible. No one can surrender her/his entire self will to God, or to the spiritual force for evil, or to anyone else.

A soul shift from unbelief to faith or vice versa is a major change - a *leap* to faith or away from faith. One can take small or large steps towards that leap, all of which will reflect baseline shifts towards new soul states and exchanges of soul territory. In the end, however, the soul shift to or from faith is always a leap.

Søren Kierkegaard wrote much about the leap to faith. Alastair Hannay and Gordon Marino's multi-authored guide to Kierkegaard's life and works includes a fine contribution by M. Jamie Ferreira[31] on "*Faith and the Kierkegaardian leap.*" According to Kierkegaard,[32] those who do not leap remain incomplete as selves and in a state of despair until they die. He emphasized the self's free choice in the quest for truth and held that a leap to faith was always made at the leaper's own peril. The same applies for leaps to unbelief and for decisions not to leap at all.

---

29  Havergal, F.R. 1874. From the hymn: *Take My Life and Let It Be Consecrated, Lord, to Thee.*
30  Van Deventer, J. 1896. From the hymn: *I Surrender All.*
31  Ferreira, J.M. 1998. Faith and the Kierkegaardian Leap, p.207-234. In A. Hannay and G.D. Marino (eds.) *The Cambridge Companion to Kierkegaard.* Cambridge University Press: Cambridge.428p.
32  Kierkegaard, S.A.1989. *The Sickness Unto Death.* Translated by A. Hannay from the Danish original, which was first published in 1849. Penguin Books: London.179p.

# FAITH

Kierkegaard also held that the leap to faith and the Christian's life of faith entail the subjugation of reason. In his commentary on Kierkegaard, Harold Blackham[33] wrote the following: "*the God-man of history which Christianity claims as the truth is conceptually absurd...The intelligible God of reason and the immanent infinity of the individual are abandoned for the God of an historical Incarnation and the conviction of sin...Once the total decision is taken, the tension is not relaxed but increased...faith and reason remain discontinuous.*"

I disagree. The Free Thought that brings about a leap to faith or to unbelief *always* includes reasoning. I know that from my own experience and from what others have shared with me. The leap to faith and the life of faith that follows might appear to require belief and trust in the absurd. However, in leaping to faith and living in faith, the believer has recognized the *truth* of the existence of her/his soul - as a little bit of absurdity in material realm terms, but as the necessary receiver and transmitter of information in the spiritual realm, of which it is part.

It is perfectly legitimate to reason that the spiritual realm might exist, with or without assuming any particular connections to the material realm. It is then perfectly legitimate to reason that a leap to faith looks like the best way to connect one's soul to God. It is then perfectly reasonable to make that leap. Thereafter, it is perfectly legitimate to continue mental reasoning about the material realm and soul processing of spiritual revelations, combined in Free Thought. A believer who has recognized the existence of the soul and her/his spiritual life and connections has *not* abandoned reason.

Belief in the Incarnation of God as Christ *is* entirely compatible with reason. If one has reasoned that God and the spiritual realm exist, then one can also reason that God is well capable of inserting His presence into the material realm in human form. The once and

---

[33] Blackham, H.J. 1961. *Six Existentialist Thinkers*. I. Søren Kierkegaard. Routledge and Kegan Paul: London.179p.p.17.

forever nature and intent of that miraculous intervention make it a reasonable proposition.

For anyone who can conclude, reasonably, that she/he has a non-material soul, as the spiritual core of self, it is also reasonable to conclude that the human soul can connect to God and vice versa and that God could insert Himself in a miraculous human body-mind as Christ, for a limited period in the material realm. It is eminently reasonable to assume that the One who made the material realm had the power to take that form, if He so wished.

We all employ reasoning in basic thought and Free Thought, including the Free Thought that leads to our choices to take leaps from unbelief to faith or vice-versa. To any readers who think that I have argued wrongly in countering the arguments that faith and reason are incompatible, I can say only that I did not abandon my own reason when I chose faith.

There has been more research on leaps to faith (i.e., conversions) than on leaps to unbelief, which are usually termed losses of faith. William James'[34] great treatise supports the description of conversion as a leap to faith. James quoted John Wesley's survey in the Methodist Church as follows: *"In London alone I found 652 members of our society who were exceedingly clear in their experience, and whose testimony I could see no reason to doubt. And every one of these (without a single exception) has declared that his deliverance from sin was instantaneous; that the change was wrought in a moment...thus, I cannot but believe that sanctification is commonly, if not always, an instantaneous work."*

---

34  James, W.2002. *The Varieties of Religious Experience: A Study in Human Nature*. The Modern Library: New York.602p.p.250-251, 276-277, 445-450.First Published in 1902 as the Gifford Lectures on Natural Religion, delivered at Edinburgh in 1901-1902. James cited p.462-463 in Tyerman, L. 1872. *The Life and Times of the Rev. John Wesley, M.A., Founder of the Methodists*. Volume I.564p. First published in 1872 by Harper & Brothers: New York.

# FAITH

James also drew on Edwin Diller Starbuck's[35] painstaking research on conversion, which I discuss in the final chapter on Unity. From such work it is clear that many who make the leap to faith have difficulty in finding words to describe their experience and its aftermath. Some personal testimonies seem transparently honest and can be extremely moving. In others, the convert seems to be communicating what she/he thinks fellow believers would like to hear.

Concerning dramatic conversion experiences, James stated: "*unconsciousness, convulsions, visions, involuntary vocal utterances, and suffocation, must be simply ascribed to the subject's having a large subliminal region, involving nervous instability.*" In other words, the claim of a comparatively unstable person to have experienced any or all of the above symptoms at her/his conversion can be taken as true and can be explained as the kind of impacts that typically accompany very large decisions and changes in that type of person.

James wrote about Colonel Gardiner, who claimed to have had the following experience: "*All at once the glory of God shone upon and round me in a manner almost marvelous…A Light perfectly ineffable shone into my soul, that almost prostrated me to the ground… This light shone like the brightness of the sun in every direction. It was too intense for the eyes…I think I knew something then, by actual experience, of the light that prostrated Paul on the way to Damascus.*"

To me, that smacks too much of well-known biblical texts and taking care not to upstage the Apostle Paul. When the sharing of real spiritual revelations includes role-playing and/or the parroting of words that others expect to hear, its impacts on nonbelievers are greatly reduced.

---

35 Starbuck, E.D. 2010. *The Psychology of Religion: An Empirical Study of the Growth of Religious Consciousness.* General Books: Memphis TN. 226p.p.224. Starbucks' study was first presented as lectures, in 1894 and 1895 to the Harvard Religious Union, and was published as two articles in the American Journal of Psychology in January and October 1897.First published as a complete study in 1899.

## Free Thought, Faith, and Science

James encountered more of the same in the experiences reported by mystics; for example, Saint Teresa's claims of having attained an *"orison of union"* with God, in which her soul was: *"fully awake as regards God but wholly asleep as regards things of this world and in respect of herself."* She allegedly saw God as *"a limpid diamond"* and understood *"how three adorable persons form one God"* and how: *"the Mother of God had been assumed into her place in Heaven."*

Some who *lose* their faith claim to have experienced a feeling of liberation, especially after bad treatment by religious authorities and other believers. When so-called 'faith' has become mainly or exclusively an adherence to requirements of an organized religion, from which the adherent longs to break free, the leap to unbelief and the escape from religious strictures must feel marvelous, at least for a while. However, the nonbeliever will not find anything to compare with what the Apostle Paul called: *"the peace of God, which passeth all understanding."*[36]

André Comte-Sponville[37] stated: *"The loss of faith brings about no transformation in knowledge and almost none in morals. But it considerably changes the degree of hope – or hopelessness – in human existence."* I agree with Comte-Sponville concerning morals. Nonbelievers and believers are equally capable of leading moral lives. I agree also that a loss of faith has an impact on hope. After a loss of faith, cynicism can become increasingly dominant over hope.

Knowledge about the material realm is not transformed by having found or lost faith. God's truthful disclosures through science are made to believers and nonbelievers. The speed of light, acceleration due to gravity, and rates of decay of radioisotopes are the same for everyone. However, loss of faith lessens knowledge of God and the spiritual realm. Faith brings ever-greater knowledge about God and His ongoing war with the spiritual force for evil.

---

36   Philippians 4:7
37   Comte-Sponville, A. 2007. *The Little Book of Atheist Spirituality*. Translated from the French by Nancy Huston. Penguin Books: London.212p.p.50.

# FAITH

## SOME PERSONAL EXPERIENCES OF FAITH

Many nonbelievers imagine that a life of faith must be miserable, because pleasures must be prohibited. This is nonsense. My life has been much happier and much more satisfying in faith than in unbelief. Faith transformed my life from a hectic quest for recognition and approval, to a serene but very active quest for truth. I receive guidance from God as floods of clarity on the pros and cons of decision-making. The result is deep peace of mind and a feeling of completeness. God also leads me to people, places and information that help me to follow His will.

When I wake up, I say or sing a prayer in my head, such as one of the following hymns: "*Forth in Thy name, O Lord I go;*"[38] or "*Be Thou my vision;*"[39] or "*Come down O Love Divine;*"[40] or "*Come, my soul, thou must be waking.*"[41] After that I get a God-given 'cool' that lasts through the day. However difficult things might get, I know that I can "*be careful for* (i.e., anxious or worried about) *nothing.*"[42]

In St. Jude's church Wolverhampton, where I sang as a choirboy, the Vicar's wife Lily Simmonds advised sending a "*sky telegram*" to God when things got tough. A well-known and sometimes ridiculed country-style hymn says something similar: "*We can talk to Jesus thru this royal telephone.*"[43] I used to mock that kind of claim, but I have found that it works. Everyone has a hotline to God.

The Prophet Isaiah[44] wrote: "*And thine ears shall hear a voice behind thee, saying, This is the way, walk ye in in it.*" That was clearly figurative

---

[38] *Forth in Thy Name O Lord I Go* has words by Charles Wesley (1707-1788) and the tune *Song 34*, by Orlando Gibbons (1583-1625).
[39] *Be Thou My Vision* has words from an ancient Irish piece, translated and put into verse by Mary Byrne and Eleanor Hull respectively in 1927, and is set to the tune *Slane*, a traditional Irish melody.
[40] *Come down O Love Divine* has words translated by R.F. Littledale (1833-1890) from the writing of Bianco da Siena (d.1434) and is set to the tune *Down Ampney*, by R. Vaughan Williams.
[41] *Come, My Soul, Thou Must Be Waking* has words by Friedrich von Canitz (1700), translated by Henry J. Buckoll (1841) and is set to a dance tune, by Franz Joseph Haydn.
[42] Philippians 4:6
[43] Lehman, F.M. 1919. *The Royal Telephone*.
[44] Isaiah 30:21

## Free Thought, Faith, and Science

writing, though some claim to have had direct and lengthy conversations with God. Neale Donald Walsch[45] published his at length, but with the following rider: *"It is not necessary for you to join me in my belief about the source of my replies in order to receive benefit from them."* I have not heard God's voice in the form of a conversation, but I have received clear and intense impressions during prayer about decisions to be made and about the needs of others.

Most people experience special moments in life when they seem to be connected in some near-perfect and holistic way to everything around them; for example, with nature or a work of art. In such moments, God is saying, to believers and nonbelievers alike: 'Feel and enjoy these wonders.'

My special moments have included the following: watching huge pelagic fish whizzing past a coral drop-off; being immersed in great music, in church and on stage; running with my dog among white hares on the Manx hills in winter; watching the salmon waiting to run up river at Santon Gorge; and taking my motorbike out on a cold Yorkshire morning at first light along an empty road, feeling the purring engine, with a turquoise and orange sky and black birds flying.

My experiences of God have a lasting quality that is lacking in memories of special material realm moments. By analogy, we cannot remember how a particular physical pain or pleasure actually felt. We can remember only the circumstances in which it occurred. With experiences of God, the soul keeps a much fuller record.

My leaps to and from faith are chronicled in Appendix III. The following examples of my experiences of God are from my recent life of faith. Please be as sceptical as you like about what I recount. You might agree with Richard Dawkins[46] that claiming guidance from a personal

---

45   Walsch, N.D. 1999. *Applications for Living from Conversations with God.* Hodder and Stoughton: London.334p.p.ix.
46   Dawkins, R. 2006. *The God Delusion.* Bantam Press: London.406p.p.347-352.

# FAITH

God is like a child communing with an imaginary friend, teddy bear or security blanket. In my case, you would be wrong. Like the Apostle Paul,[47] "*I know whom I have believed.*"

Every life has ups and downs. One of my downs was the end of a relationship, mostly through my own fault. When I finally realized that it had ended for keeps, I was walking alone beside Lake Geneva on a grey afternoon. In a park populated by waterfowl, I saw a card in a muddy, excreta-filled puddle and picked it up.

It had a picture of a cute kitten in a hollow log and the following text: "*Qui m'écoute repose en securité, tranquille, loin de la crainte du Malheur*". The King James Version of the Bible has: "*But, whoso hearkeneth unto me shall dwell safely, and shall be quiet from fear of evil.*"[48] My translation was: '*Whoever listens to me rests in safety and peace, far from the dread of unhappiness.*'

Did God parachute that card down to lie in my path? 'No,' a material realm person had obviously dropped it there in the material realm mud. Was that a coincidence? A nonbeliever could argue correctly that I must have walked many miles in various states of mind and failed to chance upon any card to match my mood. Please decide as you wish. I know that I was led to that spot. No laws of nature were broken. Nobody else was there.

Some years later I was having a wonderful time singing in the Union Church of Manila (UCM) Chancel Choir and contributing to the wider music ministry, but I was singing and playing mostly blues and folk-rock elsewhere. I went on a blues and soul musical pilgrimage to Memphis. I had lunch with the amazing Rufus Thomas. I met Bo Diddley, one of my guitar heroes from college days. I went to Graceland, Sun Studios and Beale Street. I went to the Sunday school

---

47  II Timothy 1:12
48  Proverbs 1:33

and Morning Service at Al Green's Full Gospel Tabernacle Church. I was saturated with great music.

Back at the Holiday Inn, Union Avenue, I picked up the obligatory Gideon's Bible. Its bookmark was at Isaiah chapter 9, from which Handel took words for "*Messiah.*" I paced around singing Handel in my head until some words on the bookmark hit me very hard: "*We are all travellers between two eternities.*" In other words, a human life is lived in a tiny sliver of space-time, between an eternity of non-being before birth and an eternity of being, from a temporal material and spiritual life on Earth to an eternal spiritual life after death.

I resolved there and then to make the most of every day left in my life. I felt a call to focus on Gospel music. I went on to write and record two albums of original Gospel songs and another of acoustic versions of hymn tunes. I published this story through the Gideon Society.[49]

Was it nothing but set of coincidences? A nonbeliever could argue correctly that I have been in many other hotel rooms and failed to find any apt text concerning my music or any other activity. Please decide as you wish. I am sure that God encouraged me to shift my directions in music. No laws of nature were broken. Nobody else was there.

Early in 2006, I had begun to think seriously about writing a book like this, but had made no decision to proceed and had collected only one source of reference.[50] In May, I went to Victoria BC, Canada for a fish genetics workshop. I felt moved to go shopping for old natural history volumes, preferably about fish. I was directed to Russell Books, a famous secondhand bookshop.

At Russell's I found a nice paperback reissue of Charles Darwin's work

---

[49] Pullin, R.S.V. 2010. A Bible, a Bookmark and a New Direction. *Gideon Testimonies from the Philippines* Volume I. Gideon's International in the Philippines: Quezon City, Philippines.33p.

[50] Mott, N. Editor. 1991. *Can Scientists Believe? Some Examples of the Attitudes of Scientists to Religion.* James and James: London.182p.

# FAITH

on corals.[51] I did not even think to look for anything about faith and science. At the checkout, on a shelf just behind the cashier's head among the newly arrived books waiting to be catalogued, I saw two green and gold bound volumes of Thomas Henry Huxley's essays.[52] I got them at a bargain price and went out walking on air. Back at my hotel I began reading Huxley and writing my outline.

Was that a coincidence? A nonbeliever could argue correctly that I must have spent much time searching for books without being surprised by any great finds. Please decide as you wish. I am sure that God led me to those sources. No laws of nature were broken.

Since finding those gems by Huxley, I have been led to many other sources. More importantly, spiritual revelations about what to write and what to check have flooded into my head - usually in bed early in the morning, but often at no particular place or time.

It has not been plain sailing. I have often been embroiled in spiritual warfare about what I should or should not write. What I have written comes from my Free Thought, in which the spiritual force for evil can be a powerful spoiler of best intentions and efforts and niggles away constantly, particularly in opposition to faith. I must have made some wrong choices in what I have written in this book. I can only hope and pray that they are not too numerous.

In September 2010, my thinking and writing ground to a halt. I went to the UK and the Isle of Man to visit family and friends and to seek some renewal, by walking the hills and fishing in the glens. I had just written a song based on the biblical advice to have no worries,[53] but I

---

51  Darwin, C. 1962. *The Structure and Distribution of Coral Reefs*. University of California Press: Berkeley CA.214p.
52  Huxley, T.H. 1899. *Science and the Christian Tradition: Essays*. 419p. *Science and the Hebrew Tradition: Essays*.372p.Two volumes. D. Appleton and Company: New York. See Appendix IV.
53  Ka Roger. 2010. *Be Anxious for Nothing*. Available from: www.song-smiths.com. The song is based on the messages in Christ's Sermon on the Mount - Matthew 6:25-34 - and on the Apostle Paul's advice in Philippians 4:6 and Romans 8:28.

## Free Thought, Faith, and Science

was not living what I was singing.

In London, I felt moved to go to the early morning Eucharist at St. Paul's Cathedral. I did not know that the sermon would be preached on exactly the same texts that I had used for my song lyrics, especially the punch line: "*be anxious for nothing*." That message rang in my head as I left the cathedral and boarded the tube at St. Paul's underground station.

I took the only available seat next to a young Asian woman, probably a Korean. She must have caught the train further east. She was immersed in a book and I could see from its running titles that it was a Christian book. I plucked up courage and asked to see its main title. It was: "*Anxious for Nothing*."[54] I got the message again, loud and clear.

Was that nothing but a set of coincidences? A nonbeliever could argue correctly that I must have been in many churches and on many tube trains and at my desk for many hours without encountering any encouraging texts. The cathedral sermon text would have been set by the Anglican lectionary for the church year. The woman on the tube could have come from another church that was using the same lectionary and might have brought her book for that particular day. I am sure that God led me to St. Paul's and to her. No laws of nature were broken. I got back to writing immediately.

In May 2011, I had another period of severe doubt that I could ever finish this book. I sat down at my computer in some despair and heard in my head: "**Soldier** *of Christ arise and put your armour on; strong in the strength that God supplies through His Eternal Son.*"[55]

With all due respect to Charles Wesley, I dislike that hymn and all militaristic hymns. I would not have chosen those words from my

---

54 MacArthur, J. 2006. *Anxious for Nothing: God's Cure for the Cares of Your Soul*. David C. Cook: Colorado Springs CO.220p.
55 Wesley, C.1749. From the hymn "*Soldiers of Christ Aris*e."

large mental database of hymns. The message was from God. I was jolted back to writing confidently. You can call that a hallucination, or a delusion, or autosuggestion from a widget surfing my mental database of hymn lyrics, or whatever you like. I am simply telling you what I experienced.

In April 2012, I was with a small choral group at UCM for an evening Service led by two Brothers from the Taizé Community in France. I greeted Michelle Ruetschle, the wife of UCM Senior Pastor Steve Ruetschle. She asked me what I was working on. I said that I was writing a book. She assumed that it was a biology book. When I said that it was about faith and science, she advised me to contact Ard Louis - her friend, fellow Christian, and Professor of Theoretical Physics at Oxford.

Ard put me in touch with Denis Alexander, Director of the Faraday Institute for Science and Religion. Denis invited me up to Cambridge to meet him and his staff, after which I attended the Christians in Science (CIS) conference, *"Science and Christian Faith in 2012: An Enduring Partnership."* I had been unaware of the existence of CIS, but became suddenly gifted with many new contacts. I joined CIS and subsequently the Science and Religion Forum and the Society of Ordained Scientists. Was that just a set of coincidences? Please make your choice. No laws of nature were broken.

## Postscript

The main aspect of many testimonies of faith is the believer's claim to feel the presence of God as an enabling power. As the Apostle Paul[56] put it: *"I can do all things through Christ which strengtheneth me."* Faith also brings the knowledge of being loved by God: *"the*

---

56   Philippians 4:13

# Free Thought, Faith, and Science

*love of Christ, which passeth all knowledge."*[57] With that love one is never alone, even when facing death.

Jonathan Sacks[58] wrote: "*What led me to examine my faith in depth was not the success of philosophy in refuting proofs for the existence of God. It was its failure to say anything positive of consequence about the big question of life. Who am I? Why am I here? To what story do I belong? How then shall I live?*"

We seek answers to those big questions by immersion in the arts, mathematics, philosophy, science, sport and other creative activities, *all* of which are compatible with faith. Moreover, as David Fergusson[59] put it: "*Despite centuries of scepticism and critical attack, the curious persistence of faith amongst philosophers, scientists, and artists, suggests its capacity to order life by a standard not of our own making and to impart wisdom from earlier ages that can still be ours today.*"

The biggest question for anyone setting out to explore faith is whether believers are a simply bunch of self-deluded fools or are blessed by having a relationship so precious and so powerful that to miss it would be the greatest loss imaginable. There are no higher stakes surrounding any other question about human life. The most important task in life for everyone is to use Free Thought to choose between faith and unbelief. My Free Thought led me to choose faith and to become a disciple of Christ. Your Free Thought can lead you anywhere.

The following statements provide an excellent defence of faith:

---

57 Ephesians 3:19
58 Sacks, J. 2011.*The Great Partnership: God, Science and the Search for Meaning.* Hodder and Stoughton: London.370p.p.83.
59 Fergusson, D. 2009.*Faith and Its Critics: A Conversation.* Oxford University Press: Oxford. 195p.p.181.

# FAITH

I. *From Henry Allon[60]- "You may demonstrate to a man that it is a mere philosophic imagination to believe in a personal God, that it is an unhistoric delusion to put faith in Christ, that is it a scientific absurdity to offer prayer, that it is a gratuitous expectation to dream of a life after this. He may be utterly unable to reply to your arguments, but his own conscious experience neutralizes them all. He knows, and feels, that there is a God who forgives sin, and hears prayers, who gives holy inspirations to his soul, and holds spiritual fellowship with him."*

II. *From Joseph Angus[61] - "My reason and my understanding - intuition and experience - demand a First Cause of all things. My conscience demands a Lawgiver and Judge. My entire nature cries out for forgiveness, for holiness, for happiness. The world 'sighs to be renewed'. Christianity meets every one of these instincts in a way peculiarly its own, and yet intelligible and complete."*

---

60  Allon, H. 1873. The argument for the supernatural character of Christianity from its existence and achievements, p. 249-289, In *Faith and Free Thought; A Second Course of Lectures Delivered at the Request of the Christian Evidence Society*. Hodder and Stoughton: London. 469p.p.252. Available at: www.kessinger.net

61  Angus, J. 1873. Man: a witness for Christianity, p. 439-469. In *Faith and Free Thought: A Second Course of Lectures Delivered at the Request of the Christian Evidence Society*. Hodder and Stoughton: London.469p.p.467. Available at: www.kessinger.net

## CHAPTER 6

# SCIENCE

*"In the beginning was the Word...All things were made by him"*[1]

### GOD AND SCIENCE

The Apostle John[2] wrote that God, the Word, was the Creator of everything. God is also the Discloser of all truthful information about His creation: *"For the Lord giveth wisdom; out of his mouth cometh knowledge and understanding."*

Victor Stenger[3] summarized science as follows: *"Science is a human endeavour that is really just an especially systematic and careful version of the process that each of us uses in our daily lives: observing the world around us and drawing conclusions from these observations."* I agree and I hold that all truthful disclosures through mathematics and science come from God.

For the purposes of this book, science is defined as any intellectual activity, observation or experiment concerning the composition and workings

---

1  John 1:1 and 3
2  Proverbs 2:6
3  Stenger, V.J. 2008. *God: The Failed Hypothesis*. Prometheus Books: New York.302p.p.265.

of the material realm and the spiritual dimensions of the human condition. Science is the quest for truthful explanations of *any* kind of human observation and experience, including alleged spiritual experiences.

Science seeks only the truth and exposes lies and nonsense. In science there is no such thing as an inconvenient truth. Through our Free Thought, God advances science and seeks its applications for good, while the spiritual force for evil perverts the use of science for evil ends; for example, in cruelty, the furtherance of greed, and killing.

Through science, we can all share in an increasing body of knowledge and understanding. Believers and nonbelievers develop and use their God-given creativity for participation in the advancement and/or appreciation of science. Human creativity is developed and used through Free Thought, which always includes spiritual revelations and responses.

Long before anything resembling formal scientific method was established, humans engaged in practical science to improve their quality of life and prospects for survival. Much of the traditional knowledge of indigenous peoples was developed through traditional agricultural and medical science and the applications of science in creative art. Modern science comprises basic or pure science, applied science and the development of technology. Science has changed our lives dramatically and rapidly - usually for the better, but sometimes for the worse, as we face new environmental hazards.

## THE PHILOSOPHY OF SCIENCE

Anthony Flew[4] included the following in his definition of the philosophy of science: "*Organized empirical science provides the most impressive result of human rationality and is one of the best accredited candidates for*

---

4    Flew, A. 1984.*A Dictionary of Philosophy. Revised Second Edition.* St. Martin's Press: New York. 380p.p.319-320.

*knowledge. The philosophy of science seeks to know wherein this rationality lies; what is distinctive about its explanations and theoretical constructions; what marks it off from guesswork and pseudo-science and makes its predictions and technologies worthy of confidence; above all whether its theories can be taken to reveal the truth about a hidden objective reality."*

Karl Popper[5] prefaced a new edition of his famous work "*The Logic of Scientific Discovery*" with the following quotation from Lord Acton: "*There is nothing more necessary to the man of science than its history and the logic of discovery...the way error is detected, the use of hypothesis, of imagination, the mode of testing.*" In other words, logic plus "*imagination*," which I prefer to call creativity, lead to hypotheses and testing for correctness or error. In science according to Popper,[6] tentative new theories or "*conjectures*" are formulated and are then either sustained or "*refuted.*"

Popper stressed that a hypothesis qualifies as part of science only if it is *testable* and *falsifiable*. His criteria for real science are tough and can be difficult to apply. Framing a hypothesis usually involves describing scenarios and making assumptions, some of which can be additional hypotheses. Moreover, an untestable hypothesis might become testable in the future, given new methods and tools, such as ever more powerful supercomputers.

Popper's[7] verdict on Darwinism, including the New Synthesis, was as follows, in which all emphases are his unless stated otherwise: " (it is) *not a testable scientific theory, but a* **metaphysical research program** *- a*

---

5 Popper, K. 2002. *The Logic of Scientific Discovery*. Routledge Classics: New York.513p.p.xvi.First published in 1935, as *Logik der Forschung*. Verlag von Julius Springer: Vienna, Austria. Popper cited David S. Nicholls and the Acton Manuscripts in the Library of Cambridge University.
6 Popper, K. 2002. *Conjectures and Refutations: The Growth of Scientific Knowledge*. Routledge Classics: London.582p.First published in 1963.
7 Popper, K. 2009. Darwinism as a Metaphysical Research Program, p. 105-115. In R.T. Pennock and M.Ruse (eds.) *But Is It Science? The Philosophical Question in the Creation/Evolution Controversy*. Prometheus Books: New York.577p.p.106, 109, 107,110. First published as Popper, K. 1976. Darwinism as a Metaphysical Research Program, p.167-179, 234-235. In *Unended Quest*. Open Court: LaSalle IL.315p.

*possible framework for testable scientific theories."* He added: *"Darwinism does not really predict the evolution of variety. It therefore cannot really explain it. At best, it can **predict** the evolution of variety under 'favourable conditions.' But it is hardly possible to describe in general terms what favourable conditions are - except that, in their presence, a variety of forms will emerge."*

Popper continued: *"I do not think that Darwinism can explain the **origin of life**. I think it quite possible that life is so extremely improbable that nothing can 'explain' why it originated; for statistical explanation must operate, **in the last instance**, with very high probabilities."* He concluded: *"Adaptation or fitness is defined by modern evolutionists as survival value and can be measured by actual success in survival: there is hardly any possibility of testing a theory **as feeble as this*** (present author's emphasis)."

Nevertheless, Popper recognized the importance and utility of the theory of natural selection, as follows: *"In trying to explain experiments with bacteria which became adapted to, say, penicillin, it is quite clear that we are greatly helped by the theory of natural selection. Although it is very metaphysical, it sheds great light upon very concrete and very practical researches. It allows us to study adaptation to a new environment...it suggests a mechanism of adaptation, and it allows us even to study in detail the mechanism at work. And it is the only theory so far which does that."*

Darwinism is not about the origin of life. It is about the evolution of life forms, through natural selection and descent with modification. I have no problem with envisaging the inorganic/organic chemical evolution of life on Earth and the subsequent biological evolution of a huge diversity of life forms as having been *natural* and *highly probable* processes, given that all the necessary ingredients and conditions were present, that there was scope for huge numbers of trials, and that complexity can build upon prior complexity.

Darwin knew that his theory required a mechanism by which small changes could be passed from generation to generation. He would have been delighted to see how molecular genetics now fits beautifully with evolutionary theory. Much of the body of evolutionary theory is 'effective theory,' because it includes parameters and processes that cannot be fully observed or measured. Nevertheless, it has great explanatory power and utility, as Popper admitted. I am disappointed that he called Darwinism a *"feeble theory."*

Popper's main goal was to separate what he saw as real science from metaphysics. John Polkinghorne[8] explained as follows the difficulties facing anyone who attempts that task: *"it is impossible to think seriously without taking a metaphysical stance, since this simply means adopting a world-view… The physical reductionist who claims that there is nothing but matter and energy, and no truth beyond the truth of science, is making a metaphysical statement as clearly as someone who looks at the world from a theistic perspective…Everyone, implicitly or explicitly, has a metaphysics."*

Polkinghorne distinguished between scientism, which he described as *"the metaphysical belief that science tells us all that can be known or is worth knowing,"* and *"science itself,"* meaning the conventionally defined science of the material realm, which has: *"bracketed out too much (meaning, purpose, beauty) from its consideration for it to be the universal source of understanding."* I agree. We need a trustworthy and scientific approach for exploring all human experiences, including alleged spiritual experiences.

Massimo Pigliucci[9] commented on Popper's legacy as follows: *"Most philosophers today would still agree with Popper that we can learn about what science is by contrasting it with pseudoscience (and non-science), but his notion that falsification is* **the** (original author's emphasis) *definitive*

---

8 Polkinghorne, J.C. 2011. *Science and Religion in Quest of Truth.* Yale University Press: New Haven CT. 143p.p.22-23.
9 Pigliucci, M.2010. *Nonsense On Stilts: How To Tell Science From Bunk.* The University of Chicago Press: Chicago IL.332p.p.3-4.

# SCIENCE

*way to separate the two has been severely criticized...Popper's solution is a bit too neat to work in the messiness of the real world."*

Pigliucci noted that, according to Popper, an observation of a few abnormal dogs that had only two legs - and there are indeed a few such abnormal dogs - would falsify an original statement that dogs are quadrupeds, whereas biology and common sense tell us that all *normal* dogs are quadrupeds.

Pigliucci continued: *"philosophers agree that scientists do not (and should not) in fact reject theories just because they may initially fail to account for some observation. Rather, they keep working at them to see why the data do not fit and what can be done to modify the theory in a sensible manner... So, on the one hand a strict criterion of falsification seems to throw away the proverbial baby with the bath water: too much science would fail the falsification test. On the other hand, if we allow ourselves the luxury to change theories whatever new observation comes our way we are not doing science at all, but rather engaging in a sterile exercise of rationalization."*

I agree. Irrespective of which parts of it are fully Popper-compliant or not, the successful march of science will continue, because it is a search for truth using honest, transparent and rigorous methods.

## Creativity in Science

All truthful disclosures through mathematics and science are outcomes of Free Thought, in which the development and application of human creativity has played a major part. God provides the necessary motivation, capacities and information. The new sparks for advances in mathematics, science and the arts are struck during the Free Thought of individuals.

## Free Thought, Faith, and Science

John Jeffers[10] wrote the following: "*Research is not done by institutions: it is a creative activity of exceptionally gifted individuals. The first requirement, therefore, for any society that wishes to do research is to find, recruit, and keep creative scientists... Their originality depends upon their being able to devote as much time as possible to doing what they most enjoy, their research, and on their being given the greatest possible freedom for their research activities.*" Highly creative researchers are signed up in science like star players joining football teams, albeit at much lower salaries.

The Free Thought outcomes of subjective creativity are explored through collective research and lead to bodies of objective evidence, but the resulting new knowledge and understanding remain *personal*; i.e., lodged in one or more individuals. Michael Polanyi[11] showed why this is not a paradox. God drives science through the creativity of individuals in their Free Thought. They and those who come after them become the living repositories for whatever has been truthfully disclosed.

David Deutsch[12] described the importance of human creativity as follows: "*Of all the countless biological adaptations that have evolved on our planet, creativity is the only one that can produce scientific or mathematical knowledge, art or philosophy.*" I agree that our creativity drives all of the above, but I do not agree that it evolved only as a biological adaptation. It requires the God-given portals of our souls and mind-soul interfaces, through which He communicates with us and changes us and changes the world through us.

Deutsch gave high importance to memes - Richard Dawkins' and others' proposed replicators and cultural transmitters of innovations, new

---

10  Jeffers, J.N.R. 1993. Managing creativity. *Biologist* 40 (2): 87-88.
11  Polanyi, M.1964. *Science, Faith and Society: A Searching Examination of the Meaning and Nature of Scientific Inquiry*. University of Chicago Press: Chicago IL.96p.First published in 1946. The 1964 edition has a new introduction by the author. See also Polanyi, M.1974. *Personal Knowledge: Towards a Post-Critical Philosophy*. University of Chicago Press: Chicago IL.428p. First published in 1958 - based M. Polanyi's 1951-1952 Gifford Lectures, delivered at the University of Aberdeen, Scotland, UK.
12  Deutsch, D.2011. *The Beginning of Infinity: Explanations That Transform the World*. Viking: New York.487p.p.398, 410-411.

ideas and habits. I am not attracted to meme theory. My model for Free Thought would allow sharing of the stuff of memes in human-to-human relationships, but I hold that spiritual revelations and responses are important contributors in all Free Thought, including the development and use of creativity.

Deutsch recognized the following *"two puzzles"* concerning creativity and memes: *"The first is why human creativity was evolutionarily advantageous at a time when there was almost no innovation. The second is how human memes can possibly be replicated, given that they have content that the recipient never observes."* He offered the same solution to both puzzles, as follows: *"what replicates human memes is creativity; and creativity was used, while it was evolving,* **to replicate memes***. In other words, it (creativity) was used to acquire existing knowledge, not to create new knowledge. But* **the mechanism to do both things is identical** (original author's emphases), *and so in acquiring the ability to do the former, we automatically become able to do the latter. It was a momentous example of reach, which made possible everything that is uniquely human."*

Deutsch's position is that human creativity has a dual role - acquiring new knowledge and then enabling innovations. I hold that Free Thought provides the wherewithal for both the development and use of human creativity, for participation in the arts, mathematics and science. Deutsch explained human uniqueness and human ingenuity without including the possibility of any information exchange with the spiritual realm. I contend that the leap in the human condition from *"almost no innovation"* to high creativity has been much too large to be explicable solely in material realm terms.

Creativity flourished even among early humans, as seen in cave art. God gave *Homo sapiens* a soul and a mind-soul interface, together with a high level (self-aware and potentially God-aware) consciousness and the Universal Moral Code. God's gifts to humans established a huge difference between the animal condition and the human condition.

Biological evolution alone cannot explain that difference. However, it is clear that our creativity, in common with our highly developed consciousness and moral knowledge, has foundations in our evolutionary legacy from animals.

## The Conduct of Science

Codes of conduct and ethics in science are not in conflict with anything in the Universal Moral Code. Scientists search for truth. Most scientists tell the truth, though some can be persuaded to misuse their talents and findings. Naomi Oreskes and Erik Conway[13] described how scientists had been hired to sow doubts about clear and compelling evidence that threatened commercial and political interests; for example, to cast doubt about the fact that smoking is a major cause of cancer.

Michael Shermer[14] reviewed deception and fraud in science, beginning with the following quotation from Richard Feynman: "*The first principle* (in science) *is that you must not fool yourself - and you are the easiest person to fool…After you've not fooled yourself, it's easy not to fool other scientists. You just have to be honest in a conventional way after that.*"

Shermer continued: "*Given the fiercely competitive nature of research funding and the hardscrabble intensity of scientific status seeking, it is surprising that fraud isn't more rampant. The reason that it is so rare (compared with, say, corruption in politics) is that science is designed to detect deception (of one's self and others) through colleague collaboration, graduate student mentoring, peer review, experimental corroboration and results replication.*"

---

13   Oreskes, N. and E.M. Conway. 2010. *Merchants of Doubt: How a Handful of Scientists Obscured the Truth on Issues from Tobacco Smoke to Global Warming.* Bloomsbury: London.355p.
14   Shermer, M. 2010.When scientists sin. *Scientific American*, July 2010: p.18. The quotation from Richard Feynman comes from his 1974 commencement speech at the California Institute of Technology.

# SCIENCE

Good conduct is indeed the norm in science and fraud is almost always exposed sooner or later. Observations and experimental results that cannot be replicated are not allowed to stand. Science thrives on peer review, criticism and competition, as well as collaboration. Science is based on truth. As Bertrand Russell[15] put it: "*In the welter of conflicting fanaticisms, one of the few unifying forces is scientific truthfulness…the habit of basing our beliefs upon observations and inferences as impersonal, and as much divested of local and temperamental bias, as is possible for human beings.*"

Science as a profession is typified by dedication and hard work. The Nobel Laureate biochemist Hans Krebs[16] quoted the following advice from the nineteenth century chemist Justus von Liebig, as repeated by Friedrich Kekulé: "*If you wish to be a chemist, you must be willing to work so hard as to ruin your health. He who is not prepared to do this will not get far in chemistry nowadays…For many years, three hours of sleep were enough for me.*"

## Scientific Method

John Polkinghorne[17] wrote the following on scientific method, omitting here his citations to other works: "*When scientific method is put under the epistemological microscope, many feel that a flimsy structure is exposed. The difficulties are at least as old as David Hume's critique of the method of induction. How can general conclusions arise from particular experiences, however finitely many of these there will be? How can you be sure the sun will rise tomorrow? How can we gain real knowledge of what the physical world is actually like?*"

Polkinghorne answered as follows: "*First, we must resist 'total account'*

---

15  Russell, B.2004.*History of Western Philosophy*. Routledge: London. 778p.p.744.
16  Krebs, H.A. 1967.*The Making of a Scientist*. Lecture delivered at the inauguration of the Department of Biochemistry of the University of Newcastle upon Tyne, UK.
17  Polkinghorne, J. C. 1998. *Belief in God in an Age of Science*. Yale University Press: New Haven CT. p.104-106.

# Free Thought, Faith, and Science

*theories of knowledge and be prepared instead to value more piecemeal achievements...Second, it has proved impossible to distil the essence of the scientific method...It has not proved possible to draw up a universal protocol for scientific research and I accept the verdict of Michael Polanyi...that this is because science is an activity of persons, drawing on tacit skills learned through apprenticeship in a community whose purpose is the universal intent to seek truth about the physical world, while also acknowledging that current conclusions must remain open to the possibility of correction."*

I agree and must draw further on Michael Polanyi's[18] insights, beginning with the following: "*...the scientist may appear as a mere truth-finding machine steered by intuitive sensibility. But this view overlooks the curious fact that from beginning to end he is himself the ultimate judge in deciding on each consecutive step of his enquiry. He has to arbitrate all the time between his passionate intuition and his own critical restraint of it.*" Exactly! Every scientist uses her/his Free Thought in science. Science is shot through with subjectivity, while in pursuit of objectively agreed findings.

Polanyi held that the "*scientific conscience*" is uppermost in the scientist's mind, as the adjudicator in mixes of "*intuitive impulses...stimulated by some of the evidence but conflicting with other parts of it*" and in solving the problem that: "*unfettered intuitive speculation would lead to extravagant wishful conclusions; while rigorous fulfillment of any critical set of rules would completely paralyze discovery.*"

Polanyi concluded: "*We recognize the note struck by conscience in the tone of the personal responsibility in which the scientist declares his ultimate claims.*" What Polanyi called the "*scientific conscience*" is part of the scientist's episodes of Free Thought - from the development and use of her/his creativity in research to the making of ethical decisions about whether or how to apply results.

---

18  Polanyi, M.1964. *Science, Faith and Society: A Searching Examination of the Meaning and Nature of Scientific Inquiry.* University of Chicago Press: Chicago IL.96p.p.15,40-41. First published in 1946. The 1964 edition has a new introduction by the author.

# SCIENCE

Creativity and personal judgement in science might seem to have taken us far from objectivity and even farther from being fully Popper-compliant, but we must acknowledge their importance. Objectivity, creativity and personal judgement are all essential in science.

Despite all of the complications mentioned above, scientific method can still be expressed in terms that are generally acceptable for most of the practical conduct of science. Such a synthesis will not allow us to tick all of the boxes of universality and compliance with the philosophy of science, but it will suffice to indicate how most scientists go about their work. Recognizing those qualifications, scientific method can be said to have the following steps:

I. *Ask a question about a phenomenon.*

II. *Gather relevant information and make relevant observations.*

III. *Formulate one or more testable hypotheses to explain the phenomenon.*

IV. *Test each hypothesis by well-designed experiments.*

V. *Analyse the results by sound statistical procedures.*

VI. *Draw conclusions.*

VII. *Communicate the results through peer-reviewed publication.*

Scientists go through some or all of those steps in their daily work. Most do not go to their laboratories or field sites every day musing about whether they are being fully Popper-compliant in terms of the testability and falsifiability of their hypotheses. However, in designing experiments and processing data, all scientists are constrained to follow scientific method, in order to validate their conclusions and publish in peer-reviewed outlets.

The same method and rules apply throughout science - from research on particle physics and the cosmos to solving practical problems in agriculture and medicine. Research on how the material realm actually works, from the subatomic to the cosmological level, takes us down the path of increasing knowledge to which all science contributes - from hypotheses and models to improved theory. The most important steps are the formulation and testing of new hypotheses, in attempts to improve theory.

In science, a current theory *explains* a set of observations. Those who argue against evolution on the basis that 'It's only a theory, not a fact' display their ignorance about what a theory means in science and indeed for any and all participation in the honest quest for truth about the composition and workings of the material realm and the human condition, including alleged spiritual experiences. Theology is also a body of theory.

In science and for the whole of humanity, a theory that explains all observations within its purview has a much higher status than any single observation or so-called fact. An observation that cannot be explained by a current theory puts the theory in question and creates a tension, which must be resolved as quickly as possible. A theory is modified or discarded when it cannot explain one or more well-authenticated observations within its purview.

Normal science advances slowly until new ideas, observations and theory bring about what Thomas Kuhn[19] called scientific revolutions. Following Kuhn, I define a revolution in science as a major change, resulting from the establishment and testing of new theory, with radically new perspectives and methods. Revolutions in science bring

---

19  Kuhn, T.S. 1996. *The Structure of Scientific Revolutions.* University of Chicago Press: Chicago IL.212p.First published in 1962.

# SCIENCE

paradigm shifts. During normal change *and* revolutions in science, scientists continue to formulate and test new hypotheses and attempt to improve methods.

## Indicators of Correctness

William of Occam (ca.1288-1347 CE) held that prior assumptions about anything to be explained are best limited to a minimal set based on evidence and not multiplied to include more complex conjectural options - the so-called Occam's Razor. In science, a simple answer often turns out to be the correct one, but some systems are necessarily highly complex. The best indicator of correctness in mathematics and science is often the appearance of elegance, beauty or perfection, with appropriate simplicity or complexity.

Bertrand Russell[20] put that beautifully as follows: "*Mathematics, rightly viewed, possesses not only truth but supreme beauty, a beauty cold and austere, like that of sculpture, without any appeal to any part of our weaker nature, without the gorgeous trappings of painting or music, yet sublimely pure, and capable of a stern perfection such as only the greatest art can show.*"

Richard Feynman[21] stated: "*You can recognize truth by its beauty and simplicity... When you get it right, it is obvious that it is right - at least if you have any experience - because usually what happens is that more comes out than goes in... What is it about nature that lets this happen, that it is possible to guess from one part what the rest is going to do? That is an unscientific question... and therefore I am going to give an unscientific answer. I think it is because nature has a simplicity and therefore a great beauty.*"

---

20   Russell, B. 1917. The study of mathematics, p.58-73. In B. Russell's *Mysticism and Logic and Other Essays*. George Allen & Unwin: London.234p.First published in 1910, in Russell's *Philosophical Essays*.
21   Feynman, R.P. 1992. *The Character Of Physical Law*. Penguin Books: London.173p. p.171,173.

# Free Thought, Faith, and Science

Chris Packham[22] described science well as: *"the art of understanding truth and beauty."* Beauty and elegance in nature are seen not only in butterfly wings, flowers, snowflakes, trees and the motion of cheetahs and dolphins, but also in our *representations* of the composition and workings of the material realm - in algebra and geometry, in proofs of theorems; and in molecular models, from simple methane to the complex double helix of DNA.

In mathematics and science, explanations that are beautiful and *look* right usually *are* right. God sends us *His* truthful disclosures through mathematics and science, using *our* Free Thought as His conduit. In Free Thought, our God-given creativity uncovers truth and beauty. Beauty, with appropriate simplicity or complexity, is the best indicator of correctness in mathematics and science - a divine confirmation of a truthful disclosure.

## Undervalued Science and Persistent Pseudoscience

Science gets us as close as possible to learning truths about the composition and workings of the material realm - for example, electromagnetism, genomes, gravity, pathogens, radioactivity and toxins - and about the human condition in relation to all of the above. Science can also help us to explore the higher things that are the territory of Free Thought. Through science, we obtain new knowledge and explanations that are objective, value-free and non-negotiable, because they are God's truthful disclosures.

Daniel Dennett[23] insisted that no scientist can retain credibility by arguing along any of the following lines: *"'If you don't understand my theory, it's because you don't have **faith*** (original author's emphasis) *in it'* or *'Only official members of my lab have the ability to detect these effects,'*

---

22  Packham, C. 2012. *Secrets of the Living Planet 1. The Emerald Band.* BBC TV, July 9, 2012.
23  Dennett, D.C. 2007. *Breaking the Spell: Religion as a Natural Phenomenon.* Penguin Books: London. 448p.p.363.

# SCIENCE

*or 'the contradiction you think you see in my arguments is simply a sign of the limits of human comprehension. There are some things beyond all understanding.'"* Dennett saw the *"intellectual bankruptcy"* of all those arguments. I agree.

Carl Sagan[24] distinguished between science and pseudoscience as follows: "*Science thrives on errors, cutting them away one by one. Hypotheses are framed so that they are capable of being disproved. A succession of alternative hypotheses is confronted by experiment and observation. Science gropes and staggers toward improved understanding...Pseudoscience is just the opposite. Hypotheses are framed precisely so that they are invulnerable to any experiment that offers a prospect of disproof, so even in principle they cannot be invalidated. Practitioners are defensive and wary. Skeptical scrutiny is opposed. When the pseudoscientific hypothesis fails to catch fire with scientists, conspiracies to suppress it are deduced.*"

Scientists argue on the basis of evidence. Pseudoscientists argue on the basis of wishful thinking and/or mysterious mechanisms. In a world populated entirely by persons honestly seeking truth from all available sources, truthful disclosures through science would be highly valued and communicated well to well-educated decision makers and a well-educated public, while untruthful claims of pseudoscience would be discredited and discarded. In the real world, however, science is often undervalued while pseudoscience acquires lots of publicity and many followers. Belief in pseudoscience often persists even when its claims have been debunked by real science.

Homeopathy is a good example. John Sharp[25] reviewed homoeopathy from its beginnings in 1796 and quoted as follows from the 1986 policy statement of the Council of the Royal Pharmaceutical Society of Great Britain: "*The Council of the Society...recommends members to*

---

24   Sagan, C.1997. *The Demon Haunted World: Science As a Candle In the Dark.* Ballantine Books: New York. 457p.p.20-21.
25   Sharp, J. 2010. A complementary alternative? The non-sense of homoeopathy. *Biologist* 57 (1): 28-34.

*inform any persons seeking advice on homoeopathic products that there is no scientific evidence for their efficacy, beyond that to be expected from a placebo response."*

Edzard Ernst,[26] a former practitioner of homoeopathy who became a scientific researcher on alternative medicine, also found no evidence to support any claims that homeopathic treatments are superior to placebos. He concluded: *"homeopathy is a fascinating chapter from the history of medicine - but is not an evidence-based, ethical therapeutic approach."* Despite these clear findings from science, the mass marketing and mass consumption of homoeopathic products are continuing unabated, with endorsements from celebrities, the mass media and even the health services of governments.

When science is ignored and pseudoscientific nonsense is disseminated to support commercial, political and personal agendas, the consequences can be disastrous. Paul Torday's[27] fictional but fully credible story of a British attempt to develop salmon angling in the Yemen illustrates well the lunacy of undervaluing science while pursuing political ends. I was told once of a proposal to grow tea on the Isle of Man, among the hill sheep and bilberries.

Many introductions of alien species have gone ahead with inadequate prior appraisal by scientists and/or with politics and false economics overruling scientific advice. The golden snail (*Pomacea canaliculata*), a known invasive species, was introduced to the Philippines in 1983 - on the premise that it could be exported as escargots to Europe. The exports of snails did not materialize, but they invaded and became established and non-eradicable in most Philippine wetlands - including rice farms, where they eat young rice plants and require costly control measures.[28]

---

26   Ernst, E. 2012. Testing the water. *Biologist* 59 (1): 19-21.
27   Torday, P. 2007. *Salmon Fishing in the Yemen*. Phoenix: London.329p.
28   Acosta, B.O. and R.S.V. Pullin, Editors. 1991. Environmental impact of the golden snail (*Pomacea* sp.) on rice farming systems in the Philippines. *ICLARM Conference Proceedings* 28. 34p.

# SCIENCE

Commercial advertisements are often dressed up to appear highly scientific. A particular brand of petrol has been advertised as working "*at the molecular level.*" How could it and its rival brands *not* be working at that level? Various companies have intoned at length the chemical names of constituents of breast milk substitutes, accompanied by images implying that the lucky babies who consume them will become child prodigies.

The entertainment industry loves to portray scientists as mad and misguided meddlers with nature. Christopher Toumey[29] suggested that fictitious portrayals of scientists have led to: "*a procedure for censuring scientists and scientific knowledge*" and have been "*exercises in antirationalism.*" I have seen "*Frankenstein*" and "*Jurassic Park*" cited in a TV broadcast about research for the farming of genetically modified salmon. Science fiction and conjecture got more coverage than any scientific appraisals of risks.

Media professionals have sought to prolong reporting on controversies already settled by science. For example, Ben Goldacre[30] wrote the following about the media and the myth that the measles, mumps and rubella (MMR) vaccine might be linked to autism: "*hundreds of journalists, columnists, editors and executives…drove this story cynically, irrationally, and willfully onto the front pages for nine solid years…they overextrapolated from one study into absurdity, while studiously ignoring all reassuring data, and all subsequent refutations. They quoted 'experts' as authorities instead of explaining the science, they ignored the historical context, they set idiots to cover the facts, they pitched emotive stories from parents against bland academics (who they smeared), and most bizarrely of all they simply made stuff up.*"

---

29  Toumey, C.P. 1992. The moral character of mad scientists: a cultural critique of science. *Science, Technology and Human Values* 17 (4): 411-437.
30  Goldacre, B. 2008. *Bad Science*. Fourth Estate: London.370p.p.290-291.

# Free Thought, Faith, and Science

## A Theory of Everything

Stephen Hawking and Leonard Mlodinow[31] wrote a fine layperson-friendly account of progress towards a theory of everything called M-theory, which is a family of theories - an attempt to integrate other grand theories, including Newtonian physics, gravity, thermodynamics, electromagnetism, relativity, quantum physics, quarks, strings and p-branes.

Hawking and Mlodinow's description of M-theory included the following: "*M-theory has eleven space-time dimensions...M-theory can contain not just vibrating strings but also point particles, two-dimensional membranes, three-dimensional blobs, and other objects that are more difficult to picture...These objects are called p-branes (where p runs from zero to nine)...The mathematics of the theory restricts the manner in which the dimensions of the inner space can be curled. The exact shape of the inner space...determines the apparent laws of nature. We say 'apparent' because we mean the laws that we observe in our universe...The laws of M-theory...allow for **different universes*** (original authors' emphasis) *with different apparent laws, depending on how much the internal space is curled. M-theory allows for 10 to the power 500 different universes, each with its own laws.*"

In that same vein of mind-boggling complexity, Hawking and Mlodinow commented on the human condition as follows: "*While conceding that human behaviour is indeed determined by the laws of nature, it also seems reasonable to conclude that the outcome is determined in such a complicated way and with so many variables as to make it impossible in practice to predict. For that one would need a knowledge of the initial state of each of the thousand trillion trillion molecules in the human body and to solve something like that number of equations...would take a few billion years.*"

---

31  Hawking, S. and L.Mlodinow. 2010. *The Grand Design: New Answers to the Ultimate Questions of Life*. Bantam Press: London.199p.p.117-118, 32,180.

# SCIENCE

Hawking and Mlodinow explained as follows why no Creator is needed: "*On the scale of the entire universe, the positive energy of the matter **can** (original authors' emphasis) be balanced by the negative gravitational energy, and so there is no restriction on the creation of whole universes. Because there is a law like gravity, the universe can and will create itself from nothing... Spontaneous creation is the reason why there is something rather than nothing, why the universe exists, why we exist. It is not necessary to invoke God to light the blue touch paper and set the universe going.*"

John Lennox[32] included the following in his response to Hawking's assertion that a theory of everything can be Godless: "*Science and history are not the only sources of evidence for the existence of God. Since God is a person and not a theory, it is to be expected that one of the prime evidences for his existence is personal experience...Hawking's fusillade will not shake the foundations of an intelligent faith that is based on the cumulative evidence of science, history, the biblical narrative, and personal experience.*"

In company with Lennox and other fellow believers, I am not troubled by pronouncements from any grand theorists that God is not required to explain the existence and nature of the material realm. It is a pretty safe assumption that M-theory or one of its successors will lead to better explanations of the composition and workings of material realm, of which our body-minds are parts. However, I find the concept of the spiritual realm and the existence of the soul, God and the spiritual force for evil to be the best explanation of the human condition. M-theory is undoubtedly a stupendous work in progress. It does not preclude an overarching theory of God as the Creator of everything.

According to Hawking and Mlodinow, the 'M' in M-theory was derived originally from the 'm' in the hypothetical membranes of string theory, but now it might just as well stand for: "*master, miracle, or mystery.*" I suggest that it might as well stand for the 'm' in material

---

32   Lennox, J.C. 2011. *God and Stephen Hawking: Whose Design Is It Anyway?* Lion Books: Oxford UK.96p.p.94-95.

realm, because it has nothing to say about anything else. An integrated grand theory of everything would need the addition of spiritual realm (S)-theory.

M-theory and S-theory will include effective theories, in which it is not possible to measure, observe and analyse all variables. M-theorists can attempt to cover what might be going on in huge numbers of other universes, but they will be rooted in this particular universe, along with any S-theorists. Spiritual stuff is beyond detection by material realm means. However, S-theory might be improved through the systematic gathering of subjective evidence from within concerning faith and unbelief.

M-theory seems poised for rapid development. Perhaps some grand theorists of everything will be moved to consider the integration of M-theory with S-theory, taking wider perspectives than those commonly taken in science and theology. That might lead to some embryonic form of an integrated 'M and S theory' - with all due respect a very famous retailer. 'S and M theory' would have unfortunate connotations, though Richard Feynman, with his sense of fun, might have preferred it. 'M and S theory' would always be a work in progress. Free Thought theory might contribute something along the way.

## Postscript

The eminent chemist and atheist author Peter Atkins[33] wrote that: *"The scientific method can shed light on every and any concept, even those that have troubled humans since the earliest stirrings of consciousness and continue to do so today. It can elucidate love, hope and charity."* Atkins continued as follows: *"my scientific faith, is that there is nothing that the scientific method cannot illuminate and elucidate... The scientific method*

---

33  Atkins, P. 2011. *On Being: A Scientist's Exploration of the Great Questions of Existence.* Oxford University Press: Oxford UK.111p.p. vii, 104-105.

# SCIENCE

*is a distillation of common sense in alliance with honesty, and its discoveries illuminate the world."*

On March 22, 2011, at the Royal Institution, London, I was privileged to hear Atkins give a lecture in which he confirmed his position that science has no limits. He also stated the following: *"religious belief is open to psychological investigation which means that it is open to investigation through science."* During question time, I asked Atkins whether he thought that the analysis and clustering of subjective evidence from within individuals, as in case studies for psychology, really was admissible in science? His answer was a definite 'Yes.'

In a debate[34] with John Lennox, Atkins held that the lives of brilliant men of faith, such as St. Augustine, had been *"wasted,"* because they could have contributed more to humanity had they focused on science and not on God. Atkins considered all individual experiences of alleged encounters with God to be *"hallucinations."*

We must explore further those alleged *"hallucinations"* or alleged spiritual revelations and see how the evidence stacks up. The systematic gathering and analysis of subjective evidence from within concerning faith and unbelief can bring studies of the spiritual realm into the purview of science.

In that particular debate, Atkins signed off as follows: *"Scientific method is simple - seek reliable evidence and where possible express the results mathematically. This gives me joy."* Lennox replied: *"I share Peter's joy at the results of science and mathematics. That joy is increased in that we are thinking God's thoughts after Him."*

I share with Atkins and Lennox the same joy of science. Truthful disclosures through science bring joy to those who report them and to

---

34  From the debate between Peter Atkins and John Lennox: December 31 2010, Oxford University. Available at: www.youtube.com/watch?v=YxOCXmagQuO

those who benefit from them. God speaks to believers and nonbelievers in Free Thought. God uses human creativity as His channel for truthful disclosures about the material realm, through mathematics and science.

Steven Weinberg[35] wrote: *"Whether or not the final laws of nature are discovered in our lifetime, it is a great thing for us to carry on the tradition of holding nature up to examination."* Peter Russell[36] concluded: *"There is nothing in physics, chemistry, biology, or any other science that can account for us having an interior world."* I agree with both and plead that we use science much more to explore that *"interior world."*

---

35  Weinberg, S. 1993. *Dreams of a Final Theory: The Search for the Fundamental Laws of Nature.* Vintage: London. Total pp.? p. 220.
36  Russell, P. 2005. *From Science to God: A Physicist's Journey into the Mystery of Consciousness.* New World Library: Novato CA.120p.p.17.

## CHAPTER 7

# BATTLEFIELDS[1]

*"He that is not with me is against me..."* [2]

### No Neutral Ground

We are all combatants in a huge war zone, which spans the material and spiritual realms and contains billions of battlefields. There is no neutral ground. Every soul is a battlefield. Whether we recognize it or not, we *all* take sides in the battles of faith versus unbelief, evil versus good and organized religion versus science.

As outcomes of Free Thought, our individual soul battlefields are extended to interpersonal and group battlefields: in family relationships, friendships, clubs, churches and workplaces. Good and evil pervade all our battlefields, like nutrients and pollutants spreading within and among ecosystems.

### Faith Versus Unbelief

In 2009, the British Humanist Association put the following advertisement on London buses: *"There probably is no God. Now stop worrying*

---

1  Appendix IV, Battlefield Literature, is complementary to this chapter.
2  Matthew 12: 30

*and enjoy your life."* In Islamic societies, anything similar would have caused riots and probably deaths. In London, it just passed by. Many believers must have laughed at the implication that God brings only worry to those who believe in Him. As David Robertson[3] commented: *"try telling* (that to) *the mother who just lost her child."*

I agree with Nicholas Everitt,[4] A.C. Grayling,[5] J.L. Mackie[6] and other atheist authors that none of the so-called intellectual proofs for the existence of God - the cosmological and ontological proofs and the arguments from design and alleged miracles - actually stands up. The only evidence for the existence of God and for anything made of spiritual stuff is *our* existence and our subjective evidence from within concerning faith and unbelief.

Alister McGrath[7] discussed how those writing in the new wave of atheism define faith and religion in ways to suit their arguments, while promoting an allegedly superior world, in which there would no longer be any belief in God or memberships in organized religion. He described the New Atheism well, as being: *"in an intellectually and morally uncomfortable place...caught in a dilemma framed by two of its core beliefs (neither of which, of course, can be proved: 1. God is evil and nasty. 2. God is a delusion created by human beings."*

Most atheist authors contend that the battle of faith versus unbelief is simply a battle of unreasoning versus reasoning. Some believers are indeed basing their reasoning and decisions about everything entirely on the literal contents of one or more ancient sacred texts. Nevertheless, the atheists are wrong. The battle of faith versus unbelief is always

---

3   Robertson, D. 2007.*The Dawkins Letters: Challenging Atheist Myths*. Christian Focus Publications Limited: Fearn, Ross-shire UK.159p.p.130.
4   Everitt, N. 2004. *The Non-Existence of God. Routledge*: London.326p.p.26, 28.
5   Grayling, A.C. 2013. *The God Argument: The Case Against Religion and for Humanism*. Bloomsbury: New York.269p.
6   Mackie, J.L. 1982. The Miracle of Theism. Arguments For and Against the Existence of God. Clarendon Press: Oxford. 268p.
7   McGrath, A. 2011. *Why God Won't Go Away: Engaging with the New Atheism*. Society for Promoting Christian Knowledge (SPCK): London. 118p.p.63.

fought through episodes of Free Thought and Free Thought *always* involves reasoning.

The confirmed atheist Daniel Dennett[8] wrote the following, which I can quote as *support* for the role of reason in faith: "*Now if you want me to **reason** about faith, and offer a reasoned (and reason-responsive) defense of faith as an extra category of belief worthy of special consideration, I'm eager to play. I certainly grant the existence of the phenomenon of faith; what I want to see is a reasoned ground for taking faith seriously as a way **of getting to the truth*** (original author's emphases), *and not, say, just as a way people comfort themselves and each other (a worthy function that I do take seriously).*"

I am eager to play with Dennett and other atheists, if we can agree that our subjective evidence from within is valid evidence for study.

## Evil Versus Good

Some people react against the word evil as being archaic and/or too harsh, but it is still the only word that describes adequately the spiritual force opposing good and the lies that are its stock in trade. The spiritual force for evil harms individuals and uses harmed individuals to harm others - in the home, at school, at work, at church and everywhere else.

Free process in the material realm can harm anyone, believer or nonbeliever, through accidents, infections and natural disasters etc. Those harmful events are not acts of God or acts of the spiritual force for evil. They are consequences of nature and its laws. Some are caused or exacerbated by our abuse of nature and/or our unhealthy lifestyles, but there is no such thing as natural evil. Nothing

---

8   Dennett, D.C. 1996.*Darwin's Dangerous Idea: Evolution and the Meanings of Life*. Penguin Books: New York.586p.p.154.

in the material realm is intrinsically *evil*.

Evil among humans is the quality of actors and acts pursuant to lies and/or to the harming of others. Every time we lie we further the cause of evil. Most evil acts are based on outcomes of the Free Thought of mentally normal persons. The evil acts of psychopaths and other mentally disordered persons result in part from their nature as well as from the outcomes of their Free Thought. Genetics and brain development and function are subject to free process in the material realm.

According to the Bible, the spiritual force for evil was originally an agent of good, in harmony with God in heaven. Together with its band of rebel angel followers, it allegedly fought unsuccessfully against the good Michael and his angels and was expelled from heaven: "*the great dragon was cast out, that old serpent called the Devil, and Satan, which deceiveth the whole world: he was cast into the earth, and his angels were cast out with him.*" [9] Whether you take that to be the literal truth, or imagery, or complete nonsense, I hope that most or all of us can agree that evil and good exist and are opposites.

Terry Eagleton[10] wrote the following on human evil: "*The question for religious believers is not really why there is wickedness in the world. The answer to that is pretty obvious. There is no mystery about why a pimp should lock up thirty imported Albanian sex slaves in a British brothel. The question for believers is why human beings were created free to do such things in the first place. Some believers hold that for humans not to be created free would be a contradiction in terms…because the Creator in question is God, who himself is pure freedom. To be made in God's image and likeness is precisely not to be a puppet.*"

I agree. God gives a soul to everyone and makes everyone a free agent in the war of evil versus good. It is up to each of us to weigh His influences

---

9  Revelation 12:7-9
10 Eagleton, T.2010. *On Evil.* Yale University Press: New Haven CT.176p.p.139-140.

against those from the spiritual force for evil. Irrespective of whether we are believers or nonbelievers, we are all tossed back and forth from one side to the other, in our Free Thought. The spiritual force for evil prompts us to say unkind things, to do unkind acts and to ignore the voice of God in our consciences. Through our Free Thought, it recruits us to do evil things. Bob Dylan[11] sings about the *"fiery darts"* from the spiritual force for evil that pierce our souls.

Simon Baron-Cohen[12] explained human evil by *"empathy erosion"* and suggested measuring our *"Empathy Quotients."* I agree that lack of empathy can explain human evil in some cases, but I cannot accept it as a comprehensive explanation. The perpetrators of the Holocaust included some mentally disordered persons, including psychopaths, but most were ordinary folk. They had plenty of empathy with each other, but little or none with their victims.

M. Scott Peck[13] also related evil behaviour to lack of empathy and felt that a distinction might be drawn between that which influences ordinary folk to engage in everyday evil and extraordinary demonic possessions, which some say can be cured only by exorcism. He stated: *"I have no idea whether Satan actively recruits the commonly evil to its work. I suspect not. Given the dynamics of sin and narcissism, I suspect they recruit themselves. But until such time as we have greater knowledge of Satan, my understanding remains faint."*

Scott Peck described an exorcism as follows: *"The patient suddenly resembled a writhing snake of great strength, viciously attempting to bite the team members. More frightening than the writhing body, however, was the face. The eyes were hooded with lazy reptilian torpor - except when the reptile darted out in attack, at which moment the eyes would open wide*

---

11   Bob Dylan.1980. *What Can I Do For You?* For the original source of *"fiery darts,"* see Ephesians 6:16.
12   Baron-Cohen, S. 2011. *Zero Degrees of Empathy: a New Theory of Human Cruelty*. Allen Lane: London.190p.
13   Scott Peck, M.1985. *People of the Lie: The Hope for Healing Human Evil*. Touchstone: New York.269p.p.211, 77,196,185,196-197.

*with blazing hatred...Almost all of the team members...were convinced they were...in the presence of something absolutely alien and inhuman."*

Scott Peck concluded: "*it is God that does the exorcising. But let me amend that. Human free will is basic. It takes precedence over healing. Even God cannot heal a person who does not want to be healed...Ultimately, it is the patient herself or himself who is the exorcist.*" He added: "*all good psychotherapy does in fact combat lies.*" I agree there, but I do not believe in demonic possession. I regard exorcisms as human role-playing, on behalf of the spiritual force for evil and God. The patient and the exorcism team act out an intense drama of spiritual change.

Good is the quality of actors and acts pursuant to truth and to helping others by following the Universal Moral Code. No one is completely good except God. He implants His Universal Moral Code in every soul. Transgression of that code is evil. Compliance with that code is good. A person who behaves that way will be called good or virtuous by most other persons, though some who are under the influence of the spiritual force for evil will criticize her/his behavior and encourage compliance with evil norms.

The desire for goodness, justice and fair play is prevalent across humanity. Where that desire is low or absent, evil has the upper hand and life becomes hellish. Mercifully, that is rare. More commonly we face complex mixtures of good and evil, justice and injustice, fairness and unfairness - from the outcomes of our Free Thought and the Free Thought of others, as everyone decides when they will comply with or transgress the Universal Moral Code. Our degrees of compliance are always changing. We are all complex mixtures of evil and good.

Christians believe that God, Incarnate as Christ, overcame everything that the world and the spiritual force for evil could throw at Him and that His sacrifice on the cross empowers believers to overcome evil.[14]

---

14 John 16:33 and I John 5:4.

## BATTLEFIELDS

Most Christians believe that God has defeated the spiritual force for evil so that it can no longer tempt us or harm us beyond whatever He allows. In support of that position they cite the reported negotiation over how much God allegedly allowed Satan to test Job[15] and Christ's[16] reported assurance to the Apostle Peter: "*Simon, Simon, behold, Satan hath desired you that he may sift you as wheat: but I have prayed for thee, that thy faith fail not: and when thou art converted, strengthen thy brethren.*"

Job suffered terribly. Peter was martyred. Their stories are examples of faith prevailing, despite horrific happenings. They are also illustrations of the fact that very bad things can happen to anyone. But what do they tell us about the relationships between God and the spiritual force for evil? Why did Christ, Who was God, have ask God His Father to limit what Satan could do to Peter? Satan is described as a "*roaring lion* (that) *walketh about*"[17] and is said to be walking far and wide: "*And the Lord said unto Satan, From whence comest thou? And Satan answered the Lord and said, From going to and fro in the earth and from walking up and down in it.*"[18]

I believe that the spiritual force for evil pervades all human life. I do not believe that it is like a bad dog or a roaring lion on God's adjustable leash. Accepting that model means accepting that God adjusts His limitations on the spiritual force for evil, even as it brings about *all* of the lying, betrayal, discrimination, theft, rape, torture, murder and other horrors in this world. It means accepting that God allowed the spiritual force for evil to persuade the Inquisition's torturers and the Holocaust's mass murderers to do their abominable work.

Those who believe that God controls everything like to quote texts such as the following: "*For my thoughts are not your thoughts, neither*

---

15 Job 2: 1-6
16 Luke 22:31
17 I Peter 5:8
18 Job 2:2

*are your ways my ways, saith the Lord"* and *"Is not my way equal? Are not your ways unequal?"*[19]

There no hint in that text or its wider context that God controls everything or that He sanctions whatever the spiritual force for evil does to us. If God really controlled everything, including everything evil, then the battle of evil versus good would be a sham. He has given us our freedom and with it *our* individual obligations to resist the spiritual force for evil.

The Apostle James[20] wrote: *"Resist the devil, and he will flee from you."* I agree with James, but not with the Apostle Paul's[21] blanket assurance that: *"God…will not suffer you to be tempted above that ye are able; but will make with the temptation a way to escape, that ye may be able to bear it."* That does not fit with some of my experiences or with experiences that others have shared with me.

Evil is a powerful force. The spiritual force for evil wins some skirmishes on our battlefields of evil versus good. I do not believe that it wins those skirmishes because God organized, willed and negotiated its wins in advance. It wins when *we* let it win. Each of us bears personal responsibility for the extents to which we allow the spiritual force for evil to occupy our soul battlefields.

The Anglican Collect for Easter Day contains the following recognition of God's influences on us, plus a request to Him: *"that, as by thy special grace preventing us thou dost put into our minds good desires, so by thy continual help* **we** (present author's emphasis) *may bring the same to good effect."*

---

19  Ezekiel 18:25
20  James 4:7
21  I Corinthians 10:13

# BATTLEFIELDS

## ORGANIZED RELIGION VERSUS SCIENCE

Albert Einstein[22] expressed beautifully what he called the *"cosmic religious feeling"* that drives honest quests for truth, in organized religion and science: *"The beginnings of cosmic religious feeling already appear in earlier stages of development - e.g., in many of the Psalms of David and in some of the prophets... The religious geniuses of all ages have been distinguished by this kind of religious feeling, which knows no dogma and no God conceived in man's image; so there can be no Church whose central teachings are based on it. Hence it is precisely among the heretics of every age that we find men who were filled with the highest kind of religious feeling... Looked at in this light, men like Democritus, Francis of Assisi and Spinoza are closely akin to one another."*

Those driven to explore the universe by that *"cosmic religious feeling"* have often been in conflict with their contemporary religious authorities. David Landes[23] described as follows some of the suppression of science by the Catholic Church: *"When the Portuguese conquered the South Atlantic...A readiness to learn from foreign savants, many of them Jewish, had brought knowledge that translated directly into application...in 1497, pressure from the Roman Church and from Spain led the Portuguese crown to abandon this tolerance...the intellectual and scientific life of Portugal descended into an abyss of bigotry, fanaticism, and purity of blood...by the 1520s, scientific leadership had gone."*

Vox Day[24] discussed the history of organized Christianity opposing science, including the persecutions of astronomers, chemists, mathematicians and philosophers and the large-scale banning and destruction of literature. Most of Day's examples of conflicts between Christianity and science happened more than 250 years ago, but he pointed out rightly that the following three are still prominent today: opposition

---

22  Einstein, A. 2011. *The World As I See It*. Translated by Alan Harris. Open Road: New York. 168p.p.34. First published in 1956, as *Mein Weltbild*.
23  Landes, D.1999. *The Wealth and Poverty of Nations: Why Some Are So Rich and Some are So Poor*. W.W. Norton & Company Incorporated: New York.658p.p.133-134.
24  Day, V. 2008. *The Irrational Atheist*. BenBella Books Incorporated: Dallas TX.305p.pp.29-30.

to the theory of evolution; ecclesiastical monopolies on lay-education; and rejection of modern medicine by some sects.

In particular, religious schools and fundamentalist believers who home-school their children are unlikely to emphasize that science should be embraced as a path to truth. Some believe that by de-emphasizing science they are doing God's will and protecting the faith of those under their care. This has a long history.

Owen Chadwick[25] quoted Cardinal John Henry Newman (1801-1890 CE) as follows: *"When science crosses and breaks the received path of Revelation, it is reckoned a serious imputation on the ethical character of religious men, whenever they show hesitation to shift at a minute's warning their position and to accept as truths shadowy views at variance with what they have been taught."* In this context, *"shadowy views"* refers to disclosures through science and *"taught"* refers to church dogma and doctrines.

Chadwick further explained Newman's position as follows, in which the quotations are from Newman: *"A scientist can throw off a startling theory and no one cares if he is wrong. If the Church adopts a theory which is later proved wrong, it matters, because its teaching affects the moral convictions of ordinary folk...To be slow in accepting what turns out in the end to be true, may be not dishonest but responsible...Sometimes we have to be silent before excess. If a people is convinced that a picture or crucifix works miracles, he is a crude irreverent pastor who destroys the picture or crucifix, for he needs to think of the simple faith and to bide his time. If the Church treated popular superstition so roughly, it would endanger 'the faith and loyalty of a city or district, for the sake of an intellectual precision which was quite out of place.' If the bishop knows the cult to be false, he must try to quell. But 'errors of fact may do no harm, and their removal may do much.' 'We may surely concede a little superstition, as not the worst of evils, if it be the price of making sure of faith.'"*

---

25   Chadwick, O. 2010.*Newman: A Short Introduction*. Oxford University Press: Oxford.83p.p.50, 65 - which contain, respectively, quotations from the following of the Charles Stephen Dessain volumes on Newman's papers: V 1. liii to liv and V 1.lxv-lxix.

# BATTLEFIELDS

I disagree. All avoidance of truthful disclosures through science and all tacit acceptances of baseless superstitions are dishonest and irresponsible. All moral convictions in any person, ordinary or extraordinary, must be based on truth. What Newman endorsed continues today across much of organized religion and is a huge disincentive to nonbelievers, especially scientists, exploring faith.

A more difficult problem arises when both sides make the mistake of agreeing that a particular view on a particular issue must always keep them apart, when in truth it can unite them. For example, many believers and many nonbelievers claim that one cannot believe in God *and* in the evolution of life forms. By pursuing their chosen and allegedly exclusive paths to truth, both sides are missing God-given opportunities to find more of His truth about His creation.

Ronald Britton[26] cited as follows F.R. Leavis' description of this kind of folly in his work on John Stuart Mill's essays: *"the besetting danger is not so much of embracing falsehood for truth, as of mistaking part of the truth for the whole. It might be maintained that in almost every one of the leading controversies, past or present, in social philosophy, both sides were in the right in what they affirmed, though wrong in what they denied."* Creation is true. So is evolution.

## EVOLUTION

Much of the history of the battle about evolution is available on a website that tracks anti-evolution activities and the law.[27] It is a history of bigotry, lying and depriving children of an education through which they can appreciate fully the God-given wonders of nature. I quote here at length from a source on the early history of the battle to show

---

26  Britton, R. 2009.Mind and matter: a psychoanalytic perspective, p.1-14. In R. Doctor and R. Lucas (eds.) *The Organic and the Inner World*. Karnac Books Limited: London.115p.p.9. Britton quoted from Leavis, F.R.1950.*Mill on Bentham and Coleridge*. Cambridge University Press: Cambridge. p105.

27  www.antievolution.org

how little things have changed.

In 1873, the Reverend J.H Gladstone[28] wrote the following: *"When Darwin's book on "The Origin of Species" made its appearance, I read it with great pleasure and interest... Though Darwin in that work treats only the lower animals, it was perfectly plain that the argument must also include the genus Homo, as far as his bodily frame and instincts are concerned. Nevertheless, I felt no shock to my religious faith: indeed the progressive development of animated nature seemed to harmonize with that gradual unveiling of the Divine plan which I had loved to trace in the Bible, while it offered a satisfactory explanation of those rudimentary organs which had puzzled me as a student of natural theology. But presently I heard around me many voices opposing the theory, not only as untrue, but as irreligious, while some of the other voices were loud in its praise because it was reputed anti-Christian. On listening, I seemed to distinguish two principal grounds of supposed antagonism between the development theory and Scriptural theology: 1st It cannot be true that God created all the different plants and animals if they only descended from other pre-existent forms. 2nd This view removes God further from His universe, and only allows His operation in the primitive forms or form at some incalculably remote epoch."*

Gladstone continued: *"Now the first of these objections turns on the meaning of the Hebrew word* Bara. *I failed to discover any philological reason for supposing this word means necessarily to make out of nothing, and I examined all the places - about fifty in number in which it occurs in the Old Testament. In each case it refers to a Divine act, but in not one is there any suggestion that the Divine action was exerted on nothing...As to the second objection...If we believe in the God of Epicurus, who set the world a-spinning, and then retired into inactivity, we certainly lessen the interest we can have in such a Being by widening the distance that separates us from the period when he handed over His creation to the guidance of*

---

28 Gladstone, J.H. 1873. Points of Supposed Collision Between the Scriptures and the Natural Sciences, p. 133-171. In *Faith and Free Thought: A Second Course of Lectures Delivered at the Request of the Christian Evidence Society*. Hodder and Stoughton: London.469p.p.155-158. Available from www.kessinger.net

*physical laws. If, however, we believe in the God of St. Paul, in whom 'we live and move and have our being*[29] *and 'by whom all things consist,*[30] *the sustainer as well as the giver of life,* ***it becomes a matter of no theological importance in what way He created each species, and development or evolution, if established, becomes merely the gradual carrying out of His mighty scheme of creation*** (present author's emphasis)."

I invite all anti-evolutionist believers to note the phrase: "***of no theological importance.***" Right there, back in 1873, it was made clear that the battle over evolution is completely unnecessary. God exists. Evolutionary theory explains much of the history and state of His creation. There is abundant evidence for God and evolution. The evolution of humans is evidence for God's existence.

This battle thrives on ignorance and lies, from both sides. The Christian philosopher Stephen Clark[31] wrote: "*I have no principled objection to the notion that all terrestrial life is genealogically related…But one of the most annoying features of current militant atheism is the habit of equating evolutionary theory solely with the most atheistical form of Neo-Darwinism…It is also, on the other hand, annoying to find that some American Protestants - and strangely enough, some Muslims - imagine that any benevolent interest in non-human animals, any insistence that we are all 'God's creatures', is a sign of Satanic influence. The only (bad) excuse for this is ignorance.*"

Kenneth Miller,[32] who explained evolutionary theory to those present during the famous court case in Dover Pennsylvania in 2005 about the teaching of Intelligent Design (see below), wrote the following about anti-evolutionists: "*For them, evolution is far more than a mistaken scientific theory. It is the cutting edge of a dangerous and destructive materialism that threatens the heart and soul of our civilization and culture, and*

---

29  Acts 17:28
30  Adapted from the Nicene Creed's: "*by whom all things were made.*"
31  Clark, S.R.L.2009. *Understanding Faith: Religious Belief and Its Place In Society*. St. Andrew's Studies in Philosophy and Public Affairs, Imprint Academic: Exeter, U.K.280p.p.2.
32  Miller, K.R. 2008. *Only a Theory: Evolution and the Battle for America's Soul*. Viking: New York. 244p. p.x , 221.

*it must be opposed at all cost."* He asked: *"Are we willing to allow science to work? Do we have the strength and wisdom to allow science to discard the ideas that don't work, and to search for genuine truth about the natural world?"* All honest seekers of truth, believers and nonbelievers, must answer 'Yes.'

Jerry Coyne[33] wrote the truth about evolutionary theory as follows: *"Evolution is simply a theory about the process and patterns of life's diversification, not a grand philosophical scheme about the meaning of life. It can't tell us what to do or how we should behave."* He told of a Chicago businessman who heard him lecture on evolution and then said: *"I found your evidence for evolution very convincing - but I still don't believe it."* That man chose to ignore what was for him an inconvenient truth - probably for the sake of maintaining his relationships with others who see rejecting evolution as politically and socially correct.

Some Christian apologists tell the truth that accepting evolution is compatible with faith, but many continue to bash Darwin and Darwinism, as defined for their purposes. Dinesh D'Souza[34] wrote: *"Evolution is a scientific theory. Darwinism is a metaphysical stance and a political ideology. In fact, Darwinism is the atheist spin imposed on the theory of evolution. As a theory, evolution is not hostile to religion. Far from disproving design, evolution actually reveals the mode by which design has been executed. But atheists routinely use Darwinism and the fallacy of the blind watchmaker to undermine belief in God."* D'Souza was right that evolution as a theory is not hostile to religion or to faith, but he was wrong about *"Darwinism,"* if it is defined properly as studying and attempting to improve on Darwin's body of theory, which is based on natural selection and descent with modification.

---

33   Coyne, J.A. 2009. *Why Evolution is True*. Viking: New York.282p.p.225, 221.
34   D'Souza. D. 2007. *What's So Great About Christianity?* Tyndale House Publishers: Carol Stream IL. 348p.p.157.

# BATTLEFIELDS

Charles Colson and Nancy Pearcey[35] began their chapter *"Darwin in the Dock"* with the following quote from Phillip Johnson: *"As long as Darwinists control the definition of key terms [such as science] their system is unbeatable, regardless of the evidence."* Statements like this dissuade nonbelievers, especially scientists, from exploring faith. Darwinists do not control how science is defined. The more important issue is to define Darwinism itself accurately, or better still not to define it at all but to study Darwin's works thoroughly. All are readily available as books and online.[36] Believers who cannot stomach reading Darwin might consider reading works about evolution by fellow believers who are also scientists; for example, Denis Alexander,[37] Sam Berry,[38] Francis Collins[39] and Karl Giberson.[40]

Colson and Pearcey recounted events from the life of Dave Mulholland and his young daughter Katy, as follows: *"With the help of his pastor and several friends he* (Dave)*…could now tell Katy what the big bang really means…He could put into words, without too much trouble, how the anthropic principle draws together overwhelming evidence for design running through the physical universe at every level. He could even field most of Katy's questions about where life came from - how laboratory experiments* **often 'cheat' to get even the flimsiest results** (present author's emphasis) *and how the discovery of DNA gives positive evidence for a creator."* Dave was already planning to poison Katy's view of science, by portraying experimental scientists as habitual cheats.

---

35   Colson, C. and Pearcey, N. 1999. *How Now Shall We Live?* Tyndale House Publishers: Carol Stream IL.641p.p.81-82.
36   Darwin's major works were compiled in 2005 as one volume, with commentary and editing by Nobel Prize Laureate James D. Watson, as *Darwin. The Indelible Stamp: The Evolution of an Idea.* Running Press: Philadelphia PA.1260p. Darwin's complete works are freely available online at: http://en.wikipedia.org/wiki/The_Complete_Works_of_Charles_Darwin_Online
37   Alexander, D.R. 2008. *Creation or Evolution. Do We Have to Choose?* Monarch Books: Oxford, U.K.382p.
38   Berry, R.J. 2007. *Creation and Evolution. Not Creation or Evolution.* Faraday Paper 12. 4p. Available at www.st-edmunds.cam.ac.uk
39   Collins, F. S. 2006.*The Language of God; A Scientist Presents Evidence for Belief.* Free Press: New York.294p.
40   Giberson, K.W. 2008. *Saving Darwin: How to Be a Christian and Believe in Evolution.* HarperCollins: New York. 248p.Giberson, K.W. and F.S. Collins. 2011. *The Language of Science and Faith: Straight Answers to Genuine Questions.* IVP Books: Downer's Grove IL. 250p.

## Free Thought, Faith, and Science

The story continued: "*Dave grabbed a book at random...'Anyone who says some natural force in the chemicals 'wrote' the DNA code - well, that's like saying the chemicals in the paper and ink wrote the words on this page. He* (Dave) *was gaining confidence, and at times Katy dropped her defensive attitude...But today she was back to her old combative mode. She tossed her head. 'Oh, I don't have any problem with God being at the start of it all,' she said airily. 'Maybe he did kick everything off, back at the very beginning.' Dave...knew this admission represented progress, even though Katy refused to show it.*"

Katy continued: "'*But everyone knows that once life was here, it evolved just as Darwin said it did. I saw it in a textbook at school.'...Dave took the book from her hands and felt his stomach tighten. He hadn't covered these issues yet. Katy gazed at him triumphantly. 'It's right in the book, Dad.'...Dave clenched his hands.* **God**, *he prayed,* **no wonder she keeps fighting me - fighting you. Everything she gets at school is saying that nature can do it on its own, that you are irrelevant** (original authors' emphasis)."

A father condemned the efforts of an honest teacher and an honest textbook author to share what was known about evolution. That same father portrayed his daughter's honest quest for truth as a rebellion against him and against God and assumed that God would be pleased with him. Science was trashed. Family strife and parent-teacher strife were increased. The spiritual force for evil could thank those who had endorsed anti-science nonsense. The only actors who emerged with honour and with any record of having honoured God, whether they believed in Him or not, were the teacher, the textbook author and Katy.

Pearcey[41] included a very similar story in another publication: "*A first grader came home from school one day and asked: 'Who's lying, Mom - you*

---

41   Pearcey, N. 2005.*Total Truth: Liberating Christianity from Its Cultural Captivity*. Study Guide Edition. Crossway: Wheaton IL.511p.p207-208.

# BATTLEFIELDS

*or my teacher?' That day, it turned out, the teacher had informed the class that humans and apes are descended from a common ancestor. Little Ricky was bright enough to figure out that this didn't square with what his mother had taught him from the Bible, so he figured one of them must be making things up. Surely, it couldn't be the teacher; after all, in his young eyes she was the expert, the professional. No, the person he decided to doubt was his mother. With sorrow, she realized that she had better start on a long process of counter-education."*

That particular long process will damage everyone involved. The damage will be needless and contrary to the best interests of all and to the will of God, which is for us to be set free by getting as close as we can to *truth*.

Pearcey continued: "*It is because of incidents like this...that the controversy over teaching evolution refuses to die...the public is right to be concerned...Darwinism functions as the scientific support for an overarching naturalistic worldview, which is being promoted aggressively far beyond the bounds of science. Some even say we are entering an age of 'universal Darwinism,' when it will no longer be just a scientific theory but a comprehensive worldview. In order to have a redemptive impact on our culture, Christians need to engage Darwinian evolution not only as science but also as a worldview.*"

No one needs to "*engage Darwinian evolution,*" as if it was an enemy. Rather, everyone needs to become fully and honestly acquainted with the works of Darwin and current evolutionary theory.

In 2009, a Christian bookshop in a shopping mall in Makati City, Philippines filled the air with sounds of a bouncy children's choir singing about Bible verses to simple carousel-like melodies.[42] Track 1 on their CD was "*In the Beginning,*" with a rousing chorus

---

42  Elkins, S. *20 Bible Verses Every Kid Can Sing and Know.* Wonder Kids CD W-0760. Copyright held by Wonder Workshop: Mount Juliet TN. www.wonderkids.com

shouting about Genesis 1:1. I waited for the inevitable coverage of marine life. The kids duly sang out that God made all of it on day five. Now try teaching such kids and their listeners the truth about marine biodiversity.

Marvin Lubenow[43] portrayed scientific interpretations of the human fossil record as a conspiracy against the truth that can be found in Genesis. He alleged that conspiratorial scientists have prevented all but their like-minded fellow evolutionists from examining the original specimens and: *"will stop at nothing in attempting to influence the public toward human evolution and against creation."* He praised the so-called *"Creationist Dating Revolution"* and rejected scientifically accepted methods for dating human fossils.

Rapture expert Tim Lahaye wrote that Lubenow's book: *"exposes the current frauds the evolutionists are using to deceive yet another generation of young people that human evolution is a fact."* Ken Ham, the President of *"Answers in Genesis,"*[44] praised Lubenow's book as follows: *"After reading this masterful book, only a hardened heart would still accept human evolution and reject divine creation."* I once saw Ham on BBC World TV explaining that Noah took all the dinosaur species on board the ark, as eggs.

Lubenow endorsed Young Earth Creationism as science, but criticized science in general, as follows: *"we can never know if trusting science to lead us into truth is a very wise worldview or a very foolish one...the track record for self-correcting by scientists is very poor."* He continued: *"Human evolution has been falsified...by the recent evidence that the dating methods are a fiction and the earth is young...with a young earth, all evolution is dead."* The track record of self-correction in science is actually very good. The age of the universe is about 13.7 billion years. The age of the

---

43  Lubenow, M.L. 2004. *Bones of Contention: A Creationist Assessment of Human Fossils.* Revised and Updated. Baker Books: Grand Rapids MI.400p.p.38, 51-52, 326.
44  www.answersingenesis.org

Earth is about 4.5 billion years.[45]

Some believers who oppose evolution seem comfortable with what they can agree to call 'microevolution.' They seem able to accept that bacteria have evolved into other bacteria and maybe fish into other fish etc., but they deny that there could be an evolutionary path from unicellular life, through a vast diversity of invertebrates, to fish, amphibians, reptiles, birds and mammals, including us. They insist that one so-called 'kind' cannot evolve into another kind, because 'macroevolution' like that would contradict the Bible.

There is no valid basis for any distinction between so-called microevolution and macroevolution. The same biochemical, genetic and environmental drivers and mechanisms have operated throughout all evolution and are still in operation. The 'macro' versus 'micro' argument is anthropomorphism - a false 'human's eye view' of what constitute minor and major differences among organisms. For a virus or a bacterium, a new enzyme or a new protective coat can open up a whole new way of life.

Ever since life arose on Earth, its ecosystems have been occupied by biological communities in which diverse types have done very well, or well, or moderately well, or badly, or very badly, or so badly that they have gone extinct. No one knows how many species are presently extant on Earth. Estimates range from tens of millions to 100 million or more. Only about 1.9 million species have been described scientifically.[46]

As the planet has changed, its ecosystems and their species have changed. For example, a wide diversity of cichlid fish species has evolved over

---

45  See, for example: White, R.S. 2007. *The Age of the Earth*. Faraday Paper 8. 4p. Available at: www.faraday-institute.org
46  Chapman, A. 2009. Numbers of Living Species in Australia and the World. Australia Biological Resources Study: Canberra, Australia. Available at: http://www.environemnt.gov.au/biodiversity/abra/publications/other/species-numbers/index.html

only the last several hundred thousand to several million years in Lakes Malawi, Tanganyika and Victoria, occupying diverse ecological niches, which have also been evolving. There are now about 1,500 cichlid species in those lakes.[47] I have dived in Lake Malawi and seen some of their astonishing beauty and diversity. Telling the story of their evolution honours God.

I discovered the following wonderful words in an Order of Service used at Holy Trinity Anglican Cathedral, Suva, Fiji: "*For you* (God) *the earth has brought forth life in all its forms.*"[48] 'Yes.' God, the Original One, the Wisdom, the Word, wrote the codes and created the wherewithal by which everything else in the spiritual and material realms has ever existed, exists now, and will ever exist. His creativity produced the right ingredients and conditions for the formation of the Earth and for the chemistry of life and evolution of all life forms to follow - all for His pleasure.

Fellow believers who persist in denying the reality of the evolution of life forms, despite the abundant evidence, are urged to use their Free Thought to consider the possibility that the mechanisms of evolution are actually God's chosen mechanisms, engineered by Him from His creativity. Nonbelievers are invited to use their Free Thought to consider the possibility that the mechanisms of evolution, as seen throughout nature, might indicate a higher creativity. In truth, there is no need to fight about evolution.

The truth of evolution is clear. God did not add fully formed salmon to rivers or place fully formed sparrows in fully formed trees. Fully formed parasites and their fully formed vectors and fully formed hosts were not suddenly conjured up all together. Neither were fully formed flowers and their fully formed pollinators. All of the above and all other

---

47   Salzburger, W. 2009. The interaction of sexually and naturally selected traits in the adaptive radiations of cichlid fishes. *Molecular Ecology* 18: 169-185.
48   From *The Great Thanksgiving*, p.486 In *Liturgies of the Eucharist from a New Zealand Prayer Book/ Ite Karakia Mihinare o Aotearoa*. 1989. Genesis: Christchurch, New Zealand. 545p.

forms of life evolved *naturally*. Faith and evolutionary science are fully compatible. Both are based on truth. Unfortunately, the battle over evolution continues and has been intensified by the anti-evolutionists' promotion of the so-called theory of Intelligent Design.

## INTELLIGENT DESIGN

Allene Phy-Olsen[49] summarized the political and religious agenda of the Intelligent Design movement as follows: "*to defeat scientific materialism, with its 'destructive moral, cultural and political legacies,' and to replace it with a theistic understanding of human existence.*" John MacDonald Smith[50] described the Intelligent Design movement's strategy as follows: "*...a wedge is to be driven into the 'tree' of 'materialistic science' at its nearest point, which is Darwinian evolution.*" The Intelligent Design movement's original Wedge Document[51] has the following as its stated goal: "*To replace materialistic explanations with the theistic understanding that nature and human beings are created by God.*"

On the basis of its declared goal and strategy alone, Intelligent Design fails to qualify as a scientific theory. Disclosures of truth through science cannot be constrained by or directed towards any particular political or religious agenda. Science is *free* inquiry. Intelligent Design is simply Creationism in disguise. Neither Creationism nor Intelligent Design can say anything about the mechanisms that gave rise to the world's astonishing past and present biodiversity, except that 'God did it.' Neither Creationism nor Intelligent Design has any predictive capability or testability.

Intelligent Design has a peculiar attribute to which most of its

---

49  Phy-Olsen, A. 2010. *Evolution, Creationism, and Intelligent Design*. Historical Guides to Controversial Issues in America. Greenwood; Santa Barbara CA.171p.p.17.
50  Smith, J.M. 2009.Introduction, p.3-6.In J.M. Smith and J. Quenby (eds.) *Intelligent Faith*. O Books: Winchester, UK.330p.p.4.
51  See the Wedge Document. Available at: http://www.antievolution.org/features/wedge.pdf

proponents seem blind. To call God an 'Intelligent Designer' is to put His attributes on par with human attributes. This is at best faint praise and at worst insulting. God created the means by which our intelligence evolved. He gives us our souls and our capacity for Free Thought, through which we develop and use our creativity.

It pays God no compliment at all to accord Him the capability to *design* things *intelligently*. God has capabilities far beyond any that we can conceive. It is clearly an insult to God to deny the existence of the mechanisms that He created for the chemistry of life to arise and for Earth's biodiversity to come into being and continue in a glorious state of flux.

Everything in the material realm can be said to have the 'appearance of design' - a hydrogen atom, a protein molecule, a virus, a cell, a snowflake, a leaf, a bird, an ecosystem, a planet, and a galaxy. Everything looks as if it has been made to take its place and to assume its capacity to function in the material realm.

A human perception of an appearance of design is no proof of any act of designing. In truth, we cannot tell whether any particular natural entity or process in the material realm has originated by chance, or by design, or by some combination of both. Intelligent Design offers only the following alternatives - believe or disbelieve that God was the Creator/Intelligent Designer.

Intelligent Design has nothing useful to say about how the HIV/AIDS viruses, the Ebola virus, tuberculosis bacteria, the potato blight fungus, and the parasites that cause elephantiasis, malaria and river blindness came into being. Some believers argue that this world was once perfect and free from all nasty things, but got spoiled by the spiritual force for evil and the fall of man. This explanation is of course completely untestable, but think of its implications. Are snakes evil? Are predators evil? Are species that poison and sting evil?

# BATTLEFIELDS

'No.' Life is beautiful. No living thing has ever been intrinsically evil or intrinsically good. For example, an important farmed species in one country can be a harmful alien invasive species in another. In any case, if God did not create and design all the lethal and sub-lethal genes, pests, parasites, pathogens and predators, then who did? Was it the spiritual force for evil? 'No,' those amazing expressions of life evolved naturally, like all others.

Proponents of Intelligent Design claim that the Intelligent Designer's work is obvious in the high complex natural world in us and around us and that highly complex forms and systems must have been brought into being fully formed, or substantially so, because of the high improbability and lack of time for their stepwise development.

Those who argue thus miss the point that systems with low complexity can combine with each other and can change each other markedly and rapidly, sometimes yielding further systems of much greater complexity. One cannot prove that systems such as DNA and living cells bear the signature of God, as Stephen Meyer[52] attempted.

William Dembski[53] and other proponents of Intelligent Design hold that "*complexity is crucial for inferring design.*" Denis Alexander[54] quoted Dembski as follows: "*There has to be a **reliable way*** (present author's emphasis) *to distinguish between events or objects whose emergence additionally requires the help of a designing intelligence…the whole point of the design inference is to draw such a distinction between natural and intelligent causes.*" Alexander summarized as follows the ensuing theological problem for the proponents of Intelligent Design (ID): "*ID promotes the idea of a two-tier universe, part ruled by natural forces and*

---

52   Meyer, S.C. 2009. *Signature in the Cell: DNA and the Evidence for Intelligent Design*. HarperOne: New York.611p.
53   Dembski, W.A. 1998. Science and Design. Available at: http://www.firstthings.com
54   Alexander, D.R. 2009. Evolution - Intelligent or Designed? p.7-22. In J.M. Smith and J. Quenby (eds.) *Intelligent Faith*. O Books: Winchester, UK.330p.p.21. Alexander quoted from: Dembski, W.A. 2004. *The Design Revolution: Answering the Toughest Questions about Intelligent Design*. Intervarsity Press: Downer's Grove IL. 334p.

*part designed, whereas Biblical creation doctrine sees everything as being completely dependent upon the personal triune God."*

If everything came from the hand of God, then *de facto* He must have designed *everything*. If no God was involved, then there was no designer and *nothing* was designed. It is not tenable to believe in God as Creator, Intelligent Designer and Author of all creation *and* to posit also the existence of some kind of natural background that was not designed by Him. There is of course no *"reliable way"* by which Dembski and other proponents of Intelligent Design can distinguish between the apparently designed and the apparently not designed.

The main argument for Intelligent Design and against evolution is the argument from irreducible complexity, which states that the essential and interdependent components of complex structures and systems that we observe in life - such as the bacterial flagellum, the eye, the blood clotting system and the immune system - could not have come together naturally and in a stepwise manner. Rather, those structures and systems must have been put together as complete assemblages from the very start.

The proponents of Intelligent Design hold that God drew up a huge series of 'blueprints' and then made and somehow placed on Earth, either fully formed or well on the way to being so, the diverse and complex things that are in and around us today, including distinct plants and animals, the bacterial flagellum and the blood clotting system etc. They hail the defence mechanism of the bombardier beetle as a prime example of, allegedly, irreducible complexity.

When attacked by a predator, a bombardier beetle mixes small quantities of hydroquinone and hydrogen peroxide, which were formerly stored separately, and squirts out the resulting foul chemical compound at a temperature close to the boiling point of water. So, God must have put the bombardier beetle and its defence mechanism on Earth fully formed. Right? 'Sorry, wrong.' There are hundreds of different species of bombardier beetles in the large beetle family Carabidae, which also has non-bombardier members that deter their would-be predators with hydroquinone-based chemicals. The bombardier beetles evolved a way of adding an oxidizing agent, with spectacular results. Their defence mechanism is a product of evolution.[55]

The surgeon Charles Brooke,[56] who was an early believer in irreducible complexity, proposed the following examples: the *"Mechanism of the absorbent system,"* (meaning our bodily systems for pooling and voiding metabolic wastes); the human eye; and the ear. He asked: *"Can anyone seriously maintain that these perfect adaptations of means to an end formed themselves, or resulted from the blind action of atomic force?"* These systems and structures were never *"perfect adaptations."* They evolved by using whatever happened to be available to make the best of things for the time being. No intelligent designer would have designed the human eye as it has evolved, with its blind spot.

Brooke proposed that brass and steel sweepings from a workshop floor be put into a bag and shaken to see if: *"they would spontaneously evolve a first-rate chronometer."* Norman Geisler and William Roach[57] used another version of that bankrupt anti-evolution argument, as follows: *"All known natural forces, even over long periods of time, produce more randomness and more disorder. For example, dropping red, white and blue confetti from an airplane never produces the American flag on a field. And dropping from higher elevations (which give it more time), no matter how high you go, will not help to specify it into an American flag. Only intelligence can do that."*

---

55  Weber, C.G. 1981. The bombardier beetle myth exploded. *Creation Evolution Journal* 2 (1): 1-5; available at: http://nsce.com
56  Brooke, C.1873. The evidence afforded by the order and adaptations of nature to the existence of a God, p.51-85. In *Faith and Free Thought: A Second Course of Lectures Delivered at the Request of the Christian Evidence Society*. Hodder and Stoughton: London.469p.p.77. Available at: www.kessinger.net
57  Geisler, N.L. and W.C.Roach. 2011. *Defending Inerrancy: Affirming the Accuracy of Scripture for a New Generation*. BakerBooks: Grand Rapids MI.378p.p.359.

# Free Thought, Faith, and Science

There is also a famous imaginary band of monkeys banging away forever on typewriters but never making any observable progress towards producing even a single line of Shakespeare. Anthony Flew[58] described how Gerry Schroeder had convinced him that life on Earth could not have *"been achieved by chance,"* when Schroeder showed that the odds against those monkeys producing even one line of a sonnet were 10 to the power 690 and that there were not enough particles in the universe to even *"write down the trials."* Schroeder[59] argued that chance could not have produced all of the types of proteins and cell that we see, because of the impossibly large numbers of *"evolutionary trials"* needed.

With all due respect to Flew, Schroeder and others who accept those arguments based on numbers of required trials, time and improbability, I find them all invalid. They are all based upon the assumption that complexity and diversity can evolve only through random chance and single location, stepwise changes, such as one mutation or one amino acid hitching up to another.

Any system of complexity and diversity is a baseline for further evolution towards systems of greater complexity and diversity, which will then become new baselines, with new potentials for further change. Complexity and diversity can undoubtedly evolve rapidly in the living world. We have to recognize that many types of change proceed on many levels - RNA and DNA, cleavage planes in early embryos and interactions of the life history stages of species that comprise biological communities, in nested and interlinked ecosystems.

The post-Big Bang universe was buzzing with matter, energy and forces for change. Evolution was inevitable. I have no problem in accepting that there were enough interactions among the available ingredients and enough time. The first big trend in that direction on planet Earth

---

58   Flew, A. 2007. *There is a God. How the World's Most Notorious Atheist Changed His Mind*. HarperOne: New York. 222p.p.75-78.
59   Schroeder, G.L. 2009. *God According To God: A Physicist Proves We've Been Wrong All Along*. HarperOne: New York. 249p.p.43-46.

# BATTLEFIELDS

would have been towards an increasing diversity of carbon-based organic chemicals, in energy-rich environments resembling present day volcanic pools and submarine thermal vents. The material realm kitchen would have contained inorganic and organic substrates and catalysts and organic glues, thinners, thickeners and films. The presence of clay substrates was probably a significant factor.

The next big trend was towards self-replicating organic molecules, at which point chemical 'life' of the very simplest kind can be said to have begun. The steps from the first proteins to RNA, DNA and the first viruses are not yet known, but from the earliest self-replicating and information bearing molecules, the pace of change must have accelerated, with new proteins and new self-replicating systems arising, competing, co-existing and combining. Increasingly complex chemical systems led to the first primitive cells. Along the way, many unsuccessful systems must have been eliminated. Only a minority would have adapted to surrounding environments and coped with further change.

Quasi-living chemicals, primitive cells and their non-living surroundings formed the first ecosystems, in which a small change in one 'inhabitant' could have mega-consequences for others and for the whole system. Biological populations and species then evolved. Their interactions also evolved - in commensalism, parasitism and pollination; as predators and prey; and as biological communities. All forms of life are still evolving - as living components of ecosystems in which the non-living components are also evolving; for example, through human interventions, seismic activity and climate change.

In Dover Pennsylvania in 2005, Tammy Kitzmiller and others brought suit against the Board of Dover School for requiring teachers to read out a formal note that urged students to be sceptical of Darwin's theories and to consider Intelligent Design as an alternative explanation for the diversity of life on Earth. Anti-evolutionist Board members and friends sought the introduction of a new textbook, "*Of Pandas and*

*People,*" which was said to be about Intelligent Design as a scientific theory. It was clearly based on earlier drafts about creationism. The words "*creation,*" "*intelligent creator*" and "*creationists*" had been replaced by "*intelligent design,*" "*design proponents*" and "*intelligent agency.*" Sloppy editing had left in the phrase "*cdesign proponentsists.*"

Lauri Lebo[60] wrote a gripping account of the case. Gary Johnstone and Joseph McMaster[61] put together a stunning video. Robert Pennock and Michael Ruse[62] published discussions of the case by various authors and Pennock's own Expert Report, together with Judge John E. Jones III's judgement and part of the testimony of Michael Behe, a leading proponent of Intelligent Design.

Supporters of the book on Intelligent Design were caught lying about their creationist motives. Judge Jones wrote in his judgement: "*The citizens of Dover area were poorly served by the members of their Board who voted for the ID policy. It is ironic that several of these individuals, who so staunchly and proudly touted their religious convictions in public, would time and again lie to cover their tracks and disguise the real purpose behind the ID policy.*"

Behe[63] had long been arguing that structures such as the bacterial flagellum are irreducibly complex. Various reviewers[64] have demolished his arguments. Some of the components of the bacterial flagellum have even been described as "*second-hand parts.*"[65] In the Dover case, Behe

---

60   Lebo, L. 2008. *The Devil in Dover: An Insider's Story of Dogma v. Darwin in Small-Town America.* The New Press: New York.238p.
61   WGBH/NOVA and Vulcan Productions. 2007/20008. *Judgment Day: Intelligent Design On Trial.* DVD. Available through: www.shop.wgbh.org
62   Pennock, R.T. and M.Ruse. 2009. *But Is It Science? The Philosophical Question in the Creation/Evolution Controversy.* Prometheus Books: New York.577p.
63   Behe, M.J.1996. *Darwin's Black Box: The Biochemical Challenge to Evolution.* Touchstone: New York.307p.
64   Atkins, P. Review of Darwin's Black Box: The Biochemical Challenge to Evolution. Michael J. Behe. Available at: www.infidels.org/library/modern/peter_atkins/behe.html See also the extensive review by David Ussery; updated August 10, 2000. Available at: www.cbs.dtu.dk/staff/dave/Behe_text.html
65   Leake, M., Wadhams, G.H. and J.P. Armitage. 2008. Parts exchange: why molecular machines are like used cars. *Biologist* 55 (1): 33-39.

# BATTLEFIELDS

used again his published examples of alleged irreducible complexity, including the bacterial flagellum, and was met again with much published evidence to the contrary.

Behe also proposed that Intelligent Design was open to experimental rebuttal, by growing bacteria with no flagellum for many (ca. 10,000) generations in an environment exerting selective pressure for mobility. According to Behe, if any of those bacteria produced a flagellum, Intelligent Design would be disproved. The problem for Behe's approach is that there are so many potential combinations of diverse non-motile bacterial species and strains and the stressful environments in which they can be grown. Even if many combinations were tried and all *failed* to produce a flagellum, nothing could be concluded about what might happen with any other combination.

Readers who are not used to thinking about the planning and conduct of scientific research can compare Behe's proposal with research done by Kai Lohbeck and others[66] on *Emiliana huxleyi* - a marine phytoplankton species, which suffers impaired calcification in acidic waters. They found that some populations adapted after 500 generations (one year) to become 50% better at calcifying than non-adapted ones. They demonstrated *evolution* in action, by reproducing in the laboratory the acidification of seawater that climate change is causing globally. Marine calcifying organisms are evolving to meet this challenge, as far their genetic diversity and the chemistry of calcification will permit.

## Interpersonal, Group and Institutional Battles

Interpersonal battles begin at the family level, with sibling rivalry and parent-child conflicts that last for life in some cases. Work and play become our main interpersonal, group and institutional battlefields

---

[66] Lohbeck, K., Riebesell, U. and T. Reusch. 2012. Adaptive evolution of a key phytoplankton species to ocean acidification. *Nature Geoscience*. April 8, 2012; doi: 10.1038/NGEO 1441.

as we try to impress and outcompete others. We strive for recognition and success in academia, the arts, commerce and sport etc. Interpersonal rivalry, professional jealousy and institutional conflict pervade most fields of human endeavor, including organized religion and science. We also balance competition and conflict with collaboration and peacemaking.

The same applies for some animals. Frans de Waal[67] described how our close relatives the bonobos defuse potential conflicts: "(They) *can be unfriendly to their neighbours, but soon after a confrontation has begun, females often rush to the other side to have sex with both males and other females…the scene rapidly turns into a sort of picnic* (and) *ends with adults from different groups grooming each other while their children play."*

Domesticated animals provide their owners with various goods and/or services, while surrendering their freedom and sometimes their longevity in return for guaranteed food, shelter, health, security and (for the lucky few) mating opportunities. These complex deals require stable social environments. Beth, a much-missed family Labrador, would move between two fractious family cats and entreat for peace, by posturing and whimpering. She sought stability in the household for the continuation of its supply of goods and services to her. For the same reason, dogs are sensitive to conflicts between human family members.

God is not urging bonobos to 'make love not war' or domesticated animals to keep their deals sweet by lessening conflict in the home. All conflict reducing behavior in animals has been *selected* - naturally and/or by human breeders. It is all a product of basic thought about material realm stuff and expediency. Humans lessen conflicts by social activities - talking, eating, drinking, dancing and giving gifts. In humans, basic thought about expediency and the practicalities of life merges with Free Thought outcomes about morality, justice and peace

---

67  De Waal, F. 2009. *The Age of Empathy: Nature's Lessons for a Kinder Society*. Harmony Books: New York.291p.p.24.

of mind.

Battles within and among organized religions are mostly battles of 'us' versus a particular group of 'them,' or 'us' versus everyone else. Most brands of organized religion have staked claims to be the only true path to God; for example, Shiite and Sunni Islam; Roman Catholic and Protestant Christianity. On that basis, some fundamentalists feel spiritually commissioned to discriminate against outsiders and even to kill them. Those feelings do not come from God. They are outcomes of Free Thought dominated by the spiritual force for evil.

Intolerance and violence among members of different organized religions, sects and churches are continuing, but dialogue and peacemaking appear to be increasing. Interfaith websites are fostering ecumenism, tolerance and joint action.[68] The Alliance of Religions and Conservation[69] is working for nature conservation and responsible use of natural resources. Since the Second Vatican Council, the Catholic Church has increased its efforts to engage with other organized religions, particularly Islam and Judaism. In 1964, Pope Paul VI established the Office of Non-Christian Affairs, which was later renamed as the Pontifical Council for Interreligious Dialogue.

Christ reportedly called peacemakers blessed[70] and told us to serve one another and not seek to be the first in line.[71] He commanded us to love our enemies[72] and to love everyone as much as we love our selves.[73] Interpersonal, group and institutional battles reflect our failures to love enough. On all intrapersonal, interpersonal, group and institutional battlefields, the only good fight to be fought is the fight for *truth*, in love. Expansion of the faith-science quest for truth would lessen conflict between and within organized religion and science.

---

[68] See, for example: www.interfaithorganizations.net; www.interreligiousinsights.org
[69] www.arcworld.org
[70] Matthew 5:9
[71] Mark 10:31 and 37-44
[72] Matthew 5:44
[73] Matthew 22:39 and Mark 12:31

# Free Thought, Faith, and Science

## A Few Battles At Church

I am a member of the Union Church of Manila (UCM), which is an interdenominational Protestant church, attended also by some Catholics. UCM has Sunday congregations of about 1,400 from about 40 different nationalities. I have had few battles so far at UCM about how faith fits with science, but there would probably have been more had this come up for more discussion. Biblical literalism is well represented at UCM, as shown in the following examples.

I once heard an invited lay preacher state in his sermon that evolution was *"only a theory, not a fact"* and must be wrong because it is contrary to the truth of the written word of God. Had I invited to UCM that day a fellow biologist who was exploring faith, his explorations would almost certainly have ceased.

I heard a Pastor preach on creation. He compared the ancient Babylonian creation myth with evolution according to science and with Genesis. In the Babylonian myth, the god Marduk defeats other warring gods, and humans are created from the bones of sea monsters, to work as slaves. The Pastor's version of evolutionary science was laced with sarcasm about particles having *"tiny minds,"* *"the glory of the amoeba,"* and the dog, the monkey and humans being formed by accident. He continued with creation according to Genesis, with everything spoken into being by God. After each creation story, the congregation was asked to respond. The Babylonian myth and evolution according to science were equally ridiculed. Genesis was praised.

I went to see the Pastor. I learned that he had been unaware of much of the evidence for evolution, current evolutionary theory, and current thinking on how the chemistry of life might have evolved, especially the likely key role for RNA. He was very gracious and said that he was open to all evidence.

On the following Sunday, he began preaching with a recap from his

# BATTLEFIELDS

pro-Genesis sermon and stated rather bravely that: *"Genesis I is a poetic song which leaves room for God to create quickly or over long periods of time."* In many churches, even that mild and truthful statement implying the possibility of evolution would have been unwelcome and could have brought much disunity. The quest for truth leads to unity, but telling the truth along the way can alienate some who find it inconvenient.

I heard an Elder preach on how Christ had recruited fishermen as His disciples[74] and how we should strive to catch people for Christ, following His great commission.[75] I enjoyed his sermon very much up to its last few minutes, when he stated: *"Two billion people in this world of ours are still unreached. They have never once had a credible presentation of the good news of Jesus. Not once. Some 2,400 of them die **and go to hell*** (present author's emphasis) *every hour; forty each minute; almost one every second. In the time that I just spoke this last sentence, five people have died* (he then clicked his fingers 5 times at 1 second intervals) *forever lost."*

After the service, I asked him whether he really believed that all who had never heard the Gospel of Christ were either already in hell or bound to go there? His response was that the scriptures tell us that. I asked him how that could be seen as justice from a loving and merciful God and how, in any case, could any human pronounce on who must be in hell or bound for hell - that being the prerogative of God alone? He agreed that Jesus could show mercy to some.

When the apologist David Geisler visited UCM to give a seminar on evangelism and to distribute copies of a new book,[76] I explained to him over lunch why I could not ask my fellow scientists who are nonbelievers to accept that Jonah survived inside the digestive system of a fish for

---

[74] Mark 1:14-19 and Luke 5:1-11
[75] Mark 16:15
[76] Geisler, N. and D. Geisler. 2009. *Conversational Evangelism.* Harvest House Publishers: Eugene OR.223p.

three days and nights.[77] Geisler told me he thought that scientists had now shown that Jonah's alleged stay inside a fish stomach was possible. I told him very gently that he was mistaken.

I also asked Geisler what he thought of the notion that the souls of all of the Jewish victims of the Holocaust must be in hell, along with those of all other non-Christians? He replied that the same question had troubled C.S. Lewis and then cautioned me that he could proceed with our discussion only according to the limits of what had been revealed to him, which meant the literal words of the Bible. He would not budge from biblical literalism. Our debate ended there.

When I shared this exchange with a member of the pastoral team at UCM, he commented that Geisler might have done better to consider the possibility that I was being used as a source of God's revelations to him. Of course, the same could be said about Geisler's words to me. Did I receive a revelation from God, through Geisler, to become a strict biblical literalist and to see the light by believing in Jonah's alleged 72-hour resistance to fish digestive juices?

I am at peace in rejecting that possibility. Those who preach biblical literalism and inerrancy are misguiding believers into rejecting God's truthful disclosures through science and are turning honest seekers away from exploring faith. I will continue to confront those doctrines and advocate the quest for truth, through faith and science. Right or wrong, we will all answer to God.

---

77  Jonah 1:17

# CHAPTER 8

# TRUTH AND MORALITY

*"And ye shall know the truth, and the truth shall make you free"* [1]

### TRUTH, MORALITY AND US

According to the Apostle John, Christ assured us that we will have access to the truth and that knowledge of the truth will liberate us. John[2] also reported Christ's promise that God, as the Holy Spirit, will lead us to everything that is true: *"Howbeit that when he, the Spirit of truth is come, he will guide you into **all*** (present author's emphasis) *truth."* The truth that we find through faith and all truthful disclosures through mathematics and science come God.

Truth is defined here as that which is in accordance with facts and reality. Truth is 'what is' as opposed to 'what is not.' Truth is the opposite of lies and falsehood. Some will find this definition naïve, but it fits well with the philosophy and conduct of science, with common sense, and with my experiences through faith.

Morality is based on truth. For moral behaviour we must recognize

---
1   John 8:32
2   John 16:13

truth and act accordingly. Pontius Pilate[3] reportedly asked Jesus: "*What is truth?*" Pilate knew very well the truth about Jesus' innocence when he stated: "*I find in him no fault at all.*" Pilate acted immorally, against the truth - though his moral failure was part of the necessary path to God's Self-sacrifice as Christ, on our behalf.

Simon Blackburn[4] summarized our ethical attributes as follows: "*Human beings are ethical animals. I do not mean that we naturally behave well, nor that we are endlessly telling each other what to do. But we grade, and evaluate, and compare and admire, and claim and justify.*" According to Blackburn: "*No god wrote the laws of good behaviour into the cosmos. Nature has no concern for good or bad, right or wrong.*" I agree for all of nature, except humans.

Humans and animals pursue expediency. For animals, what works better in material terms is always right and what works less well is always wrong. Humans have codes of ethics and a morality defined by a lasting and higher standard of right and wrong, above expediency and material wellbeing.

Blackburn asked: "*So is there such a thing as moral knowledge? Is there moral progress?*" I say 'Yes,' but only at the individual personal level and I agree with Blackburn that these questions: "*have to be answered from within our own moral perspective.*" Moral knowledge and moral progress across humanity are always temporal outcomes of Free Thought about moral choices, *within* individuals.

Leslie Weatherhead[5] made following points about truth, *inter alia*: "*Religious truth has no authority, no value in personality, until it authenticates itself, until the mind leaps up and sees it true and accepts it, not

---

3 John 18:38
4 Blackburn, S. 2002. *Being Good: A Short Introduction to Ethics*. Oxford University Press: Oxford.162p.p.4, 133, 134.
5 Weatherhead, L.D. 1989. *The Christian Agnostic*. Arthur James Limited: London.264p.p.21, 26-27. First published in 1965.

# TRUTH AND MORALITY

*because its truth is **imposed**, or its refusal dreaded, but because it is **seen** to be true…truth may certainly be true whatever my opinion may be, but it has no **authority** with me unless I perceive it to be true. When I do, reality has spoken to me. This is not conceit…My plea is not for an impossible subjectivism. But the so-called infallible church or book has no power unless I **feel** what it says is true* (original author's emphases)." I agree.

Steven Pinker[6] showed that humans have become progressively less violent. The numbers of deaths in wars have declined. Warriors do much of their killing nowadays by roadside bombs and drones. But war is always horrific. Try to imagine what actually happens to households and tank crews who are obliterated by missiles that cannot miss. The statistics of modern conflicts are still truly awful: hundreds of thousands raped so far in the Democratic Republic of Congo; 60,000 murdered so far in Mexican drug wars.

Our historical retreat from violence might not be of much use for predicting what lies ahead. No one can predict the full impacts of some weapons of mass destruction. More importantly, the spiritual force for evil continues to engineer horror shows through the Free Thought of individuals. As Pinker recognized, one evil genius, like Adolf Hitler, can move many to participate in atrocities.

Jonathan Sacks[7] commented as follows on the doctors who got involved in Hitler's horror show: "*More than half of the participants at the Wannsee Conference in 1942, who planned…the murder of all Europe's Jews carried the title 'doctor.' They had either doctorates or were medical practitioners…I have known people who lost their faith in God during the Holocaust, and others who kept it. But that anyone can have faith in humanity after Auschwitz defies belief.*"

---

6   Pinker, S. 2011. *The Better Angels of Our Nature: Why Violence Has Declined.* Viking: New York.802p.
7   Sacks, J. 2011.*The Great Partnership: God, Science and the Search for Meaning.* Hodder and Stoughton: London.370p.p.86.

J.L. Mackie[8] wrote the following: *"there are no objective values...(and) no substantive moral conclusions can be derived from either the meaning of moral terms or the logic of moral discourse...But if there is no objective moral truth to be discovered, is there nothing left to do but to describe our sense of injustice? At least we can look at the matter in another way. Morality is not to be discovered but to be made: we have to decide what moral views to adopt, what moral stands to take. No doubt the conclusions we reach will reflect and reveal our sense of justice, our moral consciousness – that is, our moral consciousness as it is at the end of the discussion, not necessarily as it was at the beginning."*

Mackie's model has us defining and redefining human morality in whatever way we choose, with no need for a God or a spiritual force for evil. There is much compelling evidence that human morality is built on the *quasi-morality* seen in some animals (see below). This provided an evolutionary legacy for our morality *"at the beginning"* of our species. I agree with Mackie that we make our own moral choices about what we think is right and wrong and how we will choose to act in any given circumstances. But I believe that there is an unchanging Universal Moral Code, implanted by God in every human soul. Our compliance and non-compliance with that code are outcomes of our Free Thought. I define the Universal Moral Code as the standard for moral conduct, given by God for recognition and compliance by all humans.

James Alexander[9] summarized and commented on Derek Parfit's[10] new theory of human morality as follows: ***"everyone ought to follow optimific principles, because these are the only principles that everyone could rationally will to be universal laws.' 'Optimific' means that 'if everyone acted on these maxims, things would go in***

---

8    Mackie, J.L. 1977. *Ethics: Inventing Right and Wrong*. Penguin Books Limited: Harmondsworth, U.K.249p.p.105-106.
9    Alexander, J. 2012. *On What Matters* by Derek Parfit. *Philosophy Now*. September/October 2012. 2p. Available at: http://philosophynow.org
10   Parfit, D. 2011. *On What Matters*. Volumes 1 and 2. Oxford University Press: Oxford. 592p. and 848p.

# TRUTH AND MORALITY

*ways that would be impartially best'...Parfit's eventual Triple Theory of morality is that '**an act is wrong just when such acts are disallowed by some principle that is optimific, uniquely universally willable, and not reasonably rejectable**'* (present author's emphases of Alexander's quotations from Parfit)."

Parfit's theory is impressive, but again the lack of access to any higher/divine standard would seem to lead to never-ending successions of different moral codes, the provisions of which would last only as long as successive global human populations agreed about them. The Apostle Paul[11] wanted us to be: *"no more children, tossed to and fro, and carried about with every wind of doctrine...but speaking the truth in love."*

Towards that end, God gives every human a soul and a mind-soul interface to link the material and spiritual realms and implants in every soul the Universal Moral Code, as His standard for our morality. This applies to nonbelievers as well as believers. In faith and unbelief, we behave morally or immorally, according to the outcomes of our Free Thought.

## Perspectives on Truth

Harry Frankfurt[12] distanced himself, as follows, from the debate about whether there is a real distinction between truth and falsehood: *"I shall not even try - at least not by any directly confrontational argument or analysis - to settle once and for all the entangled debate between those who accept the reality of a meaningful distinction between being true and being false and those who energetically represent themselves...as denying that the distinction is a valid one or that it corresponds to any objective reality. That debate seems unlikely to be ever resolved, and it is generally unrewarding."*

---

11  Ephesians 4: 14-15
12  Frankfurt, H.G. 2006. *On Truth*. Alfred A. Knopf: New York.101p.p.8, 47-48,65.

Frankfurt continued: *"Practically all of us do love truth whether or not we are aware that we do so. And, to the extent that we recognize what dealing effectively with the problems of life entails, we cannot help loving truth... The notions of truth and factuality...are indispensable even for understanding the very concept of rationality itself. Without them, the concept would have no meaning, and rationality itself (whatever it might turn out to be, if anything, in such deprived conditions) would be of very little use."*

I agree. There is little point in discussing the nature of truth with anyone who denies its very existence. Neither is there any point in seeking the truth about anything if one feels that one's findings will have no significant status or value. Without *"truth and factuality"* there can indeed be no rationality.

Simon Blackburn[13] and others have explained the fundamental choice between being an absolutist, who sees truth as one and the same for all, or a relativist, who sees truth as varying with different perspectives and agendas. I am an absolutist, but I recognize that the whole truth is often elusive and sometimes unobtainable, as seen in the famous old scenario of 'Who struck John?'

John says he was struck and shows us his injuries, which we can assume with high confidence were not self-inflicted. John and his striker(s) know that he was struck. If he was struck in a melee, there might be some ambiguity about who did what to John, but in most cases the striker(s) and John know what happened. Witnesses can say what they saw and photographic or video evidence can be consulted. The whole truth about who struck John might never emerge, but this does not mean that truth itself is variable. There is only one *truthful* account of who struck John. God saw what happened. He records everything that happens in the material and spiritual realms.

---

13   Blackburn, S. 2006. *Truth: A Guide for the Perplexed*. Penguin Books Limited: London.238p.

# TRUTH AND MORALITY

Ophelia Benson and Jeremy Stangroom[14] summarized the many constraints to finding and telling the truth and rejected relativism, as follows: "*Truth is evaded when it is inconvenient, criminalized when it is 'insulting', denied when it contradicts religious beliefs, tampered with when it is in conflict with ethnic or national self-esteem, ignored when it is irritating to the powerful. Truth is always potentially a stumbling block, because* **it is of the nature of truth that it is what it is, regardless of anyone's wishes** (present author's emphasis)...*It is surely of the nature of truth that it has to be all of a piece. Its norms have to apply here as well as there, if they are to apply at all. That's why relativism about truth is always self-undermining.*" I agree. I see no basis for relativism on truth.

Norman Geisler[15] examined at length what he called: "*six alternative tests for truth...rationalism, fideism, experientialism, evidentialism, combinationalism, and pragmatism.*" He stated: "*The one insufficiency common to all of these tests for truth is that none of them could definitively establish one world view over another. Whatever applicability they may have* **within** *a world view, none was sufficient to adjudicate* **between or among** *world views* (original author's emphases)." So what? Is truth to be characterized *solely* by its capacity to adjudicate between worldviews? I think not. God's truthful disclosures are common to the worldviews of faith and science. In any and all worldviews, truth is truth and lies are lies.

Geisler continued as follows on fideism: "*...fideism is not really a* **test** (original author's emphasis) *for truth at all, it is simply a claim to truth. It reduces to the claim 'this is true because I believe it to be true.' But contrary beliefs are possible. Hence simply believing a position is an inadequate basis for contending that it is true vis-à-vis other views.*"

I agree. Fideism, as an unswerving allegiance to an organized religion, can never be in itself a test for truth. On the same basis, though Geisler would disagree strongly, I hold that no sacred text can be deemed

---

14   Benson, O. and J. Stangroom. 2007. *Why Truth Matters*. Continuum: London.202p.p.xi, 17.
15   Geisler, N.1988. *Christian Apologetics*. Baker Academic: Grand Rapids MI.393p.p.136, 137, 141,145,147.

inerrant simply because the text itself says so.

Geisler's indicators for truth and falsehood were as follows: "*We propose that **undeniability** is the test for the truth of a world view and **unaffirmability** (original author's emphases) is the test for the falsity of a world view...all non-theistic world views are directly or indirectly unaffirmable and only theism is affirmable and, hence, only theism is true. Further, we believe not only that theism is the only affirmable world view but that it is undeniably true. In short, nontheisms are sayable but not meaningfully affirmable; they are utterable but not justifiable.*"

Geisler argued for choosing evangelical Christianity as one's worldview, over Islam, Judaism and other theistic options. I am a believer, a theist and a Christian, but I see no sound basis for assuming that God reaches, among the billions of Earth's inhabitants, only the believers who adhere to one or more organized religions.

Concerning Geisler's choice of "*undeniability*" and "*unaffirmability*" as tests for the truth and falsity of a worldview, the equations 2+2 = 4 and 2+2 = 5 are, respectively, about as undeniable and unaffirmable as things ever get. The former is as close as we get to a truth. The latter is a falsehood. Mathematics is a worldview based on truth. The same applies to science. How then could theism be the "*only affirmable*" worldview? Moreover, how could either of the two worldviews of faith (defined as belief and trust in God, not as organized religion) and science ever change anything in the other's fundamentals? Both draw from one whole body of truth.

The additional problem with Geisler's criteria of undeniability and unaffirmability in the context of faith is that they cannot be applied objectively to any questions concerning the spiritual realm. God's existence becomes undeniable and His non-existence unaffirmable *only* for a person who has experienced Him. In the experiences of a nonbeliever, God remains deniable *and* unaffirmable. This is why we must

# TRUTH AND MORALITY

share and study as thoroughly as possible our subjective evidence from within concerning faith and unbelief.

Mick Brown[16] explored alleged supernatural experiences, from ashrams in Asia to a Tennessee church. He wrote the following about his interview with George Churinoff, the American Buddhist tutor of the Spanish boy Lama Osel, who had been recognized as the reincarnation of an important Lama: *"As we ate he (Churinoff) held forth on the subject of bodhisattvas, **karma** (original author's emphasis), rebirth, Buddha nature. There were Buddhas among us now, he said, bodhisattvas who had deliberately chosen rebirth in order to help mankind... What better choice could a man make than to follow that path himself? 'This is wonderful,' I said, 'but what if none of it is true?' George took the question in good spirit. 'If it is not true, I still cannot think of a better way to lead your life.'"*

In other words: 'It might not be true, but it's become my comfort zone, to which I want to welcome you.' 'No thanks.' Life is for the pursuit of truth, wherever the quest leads. All Christians should heed the following advice from C.S. Lewis[17] to Christian apologists in training: *"One must keep pointing out that Christianity is a statement which, if false, is of **no** (original author's emphasis) importance, and, if **true** (present author's emphasis) of infinite importance."*

## Postmodernism

James Sire[18] noted that postmodernism: *"has influenced religious understanding, including that of Christian theism, but it accepts the foundation at the heart of naturalism: **Matter exists externally; God does not exist***

---

16 Brown, M. 1999. *The Spiritual Tourist: A Personal Odyssey Through the Outer Reaches of Belief.* Bloomsbury Published Plc.: London.309p.p.141.
17 Lewis, C.S.1972. Christian Apologetics. Chapter 10, p.89-103. In W. Hooper (ed.) *God in the Dock: Essays on Theology and Ethics.* William B. Erdman's Publishing Company: Grand Rapids, MI. 346p.p101.
18 Sire, J. 2009. *The Universe Next Door: A Basic Worldview Catalog.* Intervarsity Press: Downers Grove IL.293p.p.217, 215.

(original author's emphasis)." He noted also the difficulty of defining postmodernism - "*It is used by so many people to focus on so many different facets of cultural and intellectual life that its meaning is often fuzzy*" - and quoted from others who have faced the same problem, including the following:

I. *Ihab Hassan,*[19] *who found in postmodernism "fragments, hybridity,* **relativism** *(present author's emphasis), play, parody, an ironic anti-ideological stance, an ethos bordering on kitsch and camp."*

II. *Mark Lilla,*[20] *who described "academic postmodernism" as "a loosely structured constellation of ephemeral disciplines like cultural studies, gay and lesbian studies, science studies and postcolonial theory."*

I define postmodernism as a diffuse set of perspectives, aspirations and types of behaviour, varying with identity and culture and denying the possibility of related standards that are universal and lasting.

Harry Frankfurt[21] criticized postmodernists as follows: "*they emphatically dismiss a presumption that is not only utterly fundamental to responsible enquiry and thought, but that would seem to be – on the face of it – entirely innocuous: the presumption that 'what the facts are' is a useful notion, or that it is, at the very least, a notion with intelligible meaning... The point on which the postmodernists especially rely is just this: what a person* **regards** *as true either is a function merely of the person's individual point of view or is determined by what the person is* **constrained to regard** *(original author's emphases) as true by various complex and*

---

19  Hassan,I. *Postmodernism to Postmodernity.* Cited by Sire, J. 2009. *The Universe Next Door: A Basic Worldview Catalog.* Intervarsity Press: Downers Grove IL.293p.p.216. Available from: www.ihabhassan.com/postmodernism_to_postmodernity.htm

20  Lilla, M. 1998. The Politics of Jacques Derrida. *New York Review of Books* June 25, 1998.p36. Cited by Sire, J. 2009. *The Universe Next Door: A Basic Worldview Catalog.* Intervarsity Press: Downers Grove IL.293p.p.216.

21  Frankfurt, H.G. 2006. *On Truth.* Alfred A. Knopf: New York.101p.p.19, 21.

# TRUTH AND MORALITY

*inescapable social pressures."*

Frankfurt made the point that we all rely on truths that have been sought, found, held in common and used for our benefit - by professionals such as architects, doctors, engineers and musicians. I hold that Free Thought enables us seek truth and to find it, albeit in the midst of lies and nonsense. Truth is sought and found by a human self, but truth is not defined by what that self and/or like-minded persons would like it to be. Truth is self-realized, not self-made.

Millard Erickson[22] wrote: "*Postmodernism challenges the idea that our beliefs are true, in the sense of being in agreement with an objective world...I would propose that on a pre-reflective level, or in actual practice, virtually all sane persons function with what I would term a 'primitive correspondence' view of truth. By this I mean an understanding of truth as a quality of statements that correctly represents the 'state of affairs' being referred to. This is true, in actual practice, of postmodernists as well as modernists.*"

Erickson included Richard Rorty in that grouping and continued as follows: "*I am proposing a perennial view of truth, one that is assumed throughout premodern, modern and postmodern periods, but am attempting to give it a uniquely post-postmodern orientation.*" With all due respect to Erickson, Rorty, and all modernists, postmodernists and 'post-postmodernists,' I cannot see how anyone can reconcile absolutism and relativism on truth, even at some so-called "*primitive correspondence*" level.

Simon Blackburn[23] commented as follows on Rorty's perspectives on truth and freedom: "*Rorty remains fond of saying that if we look after freedom, truth will look after itself...*(but) *there is no reason whatever to believe that freedom itself makes for truth...Freedom includes the freedom*

---

22  Erickson, M.J. 2001. *Truth or Consequences: the Promise and Perils of Postmodernism*. Intervarsity Press: Downers Grove IL.335p.p.234, 237.
23  Blackburn, S. 2006. *Truth: A Guide for the Perplexed*. Penguin Books Limited: London.238p.p. 166-167.

*to blur history and fiction, or the freedom to spiral into a climate of myth, carelessness, incompetence or active corruption. It includes the freedom to sentimentalize the past, or to demonize others, or to bury the bodies and manipulate the record. It is not only totalitarian societies that find truth slipping away from them: the emotionalists of contemporary populism, or the moguls of the media and the entertainment industries, can make it happen just as effectively."* I agree.

Douglas Groothuis[24] summarized the absolutists' main charge against Rorty as follows: *"Following his (Rorty's) hero, John Dewey, he (Rorty) asserts that truth is what one's peers let one get away with."* While also opposing Rorty, Alvin Plantinga[25] wrote the following: *"Perhaps you will object that I'm just belaboring a straw man: Rorty couldn't really mean that truth is what our peers will let us get away with saying. You may be right. What Rorty actually says is: 'For philosophers like Chisholm and Bergmann, such explanations* **must** *(Rorty's emphasis, as original author) be attempted if the realism of commonsense is to be preserved. The aim of all such explanations is to make truth something more than what Dewey called 'warranted assertability': more than what our peers will, ceteris paribus, let us get away with saying'."*

With or without that wrinkle, Rorty's perspectives on truth and morality will not get us far as explanations of the human condition. In particular, I cannot see how any believer could accept any postmodernist recipe for variable truth. One who believes that God exists must surely also believe that God holds a fixed and unquestionable body of truth about everything that happens in the spiritual and material realms. In truth, no one gets away with any transgressions of the Universal Moral Code, even when the rest of humanity fails to spot them.

---

24  Groothuis, D. 2000. *Truth Decay. Defending Christianity Against the Challenges of Postmodernism.* Intervarsity Press: Downers Grove IL. 299p.p.20,145-147.
25  Plantinga, A. 2008. *Against Naturalism*, p.1 to p.69. In A. Plantinga and M. Tooley's *Knowledge of God*. Blackwell Publishing: Malden MA.p16.Plantinga quoted p.175-176 In Rorty, R. 1979. *Philosophy and the Mirror of Nature*. Princeton University Press: Princeton NJ. Douglas Groothuis cited the same work by Rorty.

# TRUTH AND MORALITY

Groothuis criticized, as follows, Philip Kenneson's[26] denial of the existence of objective truth beyond the human mind and human speech: *"Following Rorty, Kenneson identifies belief in objective truth as an impossible 'view from nowhere approach' (all the while claiming that he is not a relativist)...Kenneson anticipates the objection that if we drop the concept of objective truth from apologetics we are left with only subjective 'truth', such that we make things true by believing in them. Against this, he enlists Rorty's idea that truth is merely a property of sentences, or ways we speak...When Kenneson says that 'truth cannot be out there' because sentences are not out there, he neglects the fact that God has a mind that knows every true proposition to be true and every false proposition to be false...Therefore, truth* **does** *(Groothuis' emphasis) exist outside of human sentences; it exists in the mind of God."*

I agree with Groothuis. The truth of God's existence and the truth that comes through His revelations are shown by the believer's personal experiences of God, which can be shared, at least in part, as subjective evidence from within. Our souls and minds receive truth from the mind of God, as revelations through faith and as His truthful disclosures through mathematics and science. Truth, justice and the Universal Moral Code have always been one and the same for all. The postmodernists' perspectives of different truths, justice and morality - defined by whatever gets you and your allies through the day or the night or whatever - are at odds with common sense, experience, faith and science.

## Truth About the Material Realm

Mathematics is the universal language that describes and explains material realm states, potentials, relationships and concepts. Mathematics and science bring us as close as possible to truths about the material realm. Pure mathematics can be thrilling.

---

26  Kenneson, P. 1995. There is no such thing as objective truth and it's a good thing too, p.155- 170. In T.R. Phillips and D.L. Okholm (eds.) *Christian Apologetics in the Postmodern World.* Intervarsity Press: Downers Grove IL.238p.

# Free Thought, Faith, and Science

John Lynch wrote the following in his Preface to Simon Singh's[27] account of Andrew Wiles' proof of Fermat's last theorem: *"Proof is what lies at the heart of maths, and is what marks it out from other sciences. Other sciences have hypotheses that are tested against experimental evidence until they fail, and are overtaken by new hypotheses. In maths, absolute proof is the goal, and once something is proved, it is proved forever, with no room for change."*

In the context of truth about the material realm we must include all God's truthful disclosures through mathematics and science about what it is made of and how it works. Believers and nonbelievers alike can marvel at the truth of the Anthropic Principle, as explained clearly and concisely by John Polkinghorne[28] and Martin Rees,[29] who summarized it as follows: *"A few basic physical laws set the 'rules'; our emergence from a simple big bang was sensitive to six 'cosmic numbers'. Had these numbers not been 'well tuned', the gradual unfolding of layer upon layer of complexity would have been quenched."*

Believers can say that God made it that way and thank Him. Nonbelievers can say that we are simply observing the prevailing conditions and that if they were otherwise we would not be here. Those with the required creativity and skills can make and appraise propositions about the possibility of other universes and a multiverse, but no one will never be able to *detect* anything beyond our universe except, as believers know, revelations from the spiritual realm.

The most important truth about the material realm is that it is what it is, with its fixed laws and free process under those laws. We need not look on Earth for any supernatural signs and wonders. There are none. Religious images do not bleed or cry or go for walks. Accidents

---

27  Singh, S. 2005. *Fermat's Last Theorem*. Harper Perennial: London.340p.p.ix. First published in 1997.
28  Polkinghorne, J.C. 2007. *The Anthropic Principle and the Science and Religion Debate*. Faraday Paper No. 4: 4p. Available at: www.faraday-institute.org
29  Rees, M. 2000. *Just Six Numbers. The Deep Forces that Shape the Universe*. Phoenix: London.194p.p.178-179.

happen according to the natural laws that God made for His creation to work as He wished. Accidents do not happen according to God's pre-programmed or ad hoc decisions over who gets harmed and who is spared. God intervenes in the material realm through our souls and mind-soul interfaces and the outcomes of our Free Thought.

A.C. Grayling[30] discussed how we understand realism and focused on what he called: *"so-called 'metaphysical realism' about the spatio-temporal world."* Citing the works of various authors, he stated: *"This realism is familiar because we all, in our non-philosophical moments at least, believe it. Realism in this sense is to be characterized as an independence thesis… the world of physical objects and events exists and has the character it has, independently of any thought, talk, knowledge or experience of it. This captures our belief that the relations in which thought or experience of the world consist are external relations: any difference they make to the world is nonessential, and the world and its natural functions could and would exist, even if it had never been a relatum in such relations."*

I agree in part. The attributes of the material realm always were, are, and will be the same, with or without us observing them or thinking about them. However, the outcomes of our Free Thought can and do change the material realm; for example, by deciding to build or destroy. Moreover, truth and falsehood are distinguished in the material realm *only* when one or more humans make the call, using Free Thought. The consequences that follow for the material realm are therefore consequences of Free Thought. The scope for complex, diverse and dynamic consequences is obvious, given that our spiritual revelations come not only from God but also from the spiritual force for evil.

In reviewing the role of witnesses in convincing juries as to the truth of a matter, Grayling preferred the term *"probative force"* to *"truth."* He described the general case scenario as follows: *"The witness has been*

---

30  Grayling, A.C. 2007 *Truth, Meaning and Realism: Essays in the Philosophy of Thought* Continuum: London.173p.p.109, 167-168.

*selected for his standing in the case, he has been sworn in and is liable to sanctions if he lies, he is allowed only to report what he has proper warrant to report, his evidence is elicited by interrogation and subjected to the scrutiny of cross-interrogation, and what he says is weighed in conjunction with, and comparison to, other evidence offered. Given these and allied safeguards, as much has been done as possible to allow the jury to judge whether or not to accept the evidence offered as having sufficient probative force to induce in them the crucial change of mind about the defendant's innocence…it is not strictly speaking open to the jury to decide whether what a witness says is true, but it is open to them to decide how much credence to give to it. The threshold degree of credence…which they accord to testimony and other evidence is precisely the trigger of their preparedness to change their minds in the crucial aspect."*

I agree with all of the above. The norm in most societies has been to set up systems that will deliver justice and expose attempts to pervert it. However, such systems can go horribly wrong, because of biases, ignorance, lies and corruption. The spiritual force for evil works hard against justice, by its involvement in the Free Thought of criminals, judges, jurors, police, victims and witnesses. Nevertheless, we often succeed in proving material realm truths to one another, beyond any reasonable doubt. God, as the ultimate Judge, has a record of all truth and all falsehood in the material and spiritual realms.

## Truth About the Spiritual Realm

Is there any evidence for the existence of the spiritual realm? My subjective experiences of God enable me to answer 'Yes,' but I admit that all evidence for the spiritual realm, God, the spiritual force for evil, the soul and spiritual revelations is subjective. Throughout this book, I call for the systematic gathering and analysis of subjective evidence from within concerning faith and unbelief, in pursuit of possible clustering and separation of the seemingly credible, the doubtful and the incredible.

# TRUTH AND MORALITY

Most nonbelievers, especially scientists, seek directly measurable and objective evidence for anything that they can deem to be true. I regard our subjective evidence from within as glimpses of truth about the spiritual realm. I seek ways to study those glimpses. I take heart that many nonbelievers admit that the existence of the spiritual realm can neither be verified nor falsified objectively. I take that as a good basis for dialogue on how to expand research on subjective evidence from within concerning faith and unbelief.

Much of the truth about the spiritual realm remains inaccessible during human material realm lives. As the Apostle Paul[31] put it: *"For now we see through a glass, darkly; but then face to face* (with God); *now I know in part; but then shall I know even as also I am known."* Only then will all truth become open to us. Meanwhile, many believers find it hard to agree on what is most likely to be true about anything in or from the spiritual realm, beyond similar experiences of God through faith and similar spoiling tactics by the spiritual force for evil.

Some Christians, especially biblical literalists, hold that God cannot have made any new revelations to humans since the death of Paul, the last of the Apostles, nor can He make any more new and true revelations on Earth in the future *except* through the words of the Bible. I believe that God can send new revelations of truth to any believer or nonbeliever, at any time.

## Sources of Morality

Human morality is founded on the quasi-morality seen in some animals. Animals are concerned only with expediency, in terms of material wellbeing for themselves and/or their preferred fellows having partisan interests in common. Nevertheless, the quasi-moral behaviour of animals is seen in some of the behaviour that we associate with human

---
31   I Corinthians 13:12

morality including, *inter alia*: anguish after bereavement; aversion to unfairness; consolation of suffering fellows; avoidance of causing undue pain and injury in play; peacemaking; self-sacrificial parental care; sharing resources; and teamwork.

Frans de Waal[32] told the story of morality as the evolution of empathy. He found that: *"even our most thoughtful reactions to others share core processes with the reactions of young children, other primates, elephants, dogs, and rodents."* For example, we share with many animals, including dogs and primates, a marked aversion to injustice. De Waal cited research with Sarah Brosnan in which capuchin monkeys performing the same simple repetitive task would accept either a relatively tasteless cucumber slice or a tasty grape as a reward. When a monkey performed well and was offered cucumber but saw its no more deserving neighbour getting grape, it threw a tantrum and refused to cooperate further.

A video clip from that research and other related material are available online from de Waal's TED (Technology, Entertainment, Design) lecture of November 2011,[33] during which he summarized the twin pillars of morality, in animals and humans as: "**Reciprocity**/fairness (and) "**Empathy**/Compassion (original author's emphases)." De Waal held that: *"human morality is more than this, but if you removed these pillars there would not be much remaining."* De Waal also held that empathy (i.e., an *"ability to understand and share the feelings of another"*) requires a *"cognitive channel"* (i.e., the recognition of *"self-other distinction"*) and gave convincing evidence for empathy in apes and elephants.

De Waal showed footage of chimpanzees choosing *"prosocial"* tokens more frequently than *"selfish"* tokens, when offered both types as a reward for a set task. The choice of a prosocial token ensured that a neighbour in a separate adjacent cage, though not involved in the task, would receive the same reward as the doer, while the doer got the same

---

32   De Waal, F. 2009.*The Age of Empathy: Nature's Lessons for a Kinder Society*. Harmony Books: New York.291p.p.209, 187.
33   www.ted.com/talks/frans_de_waal_animals_have_morals.html

reward irrespective which kind of token she/he chose. He also showed two chimps joining forces to haul in a heavy box of food, after which one of them took all of it. He explained that the chimp taking no food was expecting reciprocity sometime in the future.

This kind of evidence begs many questions about whether any of the quasi-moral behaviour of animals is brought about by choices comparable to the loving and caring choices made by humans? Self-sacrificial and risky parental care of progeny is common in fish, amphibians, reptiles, birds and mammals, but can we ever use properly the word *love* to describe relationships between animal parents and their progeny? What about the pair bonds forged for mating and the cooperative bonds among team members in hunting, building, providing security and fighting off enemies? Are these bonds in animals the same as human love and loyalty? If not, what are the special attributes in the human equivalents of these animal relationships that make them significantly different? In short, is human morality simply on a *continuum* with animal quasi-morality, or are they *categorically different*?

Many believers and nonbelievers will consider it unnecessary to pursue this question, on the grounds that so much of what happens in our lives clearly differs so much from what happens in the lives of animals. I disagree. The faith-science quest for truth requires a thorough, honest attempt to answer the question - continuum or categorical difference?

My general case answer is that in all cases of apparent loving and caring by animals there is a real or potential pay-off in the material realm for the individual benefactor and/or for one or more of her/his fellows having partisan interests in common; for example, as kin and/or as members of a group holding communal territory. Only the human condition provides for acts based on *unconditional* love and with *no conceivable pay-offs* for the benefactor and/or her/his fellows having partisan interests in common. This alone suggests that the human species has attributes that make it categorically different from all others.

It is undeniable that animal quasi-morality has many commonalities with human morality, but God has made us categorically different. It seems highly unlikely that He gives to other species any of the same spiritual gifts that He gives to every human - the soul, the mind-soul interface, the possibility of a personal relationship with Him throughout life after the onset of Free Thought, and the Universal Moral Code. The human condition includes unavoidable co-option into the war between God and the spiritual force for evil. All humans use Free Thought to make sovereign-to-self decisions based on information from the material and spiritual realms. The animal condition has nothing comparable.

In one of his most recent works, de Waal[34] stated the following: "*I am a firm believer in David Hume's position that reason is the slave of the passions. We started out with moral sentiments and intuitions, which is where we find the greatest continuity with other primates. Rather than having developed morality from scratch through rational reflection, we received a huge push in the rear from our background as social animals. At the same time, however, I am reluctant to call a chimpanzee a 'moral being.' This is because sentiments do not suffice. We strive for a logically coherent system and have debates about how the death penalty fits arguments for the sanctity of life, or whether an unchosen sexual orientation can be morally wrong. These debates are uniquely human. There is little evidence that other animals judge the appropriateness of actions that do not directly affect themselves.*"

I agree with de Waal on all of the above. He continued as follows, citing the work of Edward Westermarck: "(human) *moral emotions are disconnected from one's immediate situation. They deal with good and bad at a more abstract, disinterested level. This is what sets human morality apart: a move toward universal standards combined with an elaborate system of justification, monitoring, and punishment. At this point, religion*

---

34   De Waal, F. 2013.*The Bonobo and the Atheist: In Search of Humanism Among the Primates.* W.W. Norton and Company: New York.289p.p.17-18, 239.

# TRUTH AND MORALITY

*comes in.*" Again I agree, except that I hold that it is *God*, not religion coming in - in all human lives, irrespective of surrounding religious or secular circumstances.

We must all face the truth that we are very similar to animals in many respects. There is much in the work of de Waal and others to suggest a continuum in the evolution of human morality from animal quasi-morality. However, *nothing* that has emerged so far from research on animal quasi-morality precludes the existence of God, the spiritual force for evil, the soul and the spiritual realm. I hold that there is a God-given categorical difference between animal quasi-morality and human morality, with humans as God's special companion species - set apart from all other life forms and made in His image.

I consider this a tenable position for all believers, but only if based on faith *per se*, not only on religious precepts. In this respect, de Waal has helped us further, as follows: "*Morality arose first, and modern religion latched on to it. Instead of giving us the moral law, the large religions were invented to bolster it...The big challenge is to move forward, beyond religion, and especially beyond top-down morality. Our best known 'moral laws' offer nice post hoc summaries of what we consider moral but are limited in scope and full of holes...Morality has much more humble beginnings, which are recognizable in the behaviour of other animals...Everything science has learned in the last few decades argues against the pessimistic view that morality is a thin veneer over a nasty human nature. On the contrary, our evolutionary background lends a massive helping hand without which we would never have got this far.*"

I would say that organized religions have arisen for a variety of reasons, including human quests for group identity, security, wealth and power, but I also see evidence that some organized religions reflect truthful revelations from God and/or lies from the spiritual force for evil, transmitted to the souls of their authorities and/or rank and file members. I agree with de Waal that human-made moral codes can be inadequate, but I hold that God's Universal Moral Code, implanted in our souls, together with His ongoing spiritual revelations, can make good any inadequacies for those who recognize His Code *and* listen to Him. At the same time, I agree with many of de Waal's explanations. He sees strong evidence for an evolved morality. So do I. He also sees that human morality has aspects not found in any animals. So do I.

Neil Levy[35] summarized the story of human morality from a purely evolutionary perspective, as follows: *"Evolved morality is real morality… (which) comes to us as a product of our evolutionary history… From the mindless and mindlessly selfish rose beings capable of rationality and morality."* He concluded: *"We are **moral** animals, but we are still **animals*** (original author's emphases). *Our brains allow us to engage in sophisticated reasoning, but they have built in biases and heuristics of which we are largely unaware. We are not angels, fallen or otherwise, but merely jumped-up apes."*

Levy continued: *"Does the legacy of evolution, our Stone Age minds, empty our lives of meaning, and our morality of substance?"* He was inviting the answer 'No,' from the perspective that even without the existence of anything supernatural we can still be awed by how far we have come and can live accordingly, in an evolving morality. I say 'No,' but from the perspective that we alone are God's chosen and special companion species and that our morality, while clearly built on foundations from the evolution of animal quasi-morality, is actually *His* morality, which He communicates to us alone - along with the necessary attributes for us to make *our* choices, torn between His guidance and the influence of the spiritual force for evil.

There is strong evidence for a standard and lasting code of human morality. We have been enabled by God to recognize that unfairness, injustice and lies are not merely inexpedient in many circumstances,

---

35   Levy, N. 2004. *What Makes Us Moral? Crossing the Boundaries of Biology*. Oneworld: Oxford. 237p. p.88, 199-200.

# TRUTH AND MORALITY

they are definitely and inherently *wrong*, even when they bring material payoffs. God's Universal Moral Code and His guidance for our compliance are based not on expediency, or fitness to the environment, or material wellbeing. They are based on His will and His love and His grace extended to us. He expects us to do our best to comply with His Universal Moral Code, allied with Him in the war between good and evil.

I do not believe that any of the above applies to animals. I hold that there is a huge and categorical difference between the animal condition and the human condition. In the context of morality, ape compliance and non-compliance with any temporary and still evolving ape quasi-moral codes are determined, knowingly or unknowingly, by the pursuit of material realm expediency and/or the spreading and perpetuation of genes. Human compliance and non-compliance with the Universal Moral Code are outcomes of Free Thought, which involves spiritual revelations and responses.

Nevertheless, no honest seeker of truth, believer or nonbeliever, can deny our similarities with animals or that the higher primates are beyond all doubt our very close relatives. Some of our fellow primates appear to make reasoned choices to establish and retain debts of gratitude. This is more than a simple and immediate favour for favour deal such as 'you scratch my back and I'll scratch yours,' which can easily become a norm for practical purposes.

If some of our fellow primates are really choosing to establish states of indebtedness with their fellows in respect of complex tasks and are retaining memories of their states of indebtedness or their *rights* to repayment, then in this at least they are the same as us. All around the world humans follow the strategy of favour for favour and behave accordingly. One of the best descriptors for this is the Filipino phrase *utang ng loob*, which means inner debt and signifies an inescapable obligation.

I see that strategy as nothing more than the pursuit of expediency, except where it is elevated to become what I call *genuine* altruism, which is found only in humans. So, the categorical difference remains, but the animal condition and the human condition have much in common. I urge all believers and nonbelievers to keep abreast of the wonderful results of recent and ongoing research on our close primate relatives. There will undoubtedly be more surprises about our similarities and differences with the higher apes, especially the bonobos and chimpanzees. Through science there will be many more truthful disclosures from God about our fellow primates and us.

Karl Giberson[36] wrote as follows about Christianity in this context: *"Until recently just about everyone in all cultures perceived a great* **qualitative** *(original author's emphasis) distinction between humans and the higher primates. Certainly, the biblical writers and the formative thinkers of the Christian tradition could not have anticipated what we have learned from primate studies in the past few decades. So we may suppose that they would frame their religious understanding in exclusively human terms…I find no compelling reason to think that the central message of Christianity is incompatible with humanity's kinship with the rest of the animal world."* I agree. Denying that *"kinship"* amounts to denying some of God's most important truthful disclosures through science.

## Altruism

Altrusim, in animals and humans, is the practice of seemingly selfless concern for the wellbeing of others. Altruism in animals and much of human altruism carry prospects of some present and/or future pay-off in the material realm for the altruist and/or one or more of her/his fellows having partisan interests in common - such as the furtherance of genes and the holding of territory. Only humans also practice *genuine*

---

36  Giberson, K.W. 2008. *Saving Darwin: How to Be a Christian and Believe in Evolution.* HarperCollins: New York. 248p.p.14.

# TRUTH AND MORALITY

altruism, which is acting to benefit one or more persons, at one's own disadvantage and without expectation or prospect of any present and/or future pay-offs in the material realm to oneself and/or to any of one's fellows having partisan interests in common.

Those who practice genuine altruism might or might not be expecting any *spiritual* pay-offs - as peace of mind during life on Earth and/or as some kind of heavenly reward. Most Christians believe that they are saved only by the grace of God and through their faith, *not* by doing good works.[37] However, the Apostle James[38] warned that: *"faith without works is dead."* The Apostle Paul wrote: *"And though I bestow all my goods to feed the poor...and have not charity* (i.e., unconditional love for others), *it profiteth me nothing."*[39]

Christ's reported commands about charitable giving seem contradictory: *"Take heed that ye do not your alms* (i.e., charitable giving) *before men, to be seen of them: otherwise ye have no reward of your Father which is in heaven"*[40] and *"Let your light so shine before men, that they may see your good works, and glorify your father which is in heaven."*[41]

Rather than attempting any theological explanations, I will just take all of the above as fair commentary on the complexities of the human condition and the diversity of the moral choices that we are forced to make through our Free Thought, in which we receive spiritual revelations from God and from the spiritual force for evil. In the ongoing war between good and evil we are all conscripts, with potentials for selfishness and charity, self-promotion and humility.

Some persons take altruism to extremes (see below). Others do their best to avoid all charitable acts on the false but self-believed

---

37 Ephesians 2: 5, 9
38 James 2:26
39 I Corinthians 13:3
40 Matthew 6:1, 5-6
41 Matthew 5:16

basis that: 'No one ever gave me anything, so why should I give anything to anyone?' Within those extremes, believers and nonbelievers choose whether to do their good works anonymously and in secret or for personal credit and as a matter of public record. In both cases, the beneficiaries can be known or unknown to the benefactor and can either have partisan interests in common with the benefactor or not. The same applies to prayer. Believers pray as groups, or in public, or alone and in secret - as reportedly advised by Christ.[42] The beneficiaries brought to God in intercessory prayer can be known or unknown to those praying and can either have partisan interests in common with them or not.

The human condition and God-human relationships are such that believers *and* nonbelievers alike can engage in genuine altruism - through anonymous and unreported acts of what Paul called charity, in order to benefit disadvantaged strangers, with no conceivable pay-offs in the material realm to the altruists or to any of their fellows having partisan interests in common. Anyone who gives spare change to a disadvantaged stranger on the street, unobserved and with no subsequent reporting to anyone of her/his act of charity, is practicing genuine altruism. The same applies to anyone who donates anonymously to a famine relief organization and does not report her/his act of charity to anyone beyond that organization. Such genuine altruism really does exist. If you doubt this, just ask the treasurers of organizations that depend on charitable donations.

Some nonbelievers try to explain what I call genuine altruistic acts in the same way that they try to explain all other altruism and acts of apparent kindness in animals and humans - through genetics, meme theory and biophilia (love for life). However, in cases where there really are no conceivable material pay-offs for the altruist and/or her/his fellows with partisan interests in common, and with the existence of the spiritual realm having been denied, it is hard to see how any of those

---

[42] Matthew 6:6

# TRUTH AND MORALITY

nonbelievers' explanations could work.

In truth, genuine altruism is a matter between the altruist and God, even if the altruist is a nonbeliever. Believers *and* nonbelievers can increase their peace of mind by extending charity anonymously to total strangers, with no expectations or prospects for any pay-offs. Believers can explain their increased peace of mind in terms of their relationships with God and the joy of doing His will. Nonbelievers can feel good about giving generously and anonymously - and can then wonder why they feel that way.

My explanation for all genuine altruism is that all humans have souls. Believers strive to discern and to do the will of God by complying with His Universal Moral Code. Nonbelievers possess the same code and receive spiritual revelations from God during their Free Thought, though denying His existence. It feels good to give. God made us that way.

Among the many works on the science of altruism, Matt Ridley's[43] classic remains outstanding and carries the following endorsement from Richard Dawkins: "*If my 'The Selfish Gene' were to have a Volume Two devoted to humans, 'The Origins of Virtue' is pretty much what I think it ought to look like.*" In his new introduction to the 30th Anniversary Edition of "*The Selfish Gene,*"[44] Dawkins explained that he could just as well have called it "*The Immortal Gene: The Altruistic Vehicle*" or "*The Cooperative Gene,*" thereby avoiding the word "*Selfish,*" which must have put off some potential readers.

Ridley made the same point as follows: "*Selfish genes sometimes use selfless individuals to achieve their ends. Suddenly, therefore, altruism by individuals can be understood.*" He described the history of potlatch, with an example from the Kwakiutl Nation of Vancouver: "*(they) gave each other blankets, candlefish oil, berries, fish, sea otter pelts,*

---
43  Ridley, M. 1997. *The Origins of Virtue*. Penguin Books: London.295p.p.20, 121, 249, 265.
44  Dawkins, R. 2006. *The Selfish Gene*. 30th Edition. Oxford University Press: Oxford.360p.

*canoes, and, most valuable of all, 'coppers', sheets of beaten copper decorated with figures. Not content with giving away wealth, some potlatch hosts took to destroying it instead. One chief tried to put out his rival's fire with expensive blankets and canoes; the rival poured candlefish oil on the flames to keep them burning."*

Lavish displays, profligate use of wealth and highly competitive giving are very much alive and well today, particularly in business and political circles. Ostentatious gifts enhance the givers' reputations and status. Competitive giving is also seen in some philanthropy. An American church stewardship campaign used the following slogan, without knowing how funny it would appear in other parts of the English-speaking world: "*I upped my pledge. Up your's!*"

Ridley concluded the following: "*Our minds have been built by selfish genes but they have been built to be social, trustworthy and cooperative... Humans have social instincts. They come into the world equipped with predispositions to learn how to cooperate, to discriminate the trustworthy from the treacherous, to commit themselves to be trustworthy, to earn good reputations, to exchange goods and information, and to divide labour... this instinctive cooperativeness is the very hallmark of humanity and what sets us apart from other animals."*

There must be a genetic basis for the evolution of animal altruism. Some human acts of seemingly self-sacrificial kindness carry expectations of reciprocity and must also have some genetic basis and/or some basis in learned behaviour for the pursuit of expediency. For example, mutual help is often the norm for those who live in harsh circumstances such as deserts, mountains, and tough cities.

I must repeat, however, that I can find no explanation for what I call genuine altruism in humans other than the God-human relationship and His gifts to us of souls, higher consciousness, Free Thought and the Universal Moral Code. A decision to engage in genuine altruism cannot come by genes and/or memes alone. It is an outcome of Free Thought, in which the Universal Moral Code and spiritual revelations from God have won, over lies from the spiritual force for evil.

As recounted by Ullica Segerstrale,[45] William ('Bill') Hamilton contributed hugely to the science of altruism, particularly through his insight that altruistic self-sacrifice in animals depends upon their genetic relatedness. In other words, an individual will volunteer to undergo disadvantages if that will help others who share that individual's genes to perpetuate those genes, and the amount of disadvantage volunteered will increase with the number of genes shared.

Hamilton's work became intertwined with that of George Price, who contributed an important equation concerning the frequency of altruistic genes in a global population, before and after selection. Incidentally, Price believed that truthful disclosures through mathematics and science come from God. Segerstrale quoted the following from Hamilton's papers: "*George* (Price) *believed that the discovery he had made in evolutionary theory was truly a miracle…it was clear to him that he had somehow been chosen to pass on a truth about evolution to a world that was, somehow, just now ready to receive it.*"

Oren Harman's[46] fine account of Price's life and work depicts him as an unhappy genius, a creature of extremes, and one who suffered much from unsuccessful personal relationships and from botched thyroid surgery, which had disabled his right arm, shoulder and neck. Price first became a Christian when he took some improbable coincidences in his life as having been engineered by God. He worked hard for the C.S. Lewis Society in London and underwent

---

45 Segerstrale, U. 2013. *Nature's Oracle: The Life and Work of W.D. Hamilton*. Oxford University Press: Oxford.441p.p.146. Segerstrale cited p.323 in Hamilton, W.D. 1996. *Narrow Roads of Gene Land: The Collected Papers of W.D. Hamilton. Vol.I. Evolution of Social Behaviour*. W.H. Freedman: New York.

46 Harman, O. 2010. *The Price of Altruism: George Price and the Search for the Origins of Kindness*. W.W. Norton: New York.451p.p.249, 360-361.

a second conversion, through which he recognized the limitless love of Christ and became a self-professed *"slave"* of God.

As researched by Harman, Price wrote encouragingly at first to Henry Morris, the founder of the Creation Science Research Center, but then declined to join Morris' 'Society' as follows: *"the reason that I did not apply for membership is that the Lord **commanded** not to...I am sorry that you apparently did not understand that I really am a genuine slave, and not like the usual 'evangelical Christian' who prays for 'guidance' and then makes his own decision. I do not ask for 'guidance'; I ask for **commands*** (Price's emphases, as original author).

Price became an extreme altruist, expending his resources on helping homeless alcoholics in London, until he himself became sick and destitute. Unable to help as before and not getting from God the means that he sought and expected, Price committed suicide by cutting his throat with scissors. Harman concluded that *"George might have stood somewhere on the slippery spectrum between normal and autistic behaviour"* and added that those around Price, including the last doctor who saw him, felt that: *"The combination of an unstable personality and the depression brought about by no longer taking his medication* (thyroxine) *might well have pushed him over the deep end, and ultimately to suicide."*

Some nonbelievers will sneer and ask why God did not take better care of His self-professed slave George Price? God *did* take care of Price - as far as material realm free process and Free Thought allowed. Price's deteriorating physical and mental condition must have influenced the outcomes of his Free Thought. Ever worsening depression culminated in suicide. All human altruism, including that which resembles altruism in animals and the genuine altruism that is found only in humans, takes place in the context of free-willed selves employing Free Thought, and free process in the material realm.

# TRUTH AND MORALITY

## THE UNIVERSAL MORAL CODE

C.S. Lewis[47] emphasized that what he called the *"Law of Nature"* is broadly agreed by all humanity. He argued that this Law is: *"a real law, which none of us made, but which we find pressing on us."* I agree. That Law is the Universal Moral Code. Everyone, believer or nonbeliever, receives the same Universal Moral Code from God. It cannot be changed by anyone. Lewis wrote: *"If anyone takes the trouble to compare the moral teaching of, say, the ancient Egyptians, Babylonians, Hindus, Chinese, Greeks and Romans, what will strike him will be how very like they are to each other and to our own."*

Again I agree. The commonalities among moral codes associated with a wide range of organized religions and belief systems extend far beyond those that are to be expected from the three Abrahamic religions - Christianity, Islam and Judaism. Marcus Borg[48] compiled a book of the parallel sayings of Jesus and Buddha and quoted Burnett Hillman Streeter as follows: *"the moral teaching of Buddha has a remarkable resemblance to the Sermon on the Mount."*

The existence of the Universal Moral Code is doubted or denied by those who believe that definitions of correct behaviour are norms derived solely from circumstances, expediency and tradition. They point to examples of societies that from time to time have approved of, or even required, cannibalism, infanticide, participation in atrocities, and slavery etc. The believer's counterargument is that no individual or collective transgression against the Universal Moral Code can negate its existence or change its provisions.

Individual and collective compliances with the Universal Moral Code are bound to vary, because evil can become dominant in the outcomes

---

47  Lewis, C.S. 2001. *Mere Christianity*. Gift Edition, including a Foreword by Douglas Gresham and Appendices on related letters of C.S. Lewis and the history of this work. HarperCollins: New York. 237p.
48  Borg. M. Editor. 1999. *Jesus and Buddha: The Parallel Sayings*. Ulysses Press: Berkeley CA.241p.p.13.

## Free Thought, Faith, and Science

of Free Thought. The spiritual force for evil can recruit anyone to transgress the Universal Moral Code at any time. Things fall apart when non-compliance with the Universal Moral Code becomes the norm. Human relationships and society cannot be sustained without morality. Morality cannot be something made up to suit the agenda of any particular individual or group.

In the following two scenarios, please use your Free Thought and assess whether right and wrong could be distinguished without a universal standard of morality:

I. *"Good morning, my (much hated) business rival. Your plan to set up a new operation, on my patch, is not going to be allowed. I am sending my mob of hired thugs to threaten you. If you persist with your plan, my boys will wipe you out. I'm sure you will agree that this is not a question of right and wrong. It's all about power. I'm sure that you would do the same to me if you had the resources."*

II. *"Good morning, Judge. Before we go into court, I want you to know that I am going to plead not guilty to the crimes that you and I both know that I have committed. I know that you will acquit me of all of them, because of the large bribe that has been paid to you by my family and associates. I'm sure you will agree that this is not a question of right and wrong. It's all about money. Anyone would do the same if they had the resources."*

In his classic work on justice, John Rawls[49] stated: *"My aim is to present a conception of justice which generalizes and carries to a higher level of abstraction the familiar theory of the social contract as found, say, in Locke, Rousseau and Kant... These principles are to regulate all further agreements; they specify the kinds of social cooperation that can be entered into and the forms of government that can be established. This way of regarding the principles of justice I shall call justice as fairness."*

---

49  Rawls, J. 1971. *A Theory of Justice.* Harvard University Press: Cambridge MA.607p.p.11.

# TRUTH AND MORALITY

That *"high level of abstraction"* and those *"principles"* are seen in the Universal Moral Code, which is based on truth and requires that justice be dispensed by all and to all. We all know that it is *wrong* to fabricate data, to forge signatures, to claim benefits for which one is not eligible, to engage in bribery and other corrupt practices, to score a goal with one's hand in soccer, to cheat at cards etc.

The provisions of the Universal Moral Code can be generalized into statements such as the following: love is always right/hate is always wrong; truth is always right/lies are always wrong. Some philosophers and ethicists try to demolish such simple statements, but their worth is trumpeted in truth-indicating literature throughout recorded history and all around the world - from folk tales to Shakespeare and in countless stories about goodies versus baddies, from the Wild West and city streets to outer space. The provisions of the Universal Moral Code are the same for everyone. They are not Northern or Western norms foisted on the rest of the world. The United Nations Universal Declaration of Human Rights draws on the Universal Moral Code.

I once visited a school in the UK where some 6-year olds had been asked to write down their own ten commandments. One girl had written the following, given here verbatim with my clarifications in brackets: 1."*Dont spoyule* (spoil) *peoples close*" (clothes). 2. "*Do not cille* (kill) *anemules*" (animals). 3. "*Help anemules when they are porli* "(sick). 4."*Trie not to cille plants and vegdebul*" (vegetables). 5. "*Dont swoloe hard sweets strate quicli.*" 6."*Trie not to chew on wood.*" 7. "*Dont ripe* (tear) *peoples books.*" 8. "*Do not bite people.*" 9. "*Help to tideupe* (tidy up) *rooms.*" 10. "*Don't take toyse from people.*"

Apart from numbers 5 and 6, which would have been taught rapidly as practical measures, that child's ten commandments represent the following two principles from the Universal Moral Code: **Love your neighbour as yourself** (numbers 1, 7, 8, 9, 10) and **Love all living things (biophilia)** (numbers 2, 3, 4). She probably received encouragement

## Free Thought, Faith, and Science

in those directions from family members, teachers, peers and others. However, I cannot accept those influences as a full explanation of what she wrote so freely at the age of six. I believe that she wrote from her evolutionary legacy and from the Universal Moral Code in her soul.

The Universal Moral Code has been summarized in various written forms. Kent Keith[50] prepared an excellent concise wording, listing 10 principles for avoidance of doing harm and 10 principles for pursuit of doing good. Some organizations have their own short versions. For example, the Rotarians strive to apply the Four Way Test: "*Is it the truth? Is it fair to all concerned? Will it build goodwill and better friendship? Is it beneficial to all concerned?*"[51]

The sacred texts of organized religions provide concise written forms of the Universal Moral Code, including *inter alia*: the Ten Commandments;[52] Christ's two comprehensive commandments,[53] which have Old Testament origins;[54] and the seven things[55] that are said to be "*an abomination*" to God: "*a proud hand, a lying tongue, and hands that shed innocent blood, an heart that deviseth wicked imaginations, feet that be swift in running to mischief, a false witness that speaketh lies, and he that soweth discord among brethren.*"

In much of organized religion there are lengthy and detailed prescriptions for allegedly correct human behaviour in a huge range of scenarios. Some of those prescriptions have nothing to do with the Universal Moral Code. For example, the Mosaic Law allows a victorious holy warrior to select and take home a beautiful woman from among the Lord's enemies captured in war. He must make sure first that she has her head shaved and her nails cut. After allowing her a month of mourning for separation from her parents, he can try her out sexually, and then either

---

50   www.universalmoralcode.com
51   Composed by Herbert J. Taylor during the Great Depression; adopted by Rotarians in 1942.
52   Exodus 20: 2-17
53   Luke 10: 27-37
54   Leviticus 19:18 and Deuteronomy 5:6
55   Proverbs 6:16-19

# TRUTH AND MORALITY

discard her or keep her as a wife.[56]

Geza Vermes[57] noted the following additional proviso for a war booty wife: *But she shall not touch whatever is pure for you for seven years, neither shall she eat out of the sacrifice of peace offering until seven years have elapsed."* That last indignity would probably have been the least of her worries. This whole edict reads more like an endorsement of war crimes than a recipe for a sequence of moral acts.

The Universal Moral Code prohibits the killing of any human by another human. All who seek to follow the Universal Moral Code try to avoid involvement in anything that threatens or ends life - even in highly extenuating circumstances such as justifiable abortion, removing medical support for the terminally ill, requests by pain-racked and terminally ill persons to endorse and enable their suicide, and being ordered to kill in armed conflict.

The prohibition on killing is not so absolute in some versions of sacred texts. In the King James Version of the Bible, the Ten Commandments include the words *"Though shalt not kill,"*[58] but elsewhere in the Old Testament and reportedly as endorsed by Christ,[59] the original commandment was: *"do no murder."* Religious authorities and rank and file members of religious society could thereby allow killing that was not construed as murder. The Old Testament has criteria for identifying murderers and demands their execution.[60]

Dave Grossman[61] concluded the following about the well-documented phenomenon of soldiers in battle who try to avoid killing the enemy:

---

56   Deuteronomy 21: 10-14
57   Vermes, G. 2010. *The Story of the Scrolls: The Miraculous Discovery and True Significance of the Dead Sea Scrolls*. Penguin Books: London.260p.p.138.
58   Exodus 20:13
59   Matthew 19:18
60   Numbers 35: 16-31
61   Grossman, D. 2010. Hope on the Battlefield, p.36-44. In D. Keltner, J. Marsh and J.A. Smith, (eds.) *The Compassionate Instinct: The Science of Human Goodness*. W.W. Norton and Company: New York.316p.p.43-44.

*"We may never understand the nature of the force in humankind that causes us to strongly resist killing fellow human beings, but we can be thankful for it. And although military leaders responsible for winning a war may be distressed by this force, as a species we can view it with pride."*

That force is the Universal Moral Code, which also prohibits incest, rape, sexual abuse of minors and other harmful sexual behaviour. It also prohibits adultery, which means sexual intercourse between a person who is married and someone other than her/his spouse. I believe that the Universal Moral Code permits all loving, lasting and faithful heterosexual and homosexual relationships.

God will always be the Judge of our interpretations and compliance/non-compliance with the Universal Moral Code, which is His standard for all of us. Believers who transgress the Universal Moral Code are forgiven if they confess sincerely to Him and ask for His forgiveness and help to do better.[62] All who transgress and fail to repent can only be commended to God's mercy.

God, in His perfect justice, mercy, love and entirely reasonable expectations of us, requires that we be honest seekers and tellers of the truth and act accordingly. On that basis, the Universal Moral Code can be condensed into one simple commandment - *in all circumstances,* **seek and tell the truth, act according to the truth** *and confront lies and nonsense.*

---

[62] I John 1:9

# CHAPTER 9

# IN THE MIDST OF LIES AND NONSENSE[1]

*"I wish you a world free of demons and full of light"* [2]

### LIES AND NONSENSE FROM ORGANIZED RELIGION

Organized religion has provided spiritual guidance and practical support to billions of people, but has all too often become a source of illusion and confusion, by spreading lies and nonsense. Christian churches have been significant contributors to all of the above. The Christian author Philip Yancey[3] responds as follows when strangers tell him horror stories about the Christian church: "*Oh, its even worse than that. Let me tell you my story. I have spent most of my life in recovery from the church.*"

Lies and nonsense from organized religion are spread far and wide. Maureen Dowd[4] quoted the Christian evangelist Franklin Graham as follows, from a TV interview about President Barack Obama: "*I think*

---

1. The title of this chapter is not intended as an inclusive label for its contents or as a label for any particular item or items, unless stated explicitly.
2. Sagan, C.1997. *The Demon Haunted World: Science As a Candle In the Dark*. Ballantine Books: New York. 457p., the author's dedication to his grandson.
3. Yancey, P. 2001. *Soul Survivor. How Thirteen Unlikely Mentors Helped My Faith to Survive the Church*. OMF Literature Inc., Manila, Philippines. 330p.p.1.
4. Dowd, M. Going Mad in Herds: America's National Lunacy. *International Herald Tribune* August 23,2010:p.8.

*the president's problem is that he was born a Muslim. His father was a Muslim. The seed of Islam is passed through the father, like the seed of Judaism is passed through the mother."*

The jihadists who attacked America on September 11, 2001 believed that they were doing the will of Allah. Some Christians believe that God caused those attacks, because He was angry about abortion and homosexuality in America. Many people, including self-professed Christians, rejoiced when Osama bin Laden was killed. *"Love your enemies"*[5] is forgotten in most human conflicts. The BBC reported the following from the Archbishop of Canterbury on the death of bin Laden: *"I think the killing of an unarmed man is always going to leave a very uncomfortable feeling because it doesn't look as if justice has been served."*

Jerry Mitchell[6] noted that the passage of a 1997 disaster relief bill in Arkansas took five precious days while legislators debated whether tornadoes were natural disasters or acts of God. Mitchell quoted Mike Huckabee, the Governor of Arkansas, as follows: "(It is) *a matter of deep conscience to me to attribute in law a destructive and deadly force as being an 'act of God'."* Mitchell also quoted an Alabama church member, who explained a local earthquake as follows: *"I wouldn't doubt it was a message from God. God sent Jonah to warn the people of Nineveh, and think what happened to Sodom and Gomorrah."*

Lies and nonsense are reinforced by ignorance. Armando Ang[7] reviewed the Inquisitions, anti-Semitism and other horror shows in the history of Roman Catholicism and concluded that: *"very few (Catholics) know the dark side of their Church history."* Ang noted the longstanding zeal of Catholic authorities in attempting to control the sexual behaviour of their flocks; for example, Pope Gregory I degreed that: *"The sex act had to be done for procreation; otherwise the husband*

---

5   Matthew 5: 44
6   Mitchell, J.T. 2000.The hazards of one's faith: hazard perceptions of South Carolina clergy. *Environmental Hazards* 2:25-41.
7   Ang, A. 2005.*The Dark Side of Catholicism*.A1 Publishing: Manila, Philippines.444p.The quotations are from the introductory letter and p.287, where Ang acknowledged Nigel Cawthorne as his source.

# IN THE MIDST OF LIES AND NONSENSE

*was prevented from entering the church."*

The Catholic Church confirms its alleged Christ-endowed *"unique authority"* as follows: *"For expressions of faith to be authentic, they must be in harmony with the doctrine and traditions of the Catholic Church, which are safeguarded by the bishops who teach with a unique authority."*[8] Catholic catechumens are required to accept the entire Catechism,[9] including the following:

I. *"The Pope enjoys, by divine institution, 'supreme, full, immediate, and universal power in the care of souls'."*

II. *"The infallibility of the Magisterium of the Pastors extends to all elements of doctrine, including moral doctrine, without which the saving truths of the faith cannot be preserved, expounded or observed."*

Recent reporting in the predominantly Catholic Philippines included the following:

I. *"Devotees who flocked to a chapel in Bangkulasi, Navotas City to pay their respects to a 400-year-old statue of the sleeping Nazarene on Good Friday have their own stories of how the image 'wakes up' and walks out of the chapel to help people in need... the statue 'usually' takes a bath at a well beside the chapel."*[10]

II. *"God has shown his displeasure* (against the Reproductive Health Bill) *with a flood."*[11]

---

8   See: www.usccb.org/catechism/document/protocol.shtml
9   *Catechism of the Catholic Church.1994.Definitive Version: Based on the Latin Editio Typica*. Episcopal Commission on Catechesis and Catholic Education: Catholic Bishops Conference of the Philippines. Word & Life Publications: Makati City, Philippines.828p. Paragraphs 937,2051.
10  Echeminida, P. 2010. Devotees tell tales of Navotas Nazarene's 'walkabouts'. Reported in *The Philippine Star*, April 4, 2010.
11  Avila, B.S. 2012. Shooting Straight. A great flood without a typhoon? Strange! *The Philippine Star*, August 9, 2012. Volume XXVII, No. 13: p.9-10. The Philippines Reproductive Health (RH) Bill was designed to make contraception more accessible, especially to the poor. It was passed in December 2012, 14 years after its introduction, but was then put on hold after a Catholic Church-backed petition to the Supreme Court. It finally came into force, with amendments, in 2014.

III. *On Good Friday, April 6 2012, at least 27 people were nailed to crosses in the Philippines.*[12]

An American Protestant guide to morality,[13] published in 1964 but still indicative of some current attitudes and teaching, contained the following, in which all forms of emphasis are as in the original:

I. *"Homosexuals are not born - they are MADE! Homosexuality is a perversion of the MIND. It is a psychic and mental and spiritual DISEASE."*

II. *"Among angels there is no marriage - no home life - no family life! And NO SEX!"*

III. *"One doctor, a professor in a very large university, goes so far as to recommend that the bodies of husband and wife ought to be across each other, forming a cross...but on AUTHORITY...I say dogmatically, marital coitus should be in the general position of the love embrace, face to face - since God ordained this act, for humans, to be that occasional supreme expression of love."*

## Sacred Texts

Matthew Gordon[14] translated from the Qur'an's Surah 5:48 as follows: "*To you We have sent down the book, In truth, confirming the scripture that came before it, And watching over it with care. So judge between them on the basis of what God has sent down. And do not follow their false desires And so venture astray from what has come to you as the Truth. To each among you We have revealed a Law and a Clear Path.*"

---

12   Cervantes, D. 2012. 27 nailed to crosses in Pampanga, Bulacan. *The Philippine Star*, April 8, 2012. Volume XXVI.No.254: p.1 and 8.

13   Dorothy, C.V., Hoeh, H.L., Martin, E.L., Meredith, R.C., Rea, B.L., Armstrong, H.W. and R.E. Merrill. 1964. *God Speaks Out on the New Morality*. Ambassador College Press: Pasadena CA. 324p.p.114, 145,191-192,250-251.

14   Gordon, M.S. 2002. *Understanding Islam*. Duncan Baird Publishers: London.112p.p.44, 45.

# IN THE MIDST OF LIES AND NONSENSE

Gordon continued: *"The Quran...teaches that, since the dawn of time, God has communicated to humankind through a series of prophets, among whom stand Noah, Abraham, David, Moses, Jesus, and Muhammad. Each of the 'books' sent to these figures contains the same essential message - that of divine unity and of the duty to worship. In this sense, the Quran was sent down to confirm the earlier revelations...However, the Quran makes it clear that in other respects it stands apart...while each of the earlier texts communicates the 'primordial' doctrine of divine unity, each also contains a unique emphasis - for example, the message to Moses stresses law as the basis of human life, whereas that associated with Jesus emphasizes spirituality and love of God. The genius of Islam, according to this line of argument, lies in its expression of the primordial message through an integration of both the divine law and the inner, or spiritual, commitment to the divine. Thus, the Quran represents the completion, indeed the perfection, of divine revelation."*

Christians and Jews can also find in their sacred texts[15] references to the integration of divine law and personal spiritual commitment.

Diarmaid MacCullough[16] found problems in reconciling history with the order and contents of books of the Old Testament, as in the following examples:

I. *"Using the Bible's own internal points of reference, the promises to the Patriarchs would have been made...around 1800 BCE... there is little reference to the Patriarchs in the pronouncements of 'later' great prophets like Jeremiah, Hosea or the first prophet known as Isaiah, whose prophetic words date from the eighth and seventh centuries BCE... The logic of this is that the stories of the Patriarchs, as we now meet them in the biblical text,* **post-date**

---

15   See, for examples: Deuteronomy 6:5, Leviticus 19:18 and Mark 12:30-31.
16   MacCullough, D. 2010.*Christianity: The First Three Thousand Years*. Penguin Books: New York.1184p.p.p.50-51, 61, 67, 68, 81, 89, 112. MacCullough thanked James Carleton Pager for the source of the 85% statistic and estimation of a much larger total of lost documents and cited p.32 in Markschies, C. 2007. *Kaiserzeitliche christliche Theologie und ihre Institutionem: Proglegomena zu einer Geschichte der antiken christlichen*. Tübingen: Germany.

(original author's emphasis), *rather than predate the first great Hebrew prophets of the eighth and seventh centuries, even though various stories embedded in the Book of Genesis are undoubtedly very ancient.*"

II. "*The angry, precise legislative programme of the Deuteronomic party extended beyond the book itself into a wholesale rewriting of Jewish history. In an operation of remarkable scholarly and literary creativity which probably involved many collaborators working over several decades, older documents were edited and incorporated into a series of books (Joshua, Judges, Samuel, Kings, Jeremiah) which carefully told the story of Israel's triumphs and tragedies in relation to its faithfulness to Yahweh.*"

III. "*(Twenty-four books) came to be regarded as having special status. It is difficult to say exactly when this happened: in Jewish tradition the decision is said to have been made in a 'Great Assembly' in 450 BCE, but that is a fairly typical back-projection of a process which was probably gradual and incremental...it must have been completed at a much later date, especially as some books...like the prophesies of Daniel, patently cannot be as old as the fifth century BCE.*"

MacCullough also described as follows our limited knowledge about the earliest sources of what became the New Testament: "*the vast majority of early Christian texts have perished* (and) *there is a bias among those that survived towards texts which later forms of Christianity found acceptable...around 85% of second-century Christian texts of which existing sources make mention have gone missing and the total itself can represent only a fraction of what there once was...* **The documents which do survive conspire to hide their rooting in historic contexts; this makes them a gift to biblical literalists, who care little for history** (present author's emphasis)."

# IN THE MIDST OF LIES AND NONSENSE

The Bible itself refers to 'the book' or 'this book,' as in the following passage: "*For I testify unto every man that heareth the words of the prophesy of this book, if any man shall add to these things, God shall add unto him the plagues that are written in this book. And if any man shall take away from the words of the book of this prophecy, God shall take away his part out of the book of life.*" [17] By using the words "*this book,*" John the Divine must have meant either the book that he had written or that book plus Mosaic Law; i.e., "*the book of this prophesy.*"[18] The "*book of life*" was John's image for God's registry of souls going to heaven.

All of Bible's references to "*the book*" or "*this book*" mean actual books, not God Himself. Christians believe that Jesus Christ was God Incarnate; i.e., "*the Word was made flesh.*"[19] That so-called "*Word*" was God Himself. It was not the Bible or any other book. Christians call the Bible 'The Word of God,' but Bible is not God. God as the Word cannot be equated with words in any book.

Based on computer searches and a method called Equidistant Letter Spacing, claims have been made that the Bible contains hidden codes underlying its text. A website[20] that finds links to the bizarre includes the following claim: "*In his wisdom, God implanted equidistant letter codes in a manner so meticulous it proves the King James Version of the Bible (the first English translation) to be the **inerrant, infallible and literal word of God*** (present author's emphasis)." From crossword-like diagrams, claims are made that the Bible predicted the attacks of 9/11 and the death of Michael Jackson and predicts that the Rapture is near.

The Bible's alleged hidden codes were well and truly debunked by

---

17 Revelation 22:18-19
18 Deuteronomy 4:2
19 John 1:14
20 www.linkydinky.com/EnglishBleCode.shtml

Maya Bar-Hillel and others,[21] who concluded that the original claim based on Equidistant Letter Spacing was: *"fatally defective* (and) *merely reflect*(ed)...*the choices made in designing* (the) *experiment and collecting the data."* Even if there were some hidden codes in the Bible, which seems an unlikely form of subterfuge by God, their presence would merely indicate that its authors and/or editors and/or translators played games with words and letters.

## INTERPRETATIONS

Maurice Bucaille[22] wrote the following concerning the original Arabic version of the Qur'an, which is the only version regarded as authentic in Islam: "*I knew from translations that the Qur'an often made allusions to all sorts of natural phenomena. It was only when I examined the text very closely in Arabic that I kept a list of them...the Qur'an did not contain a single statement that was assailable from a modern scientific point of view.*"

Ibn Warraq[23] took a different view. He praised the Prophet as "*one of the great men of history,*" but also wrote the following: "*Perhaps the worst legacy of Muhammad was his insistence that the Koran was the literal word of God, and true once and for all, thereby closing the possibility of new intellectual ideas and **freedom of thought*** (present author's emphasis) *that are the only way the Islamic world is going to progress into the twenty-first century.*"

The Old Testament contains much guidance that is relevant for all people in all times and places; for example, the Ten Commandments,[24] the Book of Job and the following gem from Micah: "*...and what doth the*

---

21    Bar-Hillel, M. Bar-Natan, D.,Kalai, G. and B. Mckay.1999. Solving the biblical code puzzle. *Statistical Science* 14: 150-173.
22    Bucaille. M. *The Bible, The Qur'an and Science.* Translated from the French by Alastair D. Pannell and the Author. Nd. Darulfikr Moultan: Cannt, Pakistan.252p.p.viii.
23    Ibn Warraq. 2003. *Why I Am Not a Muslim.* Prometheus Books: New York.402p.p.350.
24    Exodus 20:3-17

## IN THE MIDST OF LIES AND NONSENSE

*Lord require of thee, but to do justly, and to love mercy, and to walk humbly with thy God."* The same applies to the New Testament, in which the reported words of Christ have paramount importance.

Nevertheless, any honest Christian or Jewish believer must admit that the Bible contains passages that have little or no relevance for the present day. We live in very different societies from those who wrote the Bible. They wrote some words for all time and others for their own times and circumstances.

Procreation was the paramount concern for the early Patriarchs. Abram (Abraham) copulated with his housemaid, at the request of his then childless wife.[25] Lacking any other males with whom to copulate, the two daughters of Lot got their father drunk and became pregnant by him.[26] The Old Testament also gives detailed advice about leprosy diagnosis, avoiding contact with infectious male discharges and with anything touched by menstruating women, and how to take the required countermeasures, especially through animal sacrifices.[27]

The Apostle Paul wrote to the Corinthians: *"Let your women keep silence in the churches… And if they will learn anything, let them ask their husbands at home: for it is a shame for women to speak in the church."*[28] Paul advised Timothy: *"Let the woman learn in silence with all subjection. But I suffer not a woman to teach, nor to usurp authority over the man, but to be in silence* (and to) *be saved* (in particular as a descendent of the deceitful Eve; present author's comment) *in childbearing."*[29] Should Christians today comply with this advice? I think not.

Biblical literalists claim that *nothing* on Earth can be impossible for God. Their claims are based on misrepresentations of various texts.

---

25  Genesis 16:1-6
26  Genesis 19:31-36
27  Leviticus 13 to 16
28  I Corinthians 14:34-35
29  I Timothy 2:11-15

Here are the main examples, with my comments in brackets on their contexts and meanings:

I. *"And the Lord said unto Abraham...is any thing too hard for the Lord?"* (Using a familiar conversational device, God was encouraging Abraham to believe that it was not "too hard" for his elderly wife Sarah to have a child.)[30]

II. *"Ah Lord God! Behold thou hast made the heaven and the earth by thy great power and stretched out arm, and there is nothing too hard for thee."* (Jeremiah was praising God for "houses, fields and vineyards" to be repossessed by Israel and anticipating further material gains.)[31]

III. *"Then said Jesus unto his disciples...'It is easier for a camel to go through the eye of a needle, than for a rich man to enter into the kingdom of God'...When his disciples heard it, they were exceedingly amazed, saying, Who then can be saved? But Jesus beheld them, and said unto them, 'With men this is impossible, but with God all things are possible'."* (Jesus taught that souls are saved only though God-human relationships and never by human efforts alone. He used a metaphor to illustrate something equally as impossible as humans achieving their own salvation unaided by God.)[32]

IV. *"And behold, thy cousin Elizabeth, she hath also conceived a son in her old age...for with God nothing shall be impossible."* (An angel allegedly told Mary about Elizabeth's unexpected pregnancy, in order to encourage Mary's acceptance of her own forthcoming and truly miraculous role in bearing Jesus.)[33]

---

30  Genesis 18:13-14
31  Jeremiah 32:15 and 17
32  Matthew 19:23-26; the same words are repeated in Mark 10:24-27 and Luke 18:24-27
33  Luke 1:36-37

## IN THE MIDST OF LIES AND NONSENSE

> V. *"And Jesus answering them saith unto them, 'Have faith in God. For verily I say unto you, That whosoever shall say unto this mountain, Be thou removed, and be thou cast into the sea; and shall not doubt in his heart, but shall believe that those things which he saith shall come to pass; he shall have whatsoever he saith. Therefore I say unto you, what things soever ye desire, when ye pray, believe that ye shall receive them, and ye shall have them."* (God receives and answers all prayers, but it is obvious that He does not always provide what has been asked for. Prayer does not move mountains directly. The forces of nature and/or people move mountains.)
>
> VI. *"And whatsoever ye shall ask in my name, that I will do, that the Father may be glorified in the Son. If ye ask anything in my name, I will do it."* (Christ left that promise to His contemporary disciples, who were about to be faced with His departure and their need to ask for many things in order to spread the Gospel. Clearly, God does not accomplish everything that every believer requests in His name.)[34]

All of those writings refer either to highly specific scenarios or to general case scenarios concerning faith, with extrapolations to what is spiritually possible. There are *no* explicit confirmations of limitless possibilities for interventions by God to change material events and states. God Himself decides what is possible and not possible for Him. He allows free process in the material realm according to fixed natural laws and gives humans free will to pursue good or evil ends.

Many of my fellow believers will not agree with this analysis and will insist, on biblical grounds, that God really can do anything in the material realm that they might request of Him, with no limits, because that is what the Bible says about His promises. I ask them to consider carefully the following words from the Apostle John:[35] *"And this is the*

---

34  John 14:13-14
35  I John 5:14-15

*confidence that we have in him, that, if we ask anything **according to his will*** (present author's emphasis), *he heareth us: And if we know that he hear us, whatosever we ask* (i.e., according to his will; present author's comment), *we know that we have the petitions that we desired of him."* Moreover, the Apostle Paul[36] wrote that God empowers us *"in the inner man"* (i.e., in our souls) and is thereby: *"able to do exceedingly abundantly above all that we ask or think, according to the power that worketh **in us*** (present author's emphasis)." God empowers us in our souls - to choose faith over unbelief and thereby to remain empowered by Him, no matter what happens to us in the material realm.

We should not tempt God by asking Him change the weather, the natural courses of senility, the trajectories of bullets etc. We should ask for and expect to receive only those things that are *in accordance with His will*, which includes the necessary workings of free process in our material selves and in the rest of the material realm that He made for Him and for us. God discloses to us, through mathematics and science, the truth about the composition and workings of the material realm. We are His agents of change for good in the material realm as we respond to His revelations to our souls, but such material changes cannot override the material realm's God-given free process.

We exercise our own God-given freedom in a world of free process and free-willed fellow humans. God has made it so. If we are honest observers and recorders of what truly happens in us and around us, we can see and attest that this is so. There are limits to what we can do and God has limited Himself concerning what He can do in response to our requests.

Since the Second Vatican Council, the Catholic Church has begun to take a more flexible approach to biblical interpretation. The *"Dogmatic Constitution on Divine Revelation - Dei Verbum"*[37] contains the follow-

---

36   Ephesians 3:16 and 20
37   Second Vatican Council, 1965. *Dogmatic Constitution on Divine Revelation - Dei Verbum.* Chapter III, paragraphs 11 and 12. See also paragraph 110 in the Catholic Catechism.

ing: "*Therefore, since everything asserted by the inspired authors must be held to be asserted by the Holy Spirit, it follows that the books of Scripture must be acknowledged as teaching solidly, faithfully and without error that truth which God wanted put into sacred writings... **However, since God speaks in Sacred Scripture through men in human fashion, the interpreter of Sacred Scripture, in order to see what God wanted to communicate to us, should carefully investigate what meaning the sacred writers really intended, and what God wanted to manifest by means of their words. For the correct understanding of what the sacred author wanted to assert, due attention must be paid to the customary and characteristic styles of feeling, speaking and narrating which prevailed at the time of the sacred writer, and to the patterns men normally employed at that period in their everyday dealings with one another*** (present author's emphasis)."

William James[38] gave the following fine perspective on how to approach the Bible: "*if our theory of revelation-value were to affirm that any book, to possess it, must have been composed automatically or not by the free caprice of the writer, or that it must exhibit no scientific or historic errors and express no local or personal passions, the Bible would probably fare ill at our hands. But if, on the other hand, our theory should allow that a book may well be a revelation in spite of errors and passions and deliberate human composition, if only it be a true record of great-souled persons wrestling with the crises of their fate, then the verdict would be much more favourable.*"

John Polkinghorne and Nicholas Beale[39] described the Bible well, as follows: "*It is not some divinely dictated textbook giving all the answers, which have to be accepted without question. It is much more like a laboratory notebook, recording the unique events of divine self-disclosure made in the course of the history of Israel and in the life of Jesus and its astonish-*

---

[38] James, W. 1901-1902. *The Varieties of Religious Experience: A Study in Human Nature.* 2002 Edition. Modern Library: New York.602p.p.7

[39] Polkinghorne, J. and N. Beale. 2009. *Questions of Truth: Fifty-one Responses to Questions about God, Science, and Belief.* Westminster John Knox Press: Louisville KY.186p.p.7.

*ing aftermath. The Bible is not a book but a library, with various types of writing in it. There is much history, but there are also symbolic stories that convey truths so deep that only a story form could express them."* I agree.

### LITERAL IMPOSSIBILITIES

The following accounts of literal impossibilities reported in the Bible are given here to illustrate the difficulty in reconciling the doctrines of biblical literalism and inerrancy with God's truthful disclosures through science about how the material realm, as made by Him, actually works.

First, Jonah allegedly survived for three days and nights inside the stomach of a big fish.[40] Christ[41] reportedly endorsed the story of Jonah's ingestion and subsequent release, calling the fish a *"whale."* The Reverend J.H Gladstone,[42] who was a Fellow of the Royal Society and a believer in the inerrancy of the Bible, wrote: *"The identification of the names of plants and animals* (in the Bible) *is not always possible; but there is an unlucky translation in Matthew xii, 40, which has added a needless difficulty to the story of Jonah, for a whale's gullet is far too narrow for a man to pass."* According to Gladstone, the original Greek word in the earliest available texts meant: *"any large fish and on the shores of the Mediterranean was often applied to the tunny or the shark."*

Gladstone had no problem with believing that Jonah could have entered and survived inside the stomach of a shark or a large tuna. I beg to differ. Anyone ingested by a large fish or a whale would not survive more than a matter of minutes in its stomach, let alone three days and nights. There would be blinding, suffocation from lack of oxygen and gradual digestion into constituent amino acids, carbohydrates and fats

---

40 Jonah 1:17 and 2:1
41 Matthew 12:40
42 Gladstone, J.H. 1873. Points of Supposed Collision Between the Scriptures and the Natural Sciences, p. 133-171. In *Faith and Free Thought: A Second Course of Lectures Delivered at the Request of the Christian Evidence Society*. Hodder and Stoughton: London.469p.p.154-155. Available from: www.kessinger.net

# IN THE MIDST OF LIES AND NONSENSE

etc. The story of Jonah is obviously a metaphor for isolation from God, followed by reconciliation. When Christ endorsed the story of Jonah, He was probably hinting at His forthcoming death on the cross and His temporary separation from God, followed by His Resurrection.

Second, we have the story of Balaam beating his donkey unjustly after "*she*" was allegedly thrown off balance by an appearance of "*the angel of the Lord*" and then allegedly spoke to Balaam complaining about how badly he had been treating her, despite her faithful service as his beast of burden.[43] I take that as a metaphor for human blindness to God's oversight and His awareness of injustice.

Third, we are told that three very brave servants of God, who defied their captor Nebuchadnezzar by refusing to worship his golden image, were tied up and thrown into a "*burning fiery furnace*", which was allegedly heated: "*seven times more than it was wont to be heated.*" Then, allegedly, they simply walked through the furnace, together with a fourth man who was "*like the Son of God,*" and emerged unscathed as a threesome again.[44] That story celebrates those who go through extreme discomfort for their faith and emerge victorious because of God's presence and encouragement.

Fourth, when Pharoah's "*magicians of Egypt*" and Aaron cast down their wooden rods, all of the rods allegedly became snakes, but Aaron's rod/snake allegedly ate up all of the enemy rods/snakes.[45] Wooden rods cannot turn into snakes, or vice-versa. This story was spiced up in order to impress its immediate audiences and subsequent readers who craved magical happenings, and to emphasize God's relationship with His chosen people.

Fifth, the ages at which some of the Patriarchs allegedly commenced fatherhood and their ages at death, having allegedly continued fathering

---

43  Numbers 22:28 and 30
44  Daniel 3:11-30
45  Exodus 7:9-15

children for hundreds of years, are recorded in Genesis chapters 5 and 9 as follows: Adam, 130 (930); Enos, 90 (815); Cainan, 70 (910); Mahalel, 65 (895); Jared, 162 (962); Enoch, 65 (365, but it is alleged that Enoch did not die naturally because "*God took him*"); Methuselah, 187 (969!); Lamech, 182 (777); and Noah, 500! (950).

By Genesis chapter 11, the ages at which firstborn were allegedly fathered have become lower - from Shem at age 100 to many others in the range 29 to 35, but the male life span still seems to peaking, allegedly, at about 200. On the rare occasions when they are mentioned at all, the wives of this period are alleged to have had *shorter* lives than their husbands, which is the opposite of other human norms. The chosen people then allegedly adjusted rather rapidly to having lives of about "*three score years and ten* (70)" or sometimes "*four score years* (80),"[46] in common with most of humanity ever after.

Gerry Schroeder[47] endorsed the Patriarchs' alleged delayed fatherhood, prolonged reproductive success and great longevity as follows: "*Both Maimonides...and Nahmanides...suggest that changes in the environment following the flood favored ('selected for' in modern terms) shorter life span...(and) the pre-Noah ages for both puberty and death are tenfold higher than today, as if the whole process was slowed by a factor of ten. This could have been caused by a metabolic shift.*" Schroeder also noted a trend of shorter life spans through the generations after Noah and compared that trend with the shorter life spans and earlier sexual maturity that are seen as animals are domesticated.

No mammal has a life history strategy of waiting for over a hundred years to reproduce. No mammal remains reproductively active for hundreds of years. As far as I am aware, a 10-fold and sustained metabolic shift at normal temperatures and atmospheric pressures has never been observed in humans. Most domesticated animals are bred for sending

---

46 Psalms 90:10
47 Schroeder, G.L. 1998. *The Science of God: The Convergence of Scientific and Biblical Wisdom.* Broadway Books: New York.226p.p.15-16, 203.

# IN THE MIDST OF LIES AND NONSENSE

to market *before* they can reproduce. My best guess is that the incredible age data in Genesis have been manipulated to indicate personal, family and tribal power. Allegations of great longevity and sexual prowess were probably made to celebrate the alpha males of the period.

In all of his works, Schroeder[48] seeks to match science with the Tanakh, which is essentially the Old Testament. He considers Moses Maimonides (1135-1204 CE) and Nahmanides (ca.1195-1270 CE) as excellent sources for this task, on the following grounds: *"Being ancient, the perspectives that they bring are **not biased by the discoveries of modern science*** (present author's emphasis). *The antiquity of these sources ensures that there has been no attempt to bend the biblical text to match science."*

Some scientists are indeed biased against the Bible, but some biblical scholars are biased against science. There is bias to be addressed on both sides. If science seems at odds with an interpretation of a sacred text, then *both* need careful re-examination by scholars open to the possibility that either or both could be in error. God's truthful disclosures through mathematics and science cannot be overridden on principle by sacred text literalism. The world is what it is, whatever any sacred text says about it. God made it that way.

I believe that God, as the Holy Spirit, inspired and guided the many authors, editors and translators of the Bible, but I do not believe that they could have been prevented entirely from making errors of fact or emphasis. They were fallible humans, subject to economic, institutional and political pressures.

The Bible cannot be the literal, inerrant, infallible and total source of God's revelations for all time. It is a unique and precious source through which humans can discern God's will and seek His guidance,

---

48    Schroeder, G.L. 2009. *God According to God: A Physicist Proves We've Been Wrong About God All Along.* HarperOne: New York.249p.p.117-132, 86-87.

but it is *not* the only means to those ends.

The first and last word on everything is God: the Logos, the Wisdom. He is a Living God, in and for our times and for all times to come. His guidance comes to us not only through the Bible and other sacred texts and through those who study and interpret them, but also directly, into our souls.

The Apostle Paul said: "*For our gospel **came not unto you in word only*** (present author's emphasis), *but also in power, and in the Holy Ghost and in much assurance.*"[49] Biblical literalists please take note. Just as "*the Sabbath was made for man, and not man for the Sabbath,*"[50] the Bible was also made for man and not man for the Bible.

## ANGELS

The Catechism of the Catholic Church[51] holds that the "*existence of angels*" is a "*truth of faith*" and that: "*From the beginning until death, human life is surrounded by their watchful care and intercession. Beside each believer stands an angel as protector and shepherd leading him to life. Already here on earth the Christian life shares by faith in the blessed company of angels and men united in God.*" The United States Conference of Catholic Bishops[52] advised that Catechetical Texts should: "*present angels as spiritual creatures who glorify God without ceasing and who serve his saving plans for other creatures.*"

Richard Webster[53] catalogued hundreds of angels in Christianity, Islam, Judaism and Zoroastrianism, from A to Z. Here are extracts from a few

---

49  I Thessalonians 1:5
50  Mark 2:27
51  *Catechism of the Catholic Church.1994.Definitive Version: Based on the Latin Editio Typica.* Episcopal Commission on Catechesis and Catholic Education: Catholic Bishops Conference of the Philippines. Word & Life Publications: Makati City, Philippines.828p.paragraphs 328,336.
52  http://www.usccb.org/catechism/document/protocol.shtml
53  Webster, R. 2009. *Encyclopedia of Angels.* Llewellyn Publications: Woodbury MN.246p.p. xv, 13, 32-33, 225.

# IN THE MIDST OF LIES AND NONSENSE

of his entries:

I. *"Anuael is the angel of prosperity and commerce...He can be invoked in any matter concerning money or finance...Anuael is the guardian angel of people born between January 31 and February 4...Anuael has brown hair and wears green robes."*

II. *"Barchiel is ruler of the order of Seraphim, angel of February, prince of the Third Heaven, and a leading member of the Sarim. He looks after people born under the signs of Scorpio and Pisces... Gamblers desiring success with their wagers also invoke Barchiel, though they usually call him Barakiel."*

III. *"Zuriel is the Prince Regent of the choir of Principalities. He is also Archangel of the sign of Libra and ruler of September. Zuriel can be invoked to create harmony and accord. Zuriel is also the angel of childbirth."*

D.J. Conway's[54] writings on angels include the following:

I. *"Guides, guardians and angels are powerful, important beings that influence human lives...we need to connect with each class of entity. If our telepathic networks of communications with them are smooth, we will not have to struggle so much when we need their help."*

II. *"Angels are a special species of beings, whose vibrations are so very high and pure that they have access to the Supreme Being...The true purpose of angels in both the astral and earthly planes is rather nebulous. No one has a definite answer."*

III. *"Here is a beautiful example of how the Light and Shadow angels work harmoniously together to benefit this planet - and by doing*

---

54  Conway, D.J. 2009. *Guides, Guardians and Angels*. Llewellyn Publications: Woodbury MN. 176p.p.4,69, 75.

*so, influence and balance all levels and planets in the Multiverse. Each dawn, certain Light angels sing to the area of the earth where the sun rises. This reenergizes the planet...At each sunset, certain shadow angels sing while the sun sets and the moon rises. These two angelic choirs provide balanced energy currents that sweep around the earth."*

## Faith Healing

Faith healing is defined here as the claim by a self-professed healer and her/his patient that an adverse physical or mental condition has been alleviated or cured by the application of spiritual powers coming from or through that healer.

In the first volume of the British Medical Journal, William Osler[55] reported the views of Archdeacon Sinclair of London, who had noted the alleged gifts of healing attributed to worthy persons including, *inter alia*: "*Justin Martyr, Irenaeus, Origen, Ambrose, Chrysostom, and Augustine...George Fox, John Wesley, Prince Hohenlohe, Father Matthew, Dorothea Trüdel, Pastor Blumhardt and Father John of Cronstadt.*"

Osler's review ended as follows: "*The upshot seemed to Archdeacon Sinclair to be that there were no limits to the powers of faith and prayer; that many persons, like wise parents, had a gift of personal impressiveness; that the subconscious self was a fact in psychology; that the clergy might do more in their visitations of the sick by encouragements to faith; that the laying-on of hands and the anointing with oil might be tangible helps to faith; and that religion and medical science should always co-operate, while the ultimate responsibility must lie with the accredited physician.*"

In other words, Archdeacon Sinclair toed the party line of his church

---

55  Osler, W. 1910. The faith that heals. *The British Medical Journal* 1, Issue 2581; June 18, 1910: 1,470-1,472.

## IN THE MIDST OF LIES AND NONSENSE

by confirming that faith and prayer have allegedly limitless healing potential, but also took care to assign *"ultimate responsibility"* to doctors.

In that same volume, a review entitled *"Mental Healing"*[56] covered activities at ancient Greek and Roman temples and subsequent developments in the Christian era, as churches continued to house pagan-like shrines of healing. Various saints were allocated specific healing powers: St. Lazarus for leprosy; St. Roch for plague; and St. Dodon for rheumatism. Healing powers were attributed to places at which, allegedly, there had allegedly been divine apparitions, and/or to their so-called healing waters. The reviewer commented as follows on the grotto at Lourdes: *"the pilgrim is exploited in every possible way...the cures affected...might be reckoned at about 5 per cent* (but) *even when Lourdes fails to cure the ailing body, it comforts the soul."*

In typical fundamentalist Christian faith healing events today, a charismatic preacher-cum-healer excites the congregation to expect the miraculous and then invites selected sufferers to come forward and be healed. Some of those in wheelchairs stand up and walk and are allegedly healed. Others are struck on the brow or the chest, after which they faint or appear to have fainted. The preacher-cum-healer commands the supposed demons of AIDS, cancer, diabetes etc. to leave their supposed hosts and assures everyone that no 'demon of sickness' can stand against the power of Christ. Many are defrauded by such performances.

The whole history of medicine has been a search for whatever works to alleviate human pain and suffering and to prolong life, including how to make the most of the placebo effect. When trusted friends, relatives, pastors and mentors visit the sick, the results can be highly therapeutic. However, it is dangerous, unethical and foolish to rely on faith healing and prayer to alleviate sickness or injury, while shunning conventional medicine. God sent us all of the truthful disclosures through science

---

56   Anon. 1910. Mental Healing. *The British Medical Journal* 1, Issue 2581; June 18, 1910: 1,483-1,497.

upon which modern medicine is based. He wants us to be helped accordingly. To shun medical science is to insult God, Who disclosed it.

## Heaven

Randy Alcorn[57] wrote much about what we can expect to find and do in heaven. The following are a few examples, in which I have substituted the words of the King James Version of the Bible for Alcorn's cited texts:

> I. *"Will we literally eat and drink?" Alcorn said 'Yes,' as follows: "Our resurrected bodies will have resurrected taste buds. We can trust that the food we eat on the New Earth, some of it familiar and some of it brand-new, will taste better than anything we've ever eaten before...we may consume a wonderful array of fruits and vegetables, perhaps supplemented by 'meat' that doesn't require death - something that tastes better but isn't animal flesh... Those who for reasons of allergies, weight problems, or addictions can't regularly consume peanuts, chocolate, coffee, and wine - and countless other foods and drinks - may look forward to enjoying them on the New Earth."*
>
> II. *"Will there be marriage and family?" Concerning marriage, Alcorn said 'No,' based on the following text, inter alia: 'For in the resurrection they neither marry nor are given in marriage but are as the angels in heaven.' He continued thus: "People with good marriages are each other's best friends. There's no reason to believe they won't still be best friends in heaven... The notion that relationships with family and friends will be lost in heaven, though common, is unbiblical...It completely contradicts Paul's intense anticipation of being with the Thessalonians and his encouraging them to look*

---

57  Alcorn, R. 2004. *Heaven.* Tyndale House Publishers Inc., Carol Stream IL.533p.p.303, 301- 302, 305, 307, 309, 350 (quoting Matthew 22:30), 351-353,352, 353,392 (quoting Revelation 5:13), 393-394.

# IN THE MIDST OF LIES AND NONSENSE

*forward to rejoining their loved ones in heaven."*

III. *"Will there be sex?" Alcorn said 'No,' on the following grounds: "Because...sex is intended for marriage, then logically we won't be engaging in sex...Once we're married to him (Christ), we'll be at the destination that marital sex pointed to as a signpost."*

IV. *"Will animals praise God?" Alcorn said 'Yes,' based on the following text, inter alia: "every creature in heaven and on earth; and under the earth, and such as are in the sea, and all that are in them heard I saying, 'Blessing, and honour, and glory, and power, be unto him that sitteth upon the throne, and unto the Lamb for ever and ever.'" He added: "Although earthly animals aren't capable of verbalizing praise as these animals in heaven do, the passages of earthly animals praising God **and the story of Balaam's donkey** (present author's emphasis) clearly suggest that animals have a spiritual dimension far beyond our understanding."*

The following conversation is my Free Thought experiment about the kind of exchange that might have happened if a young couple (A and B), both nonbelievers and honest seekers of truth, had come to my church for a musical event and had talked with a church usher (U) on the way out.

A. Thank you, that was great.

U. Why don't you join us for worship on Sunday?

B. Well, we are not really churchgoers.

U. Maybe you should consider becoming Christians?

A. Maybe so, but we cannot see benefits in that.

U. Being sure of going to heaven when you die is a huge benefit.

B. But how would we know what to expect in heaven?

U. It's all in the Bible and we have other books that interpret all of it for you.

A. So, if we got married, would we be married in heaven?

U. No, but you would be best friends; everyone in heaven is married to Christ.

B. Really? Do those who got married here have no marital relations in heaven?

U. That's right, but you need to study it more; will we see you next Sunday?

B. I don't think so, but thanks again for the music.

I expect the joys of Earth to be exceeded beyond measure by the joys of heaven, but I know that we must let go of all of the things of Earth when we die. Those who claim to know what we will find in heaven are fooling themselves and others and are turning honest seekers away from exploring faith.

## Homosexuality

Humans share with many animal species the same biochemical and genetic mechanisms for gender determination as well as many of the same types of sexual behavior. Throughout the animal kingdom and among humans, gender and sexuality are variable beyond a simple female/male distinction. In 2006-2007, an exhibition at the Natural History Museum of the University of Oslo indicated homosexuality in over 1,500 species and found it well documented in over 500, including same sex pairings in birds and mammals.

Many species of fish are protogynous hermaphrodites. They start life as females and then some dominant ones become males. The clownfish portrayed on film as the little hero Nemo is a protandrous hermaphrodite. In Nemo's species, everyone starts life as a boy. The dominant few, presumably including Nemo, get to become girls. A sexually correct sequel would be interesting.

Lee Dugatkin[58] quoted as follows from E.O. Wilson's discussion of work by Hermann Spieth and Robert Trivers: "*Homosexual members of primitive societies may have functioned as helpers…Freed from the special obligations of parental duties, they could have operated with special efficiency in assisting their **close** (original author's emphasis) relatives. Genes favouring homosexuality could then be sustained at a high equilibrium level by kin selection alone.*"

The extents to which homosexuality in humans has any genetic basis remains controversial. Dean Hamer and others[59] found increased rates of same-sex orientation in the maternal uncles and male cousins of a sample of 114 gay men, but not in their fathers or paternal relatives. They concluded that there was statistically significant evidence for male sexual orientation being genetically influenced. However, regardless of whether homosexuality in humans has any significant genetic basis or not, it is obvious that the global population of humans has never been neatly divisible into strictly female and male sub-populations.

By the strict definition of possessing both ovaries and testes, human hermaphrodites are very rare, but many humans are intermediate

---

58   Dugatkin, L. 2006. *The Altruism Equation: Seven Scientists Search for the Origins of Human Goodness.* Princeton University Press: Princeton NJ. 188p.p.119. Dugatkin cited the discussion in p.555 of E.O. Wilson's *Sociobiology: The New Synthesis*, which was published in 1975 by Harvard University Press: Cambridge MA.

59   Hamer, D.H., Hu, S., Magnuson,V.L., Hu, N. and A.M.L. Pattalucci. 1993. A linkage between DNA markers on the X chromosome and male sexual orientation. *Science* 261:321-327.

between being fully masculine and fully feminine. A brilliant song by Ray Davies[60] illustrates well how some boys are naturally inclined to be girls, and vice-versa, in the muddled world of human sexuality. Robert Ritchie and others[61] reviewed the history of Olympians who appeared to be intersexes and noted that the occurrence of intersexes in the human population might be as high as 1.7%.

Homophobia is widespread in Christianity and Islam. In Uganda, following a 2009 anti-homosexual conference, a politician who boasted of *"having evangelical friends in the U.S. government"*[62] introduced the Anti-Homosexuality Bill, which contained provisions for the execution of homosexuals by hanging. In 2011, the Prohibition of Same Sex Marriage bill was passed unopposed by the Nigerian Senate. Christians, Muslims and others agreed that same sex couples should face prison terms of 14 years. Senator Ahmed Lawan[63] stated: *"We are protecting humanity and family values, in fact, we are protecting civilization in its entirety…Should we allow for indiscriminate same-sex marriage, very soon the population of the world would diminish."*

In the history of Islam, attitudes to homosexuality have ranged from tolerance to savage repression. Ibn Warraq[64] found the Qur'an unclear on the issue, but noted that the Hadith condemns sodomy and that *"The Prophet found sodomites abhorrent and asked for their execution,"* which the early Caliphs accomplished by stoning, burning, or throwing from minarets. Ibn Warraq noted that homosexuality was practiced openly by many Caliphs from the 9th to the 12th centuries and in high society in Muslim Spain. Attitudes to homosexuality in

---

60   Ray Davies.1970. *Lola.*
61   Ritchie, R., Reynard, J. and T. Lewis. 2008. Intersex and the Olympic Games. *Journal of the Royal Society of Medicine* 101 (8): 385-399.
62   Anon. 2010. Uganda considers death penalty for gays. *International Herald Tribune*, Tuesday, January 5, 2010, p.1, 4.
63   Fisher, J. 2011. Nigerian leaders unite against same-sex marriage. BBC News Africa, December 5, 2011. Available at: www.bbc.co.uk/news/world-africa
64   Ibn Warraq. 2003. *Why I Am Not a Muslim.* Prometheus Books: New York. 402p.p.342. Ibn Warraq referred to the celebration of homosexuality, including lesbianism, by poets such as Abu Nuwas, in works such as *The Perfumed Garden of the Sheikh Nefzaoui.*

present day Muslim societies range from benign tolerance to harsh persecution. Lesbians in Iran can be sentenced to death, but only after their fourth conviction.[65]

For some much needed words of good sense, I turn to the Reverend Doctors David de Pomerai and Thomas Lindell,[66] who compiled a clear and authoritative summary of what is known to date from science about the biological basis of human sexual orientation and concluded: "*the candidate mechanisms identified to date are not subject to individual choice, nor can they be readily changed. Therefore, homosexuality should not be regarded as an elective lifestyle choice* (original authors' emphasis)." Homosexuality is natural. It is part of the human condition, created by God.

## The Rapture

There is no mention of the Rapture by name in the Bible, or the Catholic Catechism, or the Anglican/Episcopalian Thirty-Nine Articles. The Rapture is part of the sequence of events known as the End Times, which some scholars claim to be able to describe in detail from their studies of sacred texts. Among and within Christianity, Islam, Judaism and Zoroastrianism there are many different versions of the sequences of events forecast for the End Times. I regard them all as claims to know the unknowable.

Rapture enthusiasts point out that Christ reportedly spoke at length on the End Times, using prophesies from Daniel and others about the coming "*tribulation,*" which would include conflicts and natural and supernatural disasters on a grand scale. Christ is reported to have said: "*And then shall all the tribes on earth mourn, and they shall see*

---

[65] Anon. 2012. Islam and Homosexuality: Straight But Narrow. *The Economist* Volume 402, No. 8770: 49-50.
[66] De Pomerai, D. and T.J. Lindell. 2005. *The Biological Basis of Human Sexual Orientation.* 10p.p.7. Available under Articles and Sermons at: www.ordainedscientists.org

*the Son of man (Christ) coming in the clouds of heaven with power and great glory. And he shall send his angels with a great sound of trumpet and they shall gather together his elect from the four winds, from one end of heaven to the other."*[67]

The doctrine of the Rapture, as currently preached and accepted, chiefly by evangelical Protestants, is based on a literal reading of the Apostle Paul's Epistle to the Thessalonians, elaborated with literal readings from the Books of Revelation and Daniel. Paul supplied the main Rapture text, as follows: *"For the Lord Himself shall descend from heaven with a shout, with the voice of the archangel, and with the trump of God, and the dead in Christ shall rise first. Then we which are alive and remain shall be caught up together with them in the clouds, to meet the Lord in the air; and so shall we ever be with the Lord."*[68]

Rapture enthusiasts believe that if they are still alive on Rapture day they will ascend bodily into the sky. They want all believers to be ready for that last free flight. How could any human be levitated to the clouds? 'No problem,' say the Rapture enthusiasts. They claim that the Bible has good examples of humans being taken up directly for Earth to heaven and they insist that those examples are true - because the Bible says so. For example, they cite the story in Genesis that *"Enoch walked with God and he was not; for God took him."*[69] Enoch was allegedly 365 years old at the time. His unexplained disappearance, in a sparsely populated land, is not strong evidence for a global ascension at Rapture time.

I do not believe in the Rapture as described in the End Times literature. If I am wrong, however, then I must go for the sequence of events favoured by Tim Lahaye,[70] who holds that the Rapture should take place

---

67  Matthew 24:30-31
68  I Thessalonians 4:16-17
69  Genesis 5:24
70  Lahaye, T. 2002.*The Rapture: Who Will Face the Tribulation?* Harvest House Publishers: Eugene OR.255p.p.39, 207.

# IN THE MIDST OF LIES AND NONSENSE

*before* the alleged Tribulation. It would seem only reasonable and just for Christ to take His faithful disciples up to heaven *before* all of the natural and supernatural mayhem to be endured, allegedly, by those who remain on the Earth.

Lahaye's following passages illustrate the alleged impacts of the Rapture on Earth and the complex speculations about its possible timetable:

I. *"I expect the Rapture to be electrifyingly sudden but not secret… millions of people will suddenly vanish from the earth…An unsaved person who happens to be in the company of a believer will know immediately that his friend has vanished…A mother will pull back the covers in a bassinet, smelling the sweet baby smell one moment but suddenly kissing empty space and looking into empty blankets."*

II. *"Of all the attacks on the Rapture, easily the most ridiculous is the one that finds it an instrument of the Illuminati (the Master Conspirators) to put the evangelical world to sleep politically so that the secularizers of the Western world are free to dominate the government, education, communications industry, and other agencies of influence. The proponents of this attack would even have us believe that Dr. C.I. Scofield, author of the study Bible that has blessed millions of Christians, was part of that plot. Robert L. Pierce, author of* The Rapture Cult, *distributed by the John Birch Society, is the first person I have ever read who called belief in the Rapture 'cultish.' Even the most zealous post-Tribbers who hate the pre-Trib position have not made such a suggestion. In fact, even the post-Millenial apologists from Tyler, Texas, in their earnest attempts to rescue their system of theology from a well-deserved grave by savaging the pre-Trib position, have not, to my knowledge made such a suggestion."*

# Free Thought, Faith, and Science

## Research on Prayer

Many who pray for their own health say that they have enjoyed beneficial results. Anne McCaffrey and others[71] reported that 35% of 2,055 Americans interviewed in 1998 said that they prayed about their own health concerns, mostly without discussion with their physicians, and 69% of those who prayed said that they: *"found prayer very helpful."*

Many who pray say that they are claiming God's much quoted (albeit often misinterpreted) promises and particularly the following reported words of Christ: *"what things soever ye desire, when ye pray, believe that ye shall receive them, and ye shall have them."*[72] Some misinterpret this as a promise that *any* request made to God in prayer *will* be granted by Him, if the one praying is strong enough to expect *no* other outcome. This text is really a statement about faith, through which the one praying trusts God to decide what is best and what He will grant. All who pray must pray responsibly and remember the following command: *"Ye shall not tempt the Lord your God."*[73] In other words, do not put God to the test by asking for anything that is unreasonable in the context of the material realm possibilities that He has given to us.

Eric Burke[74] believed that Christian and Jewish believers face the following paradox concerning intercessory prayer - only a believer is qualified to be a praying intercessor; but how can any believer become a praying intercessor without putting God to the test and therefore disobeying Him? Burke reviewed six studies on the efficacy of intercessory prayer published from 1988 to 2006, involving Buddhist, Christian, Jewish, Muslim and First Nation American praying intercessors. Only the two studies that involved Christians

---

71 McCaffrey, A.M., Eisenberg, D.M., Legedza, A.T.R., Davis, R.B. and R.S. Phillips. 2004. Prayer for health concerns. Results of a National Survey on Prevalence and Patterns of Use. *Archives of Internal Medicine* 164: 858-862.
72 Mark 11:24
73 Deuteronomy 6:16, endorsed by Christ in Luke 4:12
74 Burke, E.J. 2006. Controlling the independent variables in the clinical study of prayer: The devil is in the details. *American Heart Journal* 152 (4): e41-e42.

# IN THE MIDST OF LIES AND NONSENSE

claimed to have shown positive results for intercessory prayer and neither of those studies stands up to hard scrutiny.

The study by Randolph Byrd[75] is probably the most famous example of attempted research on intercessory prayer. Byrd randomly assigned 393 Coronary Care Unit (CCU) patients to a 'prayed for' (IP) group of 192 and a 'not prayed for' (control) group of 201. As indicators of the severity of the patients' coronary condition, he recorded the complications that they suffered and the specific treatments that they needed. He found no significant differences between the IP and control groups for 23 out of 29 indicators. His results for the other six were as follows, listed here as the numbers of 'prayed for' (IP) and control patients and in brackets the percentages of the IP and control groups that they represented and also the probabilities (p) that the IP and control data were significantly different.

I. *congestive heart failure - IP, 8 patients (4%), control, 20 patients (10%), $p < 0.03$;*

II. *diuretics - IP, 5 patients (3%), control, 15 patients (8%), $p < 0.05$;*

III. *cardiopulmonary arrest - IP, 3 patients (2%), control, 14 patients (7%), $p < 0.02$;*

IV. *pneumonia - IP, 3 patients (2%), control, 7 patients (13%), $p < 0.03$;*

V. *antibiotics - IP, 3 patients (2%), control, 17 patients (9%), $p < 0.03$; and*

VI. *intubation/ventilation - IP, no patients and control, 12 patients (6%), $p < 0.02$.*

---

75  Byrd, R.1988. Positive therapeutic effects of intercessory prayer in a Coronary Care Unit population. *Southern Medical Journal* 61(7): 826-829.

Based on those highly selected results, from only six out of his 29 indicators, Byrd claimed significant benefits for the IP group. His overall conclusion was that: *"intercessory prayer to the Judaeo-Christian God has a beneficial effect in patients admitted to a CCU."*

Gary Posner[76] pointed out that there could have been no way of knowing who else might have been praying for any of the patients - the control group might have received as many prayers, if not more, than the IP group. Byrd had admitted the same possibility in his paper. Posner also found that Byrd's statistical tests were flawed, because of obvious interrelationships among some of the six allegedly positive indicators.

Combatting congestive heart failure requires the use of diuretics and combatting pneumonia requires antibiotics. Congestive heart failure and pneumonia increase the likelihood of cardiopulmonary arrest and of requiring intubation or ventilation. In truth, Byrd's results showed no noteworthy differences between his IP and control groups. Nevertheless, his conclusions are still preached in some Christian churches - either through ignorance or in willful disregard of work that has discredited them.

The second positive study reviewed by Burke was that of William Harris and others in 1999.[77] They used a complex weighted and summed scoring system called the MAHI-CCU score - developed by doctors from Mid America Heart Institute (MAHI), Kansas City and the University of Missouri - Kansas City School of Medicine. The MAHI-CCU score is a continuous variable with which to rate patients' general outcomes.

---

76   Posner, G.P. 1990. God in the CCU? A critique of the San Francisco hospital study on intercessory prayer and healing. Originally published in the Spring1990 Issue of *Free Inquiry*. Available at: www.infidels.org/library/modern/gary_posner/godccu.html

77   Harris, W.S., Gowda, M., Kolb, J.W., Strychacz, C.P., Vacek, J.L. Jones, P.G., Forker, A., O'Keefe, J.H. and McCallister, B.D.1999. A randomized, controlled trial of the effects of remote, intercessory prayer on outcomes in patients admitted to the Coronary Care Unit. *Archives of Internal Medicine* 159 (19):2273-2278.

# IN THE MIDST OF LIES AND NONSENSE

The following passage illustrates the complexities of how patients were scored: "*For example, if, after the first day in the CCU, a patient developed unstable angina (1 point), was treated with antianginal agents (1 point), was sent for heart catheterization (1 point), underwent unsuccessful revascularization by percutaneous transluminal coronary angioplasty (3 points), and went on to coronary artery bypass graft surgery (4 points), his weighted MAHI-CCU score would be 10. Another patient might have developed a fever and received antibiotic treatment (1 point), but experienced no other problems and been discharged from hospital with a score of 1. A third patient might have suffered a cardiac arrest (5 points) and died (6 points), for a total weighted score of 11 points.*" Higher scores indicated worse courses and outcomes.

The results were as follows, given as mean MAHI-CCU weighted and unweighted scores (± Standard Error of Means), with the levels of probability (p) that the prayed for versus control group scores were significantly different: 'prayed for'-weighted, 6.35 (± 0.26) versus control-weighted, 7.13 (± 0.27), p=0.04; 'prayed for'-unweighted, 2.7 (± 0.1) versus control-unweighted, 3.0 (± 0.1), p=0.04. On this basis, intercessory prayer seems to have conferred benefits on about 10% of patients.

When Harris and others scored the same patients using the composite 'good,' 'intermediate' and 'bad' scoring method developed by Byrd (the so-called 'Byrd scores') to compare overall treatments and outcomes, there were no significant differences between the 'prayed for' and control groups. Expressed here as numbers of patients (and percentages of the 'prayed for' and control populations that they represented), the Byrd scores were as follows: good - 'prayed for,' 314 (67.4%) and control, 338 (64.5%); intermediate - 'prayed for,' 63 (13.5%) and control 71 (13.5%); bad - 'prayed for,' 89 (19.1%) and control 115 (21.9%).

The complexity of the MACCHI-CCU scoring system and the Byrd scores casts doubt even on the disappointing and questionable assessment of the power of prayer as being about 10%. Harris and others reported, carefully and correctly, that based on their data: "*Remote, intercessory prayer was associated with lower CCU course scores. This result suggests that prayer might be an effective adjunct to standard medical care.*" In truth, however, their results provide no firm support either for or against the efficacy of intercessory prayer.

Herbert Benson and others[78] also attempted to study the effects of Christian intercessory prayer on CCU patients. The intercessory inputs were not adequately defined, as will always apply in such research. There were medical complications in 59% of a group who had certain knowledge that they were being prayed for, compared to 51% in a group who had been prayed for but did not know it, and 52% in a group who were presumed not to have been prayed for at all.

The 59% statistic cannot be hailed as proof that prayer does more harm than good. However, intercessory prayer activities could have negative effects on a prayed for nonbeliever who wished strongly *not* to be prayed for, but was then prayed for and told about it and became more stressed. If these results had come in reverse order, there would probably have been some claims of scientific proof for the positive power of prayer. In truth, they yield no firm conclusions.

Mitchell Krucoff and others[79] commented as follows on the Study of the Therapeutic Effects of Intercessory Prayer (STEP): "*In STEP, the safety of patients and related ethical obligations to study subjects were conducted, like so much of the study methodology, at the very highest level. Informed consent was required, and data and safety monitoring board oversight was provided. The data, on the other hand, proved to be counterintuitive. The assumption in the analysis plan was that blinded prayer*

---

78   Benson, H., Dusek, J.A., Sherwood, J.B., Lam, P., Bethea, C.F., Carpenter, W., Levitsky, S., Hill, P.C., Clem, D.W., Jain, M.K., Drumet, D., Kopecky, S.L., Mueller, S., Marek, D., Rollins, S. and P.L. Hibberd. 2006. Study of the therapeutic effects of intercessory prayer in cardiac bypass patients: a multicenter trial of uncertainty and certainty of receiving intercessory prayer. *American Heart Journal* 151 (4): 934-942.

79   Krucoff, M., Crater, S.W. and K.L. Lee. 2006. From efficacy to safety concerns: A STEP forward or a step back for clinical research and intercessory prayer. *American Heart Journal* 151 (4): 762-764.

*would be effective and unblinded prayer even more effective, with effective complication rates of 50% in the standard* (i.e., the control, not prayed for) *group, 40% in the blinded prayer group, and 30% in the unblinded prayer group - exactly the opposite of what was actually observed.* **In the interpretation of obviously counterintuitive findings as 'what may have been chance,' the STEP investigators have allowed cultural presumption to undermine scientific objectivity** (present author's emphasis)…*mechanistically undefined 'frontier' therapy research - even well intentioned intercessory prayer - must be scrutinized for safety issues at an equal or even higher level than efficacy measures if medically important and useful knowledge in this area is to truly step forward."*

David Hodge[80] made a meta-analysis from 17 studies on intercessory prayer, which he categorized as follows: 5 with no significant effect of prayer; 5 with no significant effect of prayer, but with some *"trends in the data"* that appeared to favour the 'prayed for' groups; and 7 with some significant effects of prayer. He included the problematic Byrd study (see above) and the study by K.Y. Cha and D.P. Wirth,[81] in which Christians in Australia, Canada and the USA prayed for the success of *in vitro* fertilization (IVF) - embryo transfer in Korean women. Neither the women nor the providers of the IVF-embryo transfer knew that prayers were being offered.

Hodge summarized Cha and Wirth's experiment as follows: *"Prospective double-blind RCT* (randomized control trial)…(on) *88 women of unknown religion receiving treatment for infertility… Two-tier prayer intervention: Tier 1 - an unspecified amount of daily distant IP* (intercessory prayer) *for groups of 5 clients (the women), for approx. 4 weeks by groups consisting of 3 to 13 Christians; Tier 2 – an unspecified amount of prayer offered for the above groups by additional Christian groups."* Here, the impossibility of quantifying prayer is blindingly obvious. Moreover, the

---

80 Hodge, D.R. 2007. A systematic review of empirical literature on intercessory prayer. *Research on Social Work Practice* 17 (2):174-187.
81 Cha, K.Y. and D.P. Wirth. 2001. Does prayer influence the success of in vitro fertilization-embryo transfer? Report of a masked randomized trial. *The Journal of Reproductive Medicine* 46:781-878.

sample size seems too low for an attempt to isolate one factor among the many that might affect success rates in IVF-embryo transfer.

The Cha and Wirth study gave seemingly encouraging indications for positive effects of intercessory prayer. In the control group, only 26% of the women became pregnant, compared to 50% in the prayed for group. For those aged above 39, the corresponding data were 23% and 42%. However, there were no differences between the prayed for group and the control group for women younger than 30. For those aged 30-39, 23% of both groups became pregnant. These results are controversial[82] and probably best regarded as having yielded no firm conclusions.

Hodge's meta-analysis indicated a *"small"* positively significant effect for intercessory prayer. This small effect got even smaller if the Cha and Wirth study was excluded. Hodge concluded the following: *"at this junction in time, the results might be considered inconclusive. Indeed, perhaps the most certain result stemming from this study is the following: findings are unlikely to satisfy either proponents or opponents of intercessory prayer."*

The overall conclusion for all attempted research on prayer so far is surely that prayer is not researchable at all by scientific method. Moreover, we should not put God to the test by attempting scientific experiments to determine how He is performing in response to prayer. In any case, if He values us more than sparrows[83] and knows our needs even before we ask,[84] why do we need to pray at all for any specific needs? And if we do pray for those needs and they are fulfilled, how can we know what part our prayers have played, if any? Correlation does not prove causation. God could have done it anyway. This possibility would seem to preclude any meaningful scientific research on prayer.

---

82  See: www.reproductivemedicine.com
83  Matthew 10:31
84  Matthew 6:8

# IN THE MIDST OF LIES AND NONSENSE

I believe in the power of intercessory prayer to alleviate stress in those praying and in some of those being prayed for. I have no doubt that praying for the *spiritual* wellbeing of oneself and one's fellows, especially those who are in trouble or sick, can bring very positive results. As a believer, I *know* that all believers must pray as often as possible. Prayer is our means of dialogue with God. Prayer is not simply the means by which we can ask God to fulfil our needs. Some Christians use the acronym ACTS to prioritize and order their prayers - Adoration, Confession, Thanksgiving and, lastly, Supplication.

## Miscellaneous Superstitions

My mother was taught that the best cure for warts was to rub them with a joint of bacon and then bury it in the garden - an expensive sacrifice for a placebo effect. Most warts disappear with time, with or without the rubbing and burial of any bacon. Our immune systems and time heal all manner of maladies whatever we do about them, including doing nothing.

As Peter Medawar explained, if a disease sufferer is advised to follow a particular course of action in pursuit of a cure and is then cured, then no power on earth will persuade that person that the cure did not result from that action, though it might have had nothing to do with it.

I once worked as a volunteer diver on surveys of tin-bearing rocks off the coast of Cornwall. The skipper of the survey boat belonged to the very strict St. Ives Brethren. He would not allow the sole female member of our diving team from London on board. He prohibited use of the boat's radio for listening to music and would not allow any other radio (or black bottle) on board.

On the boat from which I caught fish for my research in the Isle of Man, the crew would spit on the trawl net for good luck. I was told:

# Free Thought, Faith, and Science

*"Well there might not be anything in it, Roger; but if there is, we might as well have it"* - the Manx fisher's version of Pasqual's wager.

While driving over the Fairy Bridge in the Isle of Man, I raise my hand to greet the 'little folk' who allegedly live beneath it. Greeting them is said to aid one's safe negotiation of the sharp bends ahead. I do not believe in or accord any protective powers to Manx fairies. I simply follow a quaint and harmless, though nonsensical, local tradition.

In Africa, I was told that a sitting President had moved from one palace to another in order to escape the spirits of his dead enemies, which were turning into rats and gnawing at his feet in bed. I was also told that a past President had been able to make himself invisible and pass through walls.

In the Philippines, some folk still believe in *aswangs*, who look like normal women by day but fly like witches at night, with their lower bodies missing and entrails dangling. An *aswang* allegedly lands on a rooftop and puts down a long, proboscis-like tongue to suck life from the foetus in a pregnant woman. A miscarriage or stillbirth can then be blamed on some poor local woman, as has happened to many alleged witches down the ages.

In early 2010, a prolonged drought was devastating Philippine agriculture. Authorities in the Catholic Church prayed for rain. Tribesmen in the northern Cordillera did rain dances and shot arrows to the sky. On March 12, a typical cold front passed through. The rains began and the prevailing El Niño climate began to change. Catastrophic floods followed. Enough said.

# CHAPTER 10

# REFORMATIONS AND REVOLUTIONS

*"It is certainly high time to look carefully at the world that God has made in order to find out how he really intended it to be."*[1]

### CONCEPTS, DEFINITIONS AND PROSPECTS

Christoph Cardinal Schönborn saw the urgency of looking *"carefully"* at the material realm *and* its spiritual context. In other words, it is urgent that we expand the faith-science quest for truth, through science-friendly reformations among believers and faith-friendly revolutions among scientists.

Believers, especially religious authorities, must become more willing to embrace science and to look beyond sacred text literalism and inerrancy. Nonbelievers, especially scientists, must become more willing to allow the possibility that the spiritual realm exists and to admit the validity of subjective evidence from within concerning

---

1 Schönborn, C.C. 2007. *Chance or Purpose? Creation, Evolution and a Rational Faith.* Edited by H.P. Weber and translated by H. Taylor from the original German. Ignatius Press: San Francisco CA. 181p.p.93. First published as: *Ziel oder Zufall? Schöpfung und Evolution aus der Sicht vernünftigen Glaubens.* Verlag Herder GmBH: Freiburg im Breisgau, Germany.

faith and unbelief.

Sam Harris[2] sought a new world devoid of faith and organized religion when he stated the following: *"Of course, one senses that the problem is simply hopeless. What could possibly cause billions of human beings to reconsider their religious beliefs? And yet, it is obvious that an utter revolution in our thinking could be accomplished in a single generation: if parents and teachers would merely give honest answers to the questions of every child."*

Any quest to eradicate faith is indeed hopeless, because faith is based on real experiences of God. God's spiritual revelations are made to whomever He chooses. They can be rejected, but not eradicated. However, I agree strongly with Harris that the future of humanity must be based solely on seeking the truth and acting accordingly. In particular, children *must* be taught truthfully about what is known and what remains unknown.

A reformation in an organized religion is a major change from an established order to new perspectives and practices. A reformation becomes essential when religious authorities are impeding the quest for truth. Reformations back to orthodoxy are sometimes needed in order to protect the truth. As G.K. Chesterton[3] pointed out: *"The orthodox church never took the tame course or accepted conventions."* Through all reformations, science-friendly or not, true Christianity will keep its core truth - the Gospel of Christ.

A revolution in science is a major change resulting from the establishment and testing of new theory, with radically new perspectives and methods. A revolution becomes essential when current theory fails to explain observations and outmoded paradigms impede the quest for

---

2    Harris, S. 2005. *The End of Faith: Religion, Terror and the Future of Reason.* Free Press: London. 336p.p.224.
3    Chesterton, G.K. 2009. *Orthodoxy.* The Moody Bible Institute: Chicago IL.239p.p.152.First published in 1908.

# REFORMATIONS AND REVOLUTIONS

truth. Through all revolutions in science, faith-friendly or not, truthful disclosures from God will remain valid.

Reformations and revolutions usually happen after long formative periods. Diarmaid MacCullough[4] quoted from Hans Urs von Baltasar as follows: *"Nothing has ever borne fruit in the Church without emerging from the darkness of a long period of loneliness into the light of community."* MacCullough asked: *"Can the many faces of Christianity find a message which will remake religion for a society which has decided to do without it."*

'Yes,' indeed it can. But the road will be long and bumpy. Christianity today has largely escaped from its centuries-long history of oppressing and killing non-Christians and doing the same within its own ranks across the Catholic-Protestant divide. However, the primary concern of some Christian churches and schools is still to resist changes that might conceivably lessen their power and influence, including science-friendly reformations.

Reza Aslan[5] wrote the following on reformations in Islam: *"When fourteen centuries ago Muhammad launched a revolution in Mecca to replace the archaic, rigid and inequitable structures of tribal society with a radically new vision of divine morality and social egalitarianism, he tore apart the fabric of traditional Arab society. It took many years to cleanse the Hijaz of its 'false idols.' It will take many more to cleanse Islam of its new false idols - bigotry and fanaticism - worshipped by those who have replaced Muhammad's original vision of tolerance and unity with their own ideals of hatred and discord. But the cleansing is inevitable, and the tide of reform cannot be stopped. The Islamic Reformation is already here. We are all living in it."*

---

4    MacCullough, D. 2010.*Christianity: The First Three Thousand Years*. Penguin Books: New York. 1184p.p.1016.MacCullough cited p.32 in von Balthasar, H.U. 1994.*The Moment of Christian Witness*. Communio Books: San Francisco CA. First published in 1966.

5    Aslan, R.2006. *No God But God: The Origins, Evolution and Future of Islam*. Arrow Books: London. 310p.p.266.

# Free Thought, Faith, and Science

I agree, but again the road is proving long and bumpy. A small but highly disruptive minority of extremist Muslim believers is oppressing and killing non-Muslims and doing the same within their own ranks across the Sunni-Shiite and other sectarian divides. Religious censorship of science education is common in Islamic schools, particularly concerning the theory of evolution.

All reformations in organized religion *and* revolutions in science begin as outcomes of the Free Thought of creative individuals, who then engage in more Free Thought and make decisions about sharing their insights and experiences with others.

Ian Barbour[6] explained as follows why sharing is essential, in faith and science: "*As there is no private science, so also there is no private religion. In both cases, the initiate joins **a particular community** (original author's emphasis) and adopts modes of thought and action. Even the contemplative mystic is influenced by the tradition in which he or she has lived. Paradigms in religion, as in science, are acquired by example and practice, not by following formal rules. Individual insights are tested against the experience of others, as well as in one's own life.*"

I agree with Barbour, though faith *per se* is not religion and is always private. The same applies to unbelief. The sovereign self has its own private set of rules. Our professions of faith or unbelief can be true or false.

Barbour also cited, as follows, Hans Küng's comparison of paradigm shifts in Christian thought and science: "*The centrality of the scriptural witness to Christ is without parallel in science. The 'biblical message,' not scripture itself, is the enduring norm. Each new paradigm* (in Christian thought) *arose from a fresh experience of the original message, as well as from institutional crises and external challenges. The gospel*

---

6   Barbour, I.G. 1997. *Religion and Science: Historical and Contemporary Issues.* HarperOne: New York.368p.p.131, 129. Barbour cited Küng, H. 1989. *Paradigm Change in Theology,* in H. Küng and D. Tracy (eds.). T&T Clark: Edinburgh.

# REFORMATIONS AND REVOLUTIONS

*thus contributed to both continuity and change...we can acknowledge the distinctive features of religion and yet find the comparison with scientific paradigms helpful in understanding processes of change in the history of a religious tradition."*

We can indeed apply the concept of the paradigm and the paradigm shift in organized religion as in science. A reformation in organized religion or a revolution in science is an escape from a truth-deficient *status quo* to a more truthful position, from which further explorations can proceed. The same applies to a shift in the *meta-paradigm* for our explorations of the totality of the spiritual and material realms.

Peter Russell[7] explained the meta-paradigm in material realm science as follows: *"All our scientific paradigms are based on the assumption that the physical world is the real world, and that space, time, matter and energy are the fundamental components of reality. When we fully understand the functioning of the physical world, we will, it is believed, be able to explain everything in the cosmos. This is the belief upon which all our scientific paradigms are based. It is, therefore, more than just another paradigm; it is a* **metaparadigm** (original author's emphasis)."

Russell called for *"a worldview that validates spiritual inquiry,"* which is pretty much the same as calling for faith-friendly revolutions across science. The faith-science quest for truth is an exploration of the whole of reality - the spiritual *and* material realms. It proceeds on personal, interpersonal, institutional and organizational battlefields. Its meta-paradigm is defined by all current theology and all current science.

A meta-paradigm shift in the faith-science quest for truth is long overdue, but progress will not be easy. Most religious authorities prefer to stay in their own dogmatic comfort zones. Scientists welcome change, but many live in a false comfort zone where they avoid thinking about

---

[7] Russell, P. 2005. *From Science to God: A Physicist's Journey into the Mystery of Consciousness.* New World Library: Novato CA.120p.p.25, 117.

anything allegedly spiritual.

Some believers, particularly sacred text literalists, hold that they have God's endorsement to lead anti-intellectual, anti-science lives and to rely entirely on the contents of a particular book and/or pronouncements from religious authorities for all that they will ever need to know. Some Christians cite the following reference to Isaiah by the Apostle Paul:[8] *"For it is written, I will destroy the wisdom of the wise, and will bring to nothing the understanding of the prudent. Where is the wise? Where is the scribe? Where is the disputer of this world? Hath not God made foolish the wisdom of this world?"*

Isaiah was *not* denigrating science. He was condemning cosmetic religion and Godless arrogance, as shown by the preceding verse: *"Wherefore the Lord said, Forasmuch as this people draw near me with their mouth, and with their lips honour me, but have removed their heart from me, and their fear toward me is taught by the precept of men: Therefore, behold, I will do a marvellous work among this people...for the wisdom of the wise shall perish, and the understanding of prudent men shall be hid."*[9]

Self-censorship and self-deception about what to explore and what to believe are widespread among present day believers. For example, Felipe Fernández-Armesto[10] described his situation as follows: *"I exasperate my family by my unwillingness to take any report on trust or any opinion on the merits of the opiner;* **yet I have no difficulty in being a Catholic and deferring to the authority of the Church, as superior to whatever my own reason or experience might tell me, on matters reserved to ecclesiastical authority** (present author's emphasis). *I do so because I count it a virtue and because - I suppose if I am honest - I find it a comfort."*

---

8   I Corinthians1: 19-20. Paul quoted Isaiah 29:14.
9   Isaiah 29:13
10  Fernández-Armesto, F. 1997. *Truth: A History and a Guide for the Perplexed*. Thomas Dunne Books: New York. 257p.p.75.

# REFORMATIONS AND REVOLUTIONS

Ecclesiastical authority cannot override the *"reason or experience"* of any honest seeker, unless the spiritual force for evil makes it so. Faith is not a divinely endorsed escape to anti-intellectualism. Christ[11] reportedly gave us the following version of what He called *"the first and great commandment"*: *"Thou shalt love the Lord thy God with all thy heart, and with all thy soul, and with all thy **mind*** (present author's emphasis)."

God is not anti-intellectual or anti-science. He uses human creativity and science as a conduit for His truthful disclosures about the composition and workings of His creation. God loves truth and hates the lies perpetrated by pseudoscience and the false, cosmetic professions of faith that pervade much of organized religion.

Carl Sagan[12] quoted as follows a *"religious leader"* seeking *"disciplined integrity in religion:"* *"Devotionalism and cheap psychology on one side, and arrogance and dogmatic intolerance on the other distort authentic religious life almost beyond recognition…Honest religion, more familiar than its critics with the distortions and absurdities perpetrated in its name, has an active interest in encouraging a healthy skepticism for its own purposes… **There is the possibility for religion and science to forge a potent partnership against pseudo-science. Strangely, I think it would soon be engaged in opposing pseudo-religion*** (present author's emphasis)."

That outcome would not be strange at all. Truthful reformations in organized religion and truth-based revolutions in science cannot be other than mutually beneficial and complementary. Believers help the quest for truth whenever they recognize that mathematics and science lead to objective evidence for that which can be taken as true. The honest quest for truth is common ground in faith and science. Seeking truth is a reasonable goal and truth is best sought in a climate of reasonableness. The same applies to seeking escapes from intolerance, lies, nonsense and the baseless polarization between faith and science.

---

11  Matthew 22: 36-38
12  Sagan, C. 1997. *The Demon Haunted World: Science As a Candle In the Dark.* Ballantine Books: New York.457p.p.20.

# Free Thought, Faith, and Science

## The Protestant Reformation

The Protestant Reformation is often known as *The* Reformation. It brought about a messy and lasting break with Roman Catholicism. When Martin Luther blew the whistle in 1517 CE, the Catholic Church was selling indulgences by which believers could, allegedly, buy remission for their past and future sins. It was also requiring pilgrimages and payments to a huge proliferation of alleged holy relics and associated art works. There were many around Luther in Wittenberg. There was also a huge barrier between the priests and the people and the Bible was inaccessible to most believers.

Alister McGrath[13] quoted as follows from the Westminster Confession of Faith (1646 CE): "*The whole counsel of God, concerning all things necessary for his own glory, man's salvation, faith and life, is either expressly set down in Scripture, or by good and necessary consequence may be deduced from Scripture; to which nothing at any time is to be added,* **whether by new revelations of the Spirit** (present author's emphasis), *or traditions of men.*"

McGrath also quoted the Princeton Reformed theologian Benjamin B. Warfield (1851-1921 CE) as follows: "'*Inspiration is that extra-ordinary, supernatural influence... exerted by the Holy Ghost on the writers of our Sacred Books, by which their words were rendered also the words of God, and, therefore,* **infallible** (present author's emphasis).'" In fundamentalist Protestantism, the doctrine of 'Sola Scriptura' (by the Bible alone) became arguably as tyrannical as the Catholic doctrine of infallible church authority.

The inevitable schism between the Protestant Church and science was bitter and remains so. Many fundamentalist Protestants today hold that God cannot reveal anything new, except through the Bible. This cannot be true. The words of the Bible cannot be God's *last and only*

---

[13] McGrath, A. 2007. *Christianity's Dangerous Idea: The Protestant Revolution; A History from the Sixteenth Century to the Twenty-first.* SPCK: London.552p.p.244, 205.

method for sending spiritual revelations. Christians worship a Living God - in personal relationships with Him that develop through Free Thought, into which He sends His spiritual revelations and from which He invites responses.

Harvey Cox[14] wrote the following: *"I am confident that it is possible to take the Bible back from its fundamentalist hijacking and make it once again a genuine support of faith, instead of an obstacle...the fundamentalist strategy was an attempt to bring the Bible back to the people, but it failed by making the Bible itself the object of a deformed and static caricature of faith."*

## Revolutions in Science

Some of the areas of science that have particular relevance for the faith-science quest for truth are undergoing revolutions right now or have revolutions pending. The quest for a theory of everything is proceeding rapidly by combining mathematical modeling with observations at all scales from fundamental particles to cosmology. Evolutionary theory has a revolution pending as the whole gamut of processes that can contribute to natural selection and descent with modification becomes better understood. Multidisciplinary research on the human brain and human consciousness is expected to lead to revolutions in neuroscience and psychology.

The biggest and most understated challenge in science is its need to engage with other worldviews, particularly with faith and organized religion. Scientists who are nonbelievers can take the easy option of claiming that there is no sound basis for any such engagement, but some will want to engage - either to demolish faith or to find out whether there might be something in it. Scientists who are believers will of course continue to pursue their faith-science quests for

---

[14] Cox, H. 2009. *The Future of Faith*. HarperOne: New York. 245p.p.167-168.

# Free Thought, Faith, and Science

truth. These diverse attitudes are found not only among individuals but also at institutional and organizational levels.

The challenge is to shift the meta-paradigm of the faith-science quest for truth, which will then expand as more believers embrace science and more scientists apply their Free Thought to explore the possibility that the spiritual realm might exist. Reformations and revolutions will be brought closer by taking small steps - spreading on both sides the truth that faith is compatible with science and that all truthful disclosures through mathematics and science come from God. Scientists who are believers already recognize that truth. Scientists who are nonbelievers can consider it a possibility.

Believers need have no fears that science-friendly reformations will threaten their faith, if indeed their faith is based on truth. God equipped all of us to be honest seekers of truth, not unquestioning followers of dogma. Believers who join science-friendly reformations will not have to abandon their books and prayers and the mutual support that comes from fellowship, in order to learn more about God's truthful disclosures through science and their complementarity with faith.

Nonbelievers need have no fears that faith-friendly revolutions across science will compromise scientific rigour and scientific method. Scientists who join faith-friendly revolutions will not have to abandon their books and instruments and base their lives entirely on Bible reading and prayer in order to learn more about faith *per se*. In the faith-science quest for truth, the key role of scientists is to uphold the philosophy of science and apply scientific rigour to the study of human experience, in faith and unbelief.

# REFORMATIONS AND REVOLUTIONS

## CHRISTIANITY IN FLUX

According to *The Economist*,[15] based on data from the Pew Forum on Religious and Public Life,[16] most of the growth of Christianity over the last century was in sub-Saharan Africa, where the Christian proportion of the population increased from 9% to 63%. In Europe and the Americas, it declined by at least 10% and 20% respectively.

David Kinnaman and Gabe Lyons[17] reported on surveys, which had been commissioned by the Fermi Project and carried out by the Barna Group,[18] to assess attitudes to Christianity among young Americans, aged from 16 to 29. The surveyed population comprised 305 Christian churchgoers (C) and 440 outsiders (O). The percentages in agreement with some negative descriptors of Christianity in the USA were as follows: anti-homosexual, C 80% / O 91%; hypocritical, C 47 % / O 85%; too involved in politics, C 50% / O 75%; out of touch with reality, C 32% / O 72%. A reformation back to the teachings of Christ - speaking the truth in love - seems sorely needed.

On November 5, 2011, the BBC World TV Series "*Our World*" contained the following items on Christianity and churches in Europe:

I. *In Germany, over one million Catholics who have refused to pay taxes to support the church are being denied Holy Communion;*

II. *In Finland, where the church is supported by public taxes, 40,000 people have resigned their church memberships after church statements against homosexuals. One young woman asked: "Why would I want to outsource my relationship with God to the church?"*

---

15 Anon. 2011. Religious Freedom: Christians and Lions. *The Economist*. Volume 401, No. 8765: p.9.
16 www.pewforum.org
17 Kinnaman, D. and G. Lyons. 2007.*Unchristian: What a New Generation Thinks About Christianity… And Why It Matters.* Baker Books. Grand Rapids, Michigan.255p.
18 www.barna.org

III. At the Exodus Church in Gorinchem, *"part of the mainstream Protestant Church in the Netherlands,"* the Reverend Klaas Hendrickse preaches as follows: *"Personally I have no talent for believing in life after death…When it happens, it happens down on earth, between you and me, between people…God is not a being at all…It's a word for experience, or human experience."*

IV. Professor Hijme Stoeffels of the Free University in Amsterdam believes that for the majority of Dutch people it is *"a bridge too far"* to label their spiritual view of *"somethingism"* as *"God"* or as *"a personal God."*

Matthew Fox[19] held that: *"When Original Blessing replaces Original Sin, relationships of awe, of passion and compassion, of love for justice and for the earth, are resurrected."* Fox advocated a *"creation-centered spirituality,"* including: *"mystical artistry, universal compassion…celebration of the divine within each human soul* (blending) *science (knowledge of creation), mysticism (experiential union with creation), and art (expression of our awe at creation)."*

Retired Episcopalian Bishop John Shelby Spong[20] has described himself as a believer in exile. He rejects the Virgin Birth, the Incarnation of God as Christ, and Christ's Resurrection and Ascension. Spong has clearly left what most people would call Christianity, but has kept the core Christian principle of love for all sorts and conditions of humans. Spong explained his position as follows: *"I do not want to be part of any faith community that says the color of one's skin, the ethnicity of one's ancestor's, the gender of one's body or the sexual orientation of one's brain determines a person's worth, holiness or access to being part of who God is."*

---

19   Fox, M. 2005. Original Blessing & Mysticism. Available at: https://rowecenter.qwknetlle.com/schedule/2005/20050408_MatthewFox.html
20   Spong, J.S. 2009. *Eternal Life: New Vision: Beyond Religion, Beyond Theism, Beyond Heaven and Hell.* HarperOne: New York.268p.p.202.

# REFORMATIONS AND REVOLUTIONS

The ministry of Tim Keller, Founding Pastor of the Redeemer Presbyterian Church, New York City, has helped many people stressed by modern life in the surrounding city and across the world. The Redeemer Presbyterian Church[21] tells all prospective new members: "*you must believe the Bible.*"[22] In his book "*The Reason for God,*"[23] which is a compelling call to faith, Keller made the following case for total belief in the Bible: "*Now, what happens if you eliminate anything from the Bible that offends your sensibility and crosses your will? If you pick and choose what you want to believe and reject the rest, how will you ever have a God who can contradict you? You won't!*"

With all due respect to Keller, I disagree. To reject literal impossibilities in Bible stories does *not* mean rejection of its main messages. A Christian must believe in the divinity and saving grace of Christ. A Christian need not believe in a talking donkey or in the resistance of Jonah to fish digestive juices.

The Union Church of Manila (UCM) requires all of its members to affirm the entire Apostles' Creed. In the liturgy for reception of new UCM members by the congregation, the officiating Pastor explains that: "*One becomes an active member of the church through faith in Jesus Christ as Savior and acceptance of his Lordship in all of life.*" He asks new UCM members the following question, *inter alia*: "*Do you receive and profess the Christian faith as contained in the Scripture of the Old and New Testaments?*" A new member need not be a biblical literalist in order to respond truthfully "*I do.*"

Leslie Weatherhead[24] wrote the following fine words about Christianity and truth: "*Not for much longer will the world put up with the lies, the*

---

21  www.redeemer.com
22  Redeemer Presbyterian Church, as distributed in 2011.*Visitor's Information*. Redeemer Presbyterian Church: New York.36p.p.21.
23  Keller, T. 2008.*The Reason for God: Belief in an Age of Skepticism*. Dutton (Penguin Group): New York.293p.p.114.
24  Weatherhead, L.D. 1989. *The Christian Agnostic*. Arthur James Limited: London.264p.p.3, 17. First published in 1965.

*superstitions and the distortions with which the joyous and essentially simple message of Christ has been overlaid...I am sure that we can only recommend* (sic) *Christianity to the thoughtful men of today by a restatement which admits a large degree of agnosticism, eliminates magic, dispenses with imposed authority, and abolishes, from our conception of God, horror and cruelty which would degrade a man, let alone God.*" Expansion of the faith-science quest for truth and science-friendly reformations would be major means towards achieving that kind of "*restatement.*"

## Reformed Epistemology

Patrick Roche[25] summarized the history and positions of the Reformed Epistemology (RE) movement, which began in the late 1960s. RE has sought to promote faith on the basis that belief in God is a proper basic belief.

In the context of faith/theistic religion and science, the leading RE theologian Alvin Plantinga[26] stated: "*there is superficial conflict but deep concord between theistic religion and science, but superficial concord and deep conflict between naturalism and science.*"

I agree that there is "*deep concord*" between faith (as the core of theistic religion) and science. Both are based on truth from God. However, I would say that the "*deep conflict*" alleged by Plantinga between naturalism and science reflects only the particular inadequacy of *material realm-only* explanations of the *human* condition. The science of the rest of nature has no conflict with anything that can be properly called naturalism.

Plantinga used evolutionary theory as an example of what he saw as

---

25  Roche, P. 2002.Knowledge of God and Alvin Plantinga's Reformed Epistemology. *Quodlibet Journal* Volume 4, issue 4. Available at: www.quodlibet.net/articles/roche-plantiga.shtml
26  Plantinga, A. 2011.*Where the Conflict Really Lies: Science, Religion and Naturalism.* Oxford University Press: Oxford.359p.p.265, 63, 313, 231,121.

# REFORMATIONS AND REVOLUTIONS

that "*deep conflict*" and concluded the following: "*The scientific theory of evolution as such is not incompatible with Christian belief...what is incompatible with it is the idea that evolution, natural selection, is **unguided*** (original author's emphasis). *But that idea isn't part of evolutionary theory as such; it's instead a metaphysical or theological addition.*"

I agree with the first statement there, but for the rest I can assume only that Plantinga discounted the possibility that a material realm, created entirely by God, could have functioned "*unguided*" for billions of years and then, also unguided and entirely naturally, led to the evolution of a wide diversity of life forms before God chose to complete us to be His special companions. This position, which is my position, is entirely compatible with faith.

Plantinga reviewed human "*cognitive faculties*" and stated the following: "*But suppose you are a naturalist: you think that there is no such person as God, and that we and our cognitive faculties have been cobbled together by natural selection. Can you then sensibly think that our cognitive faculties are for the most part reliable?*" He had assumed that a "*naturalist*" must be a nonbeliever. That is fine if one frames one's definitions accordingly, but many believers, myself included, explore the wonders of *nature* and are therefore naturalists.

Plantinga seems to have discounted the possibility that our cognitive faculties are founded on our evolutionary legacy from animals and completed by God's ongoing gifts of souls, through which we become the only links between the spiritual and material realms. Moreover, evolution through natural selection and descent with modification is *not* a cobbling together of the unreliable. It weeds out the less reliable and secures continuation of the more reliable. Human cognitive faculties would not be *de facto* unreliable even if they had indeed arisen *entirely* naturally through evolution, with no divine gifts of souls.

Still on the evolutionary front, Plantinga stated the following about

## Free Thought, Faith, and Science

Michael Behe's writings on Intelligent Design: "*Behe has not demonstrated that there are irreducibly complex systems such that it is impossible or even monumentally improbable that they have evolved in a Darwinian fashion.*" I agree and I hope to see that conclusion publicized more widely. However, Plantinga continued as follows: "*he* (Behe) *has certainly provided Darwinians with a highly significant challenge.*" I disagree, for reasons given in detail in the Battlefields chapter. Rejecting Intelligent Design is a necessary step towards science-friendly reformations in Christianity.

Plantinga also considered how God might intervene, as so-called special divine action, through quantum mechanics (QM) to make supernatural miracles; for example, by collapsing all pertinent quantum wavefunctions - the so-called "*divine collapse-causation (DCC).*"

After discussing how a "*sensible*" believer need not adjust her/his belief "*every time science changes its mind,*" on big bang cosmology and by inference everything else, Plantinga stated: "*But where Christian or theistic belief and current science can fit nicely together, as with DCC, so much the better, and if one of the current versions of QM fits better with such belief than the others, that's a perfectly proper reason to accept that version. True, this version may not win out in the long run (and the same goes for QM itself) so the acceptance in question (as of QM itself) must be provisional.*"

As Plantinga recognized, what looks like a nice fit between theology and science is not necessarily a *true* explanation. More importantly, the *misfits* need as much if not more attention as the nice fits - from both sides. There is no sound reason for any believer who is an honest seeker of truth to embrace science other than *wholeheartedly*. Moreover, there is no need to select what look like nice fits with theology, while shunning the misfits that might reveal inconvenient truths. That would be less than honest.

# REFORMATIONS AND REVOLUTIONS

Concerning Plantinga's suggested nice fit between QM and DCC, I must repeat and re-endorse here the following conclusions from Ian Barbour:[27] *"The traditional view of divine omnipotence can still be maintained without violating the laws of physics if God is the determiner of indeterminacies at the quantum level." But…**the ideas of divine self-limitation and process theology are more consistent with both scientific evidence and central Christian beliefs*** (present author's emphasis)."

In other words, all believers and nonbelievers would do well to let go of the idea that miracles can be true because of quantum indeterminacy. I have suggested that quantum indeterminacy might be an important mechanism in Free Thought, but I reject the idea that quantum indeterminacy can explain alleged miracles in the material realm that would have required contravention of its natural laws. I do not believe that such miracles have ever taken place outside the period of the Incarnation. I believe that the only ongoing miracle is that God sends spiritual information to all human souls, during Free Thought.

The RE movement thus far seems to have shied away from a wholehearted accommodation with science. It seems afraid going too far, especially in posing questions about the Bible. Beyond that sacred text constraint to reformation, which is found throughout Christianity and has parallels in other organized religions, the most difficult step for RE and most other reformist movements in Christianity seems to be to allow the possibility that all truthful disclosures through mathematics and science come from God and that He uses believers *and* nonbelievers as His channels.

---

27   Barbour, I.G. 2000. *When Science Meets Religion: Enemies, Strangers, Or Partners?* HarperOne: New York.205p.p.89.

# FREE THOUGHT, FAITH, AND SCIENCE

## CHRISTIAN APOLOGETICS

The terms Christian apologetics and apologists are still not widely known or understood, especially among non-Christians. Christian apologists do not *apologize* for the Gospel of Christ. They strive to *explain* it in order to pave the way for conversions. Their field is called Christian apologetics. Apologetics and evangelism form a continuum. Apologetics is like pre-evangelism.

Daniel Dennett[28] cited Avery Cardinal Dulles' description of apologetics as *"the rational defense of faith"* and quoted Dulles' charges against apologetics as follows: *"Apologetics fell under suspicion for promising more than it could deliver and for manipulating the evidence to support the desired conclusions. It did not always escape the vice that Paul Tillich labeled 'sacred dishonesty'."*

Dulles called for a *"renewal and reformation of apologetics"* and for apologetics to *"shift its ground."* Dennett commented: *"Cardinal Dulles is interested in getting conversions; and so are scientists. They campaign with vigor and ingenuity for their pet theories. But they are constrained by the rules of science...No such rules have yet evolved to govern the practice of religion."*

Douglas Webster[29] noted that Latin American liberation theologians spoke of *"evangelizing the church"* while rejecting: *"the Roman Catholic spiritualistic and feudalistic conception of their culture."* Relating that to North American Protestantism, he observed the following: *"The very things we take pride in exhibit our cultural captivity. Mainline Protestantism offers a highly politicized ethic controlled by the dictates of enlightened secular culture, while market-driven Protestantism offers a highly psychologized gospel conditioned by self-centered,*

---

28　Dennett, D.C. 2007. *Breaking the Spell: Religion as a Natural Phenomenon*. Penguin Books: London.448p.p.363-365. Dennett cited Avery Cardinal Dulles. 2004. The rebirth of apologetics. *First Things* 143 (May): 18-23.
29　Webster, D.1995. Evangelizing the church, p. 194-208. In T.R. Phillips and D.L. Okholm (eds.) *Christian Apologetics in the Postmodern World*. Intervarsity Press: Downers Grove IL. 238p.p. 195,208.

*consumer-oriented, media-induced felt needs."*

Webster made the following call to reformation: "*The enculturated ethic of the mainline church and the remain-as-you-were evangelism of the market-driven church expose a fundamental need for spiritual revival...If we want to evangelize our culture, we must begin by evangelizing our churches.*" His recipe for reformation was: "*begin with a biblical theology of evil* (and) *a clear proclamation of the cross and the ethic of Jesus Christ.*" I would add the need to embrace God's truthful disclosures through mathematics and science. Reformation in Christian apologetics and evangelism must be based on a theology that is at one with science.

One of the saddest aspects of much of present day Christian apologetics is the lauding of Intelligent Design, while denying evolution as disclosed to Darwin - natural selection and descent with modification. Some Christian apologists are still inviting belief in God's existence based on what the bombardier beetles squirt from their rear ends.

Miguel Angel Endara[30] included the following in his essay on "*The Design Argument*":

"*1) Specified complexity is a reliable detector of design.*

*2) The bombardier beetle's defense mechanism is an example of irreducible complexity, which is a species of specified complexity.*

*3) The naturalistic mechanism of natural selection, working on random gene mutation, reproduction, and competition, cannot account for the information-rich specified complexity of the defense mechanism.*

---

30   Endara, M.A. 2007. The Design Argument, p. 217-229. In N.L. Geisler and C.V. Meister (eds.) *Reasons for Faith: Making a Case for the Christian Faith.* Crossway Books: Wheaton IL.416p.p.229.

*4) **We infer an intelligent and transcendent designer as the source for the existence of the bombardier beetle*** (present author's emphasis) *and other biological organisms who* (sic) *exhibit specified complexity.*

*5) This intelligent and transcendent designer we call* **God** (original author's emphasis)."

Norman Geisler[31] made the case for exclusive reliance on the Bible as follows: "*Jesus taught that the thirty-nine books of the Old Testament are the authoritative, written word of God. Likewise, Jesus, who is God's full and final revelation, promised that his twelve apostles would be guided by the Holy Spirit into 'all truth.'…**Hence, the canon of God's revelation to man is closed. With these sixty-six books we have the complete and final revelation of God for the faith and practice of believers*** (present author's emphasis)…*the Bible and the Bible alone contains all doctrinal and ethical truth God has revealed to mankind. And the Bible alone is the canon or norm for all truth. All other alleged truth must be brought to the bar of Holy Scripture to be tested.*"

According to Geisler, God cannot send us any new revelations apart from those that can be gleaned through the Bible. Moreover, all scientific findings must be wrong if they do not fit with whatever can be found in the Bible about the same topic. I disagree strongly on both counts.

Norman Geisler and William Roach[32] defended further the view that Christian apologists and indeed all Christians must believe in total biblical inerrancy, albeit based solely on the Bible's own internal claims. They castigated all who stray from that position, including those who, like the present author, claim to be believers and accept the truth of evolution - past, present and anticipated.

---

31    Geisler, N.1988. *Christian Apologetics.* Baker Academic: Grand Rapids MI.393p.p.376-377.
32    Geisler, N.L. and W.C.Roach. 2011. *Defending Inerrancy: Affirming the Accuracy of Scripture for a New Generation.* BakerBooks: Grand Rapids MI.378p.p.363.

# REFORMATIONS AND REVOLUTIONS

Geisler and Roach attacked so-called *"theistic evolution"* as follows: *"The problem with theistic evolution is that it is an 'easy' solution to the problem of the conflict between prevailing contemporary scientific views and a serious literal, historical understanding of Scripture.* **Rather than challenging the philosophical and scientific basis of the evolutionary hypothesis, it is a whole lot easier to save one's academic reputation and job by just agreeing with it.** *However, the price paid for this 'easy' solution is too high. And it is unnecessary. For there is a credible…alternative that is credible science without forsaking the inspiration and inerrancy of Scripture:* **the intelligent design movement** (present author's emphases)."

Geisler and Roach went on to attack the BioLogos movement[33] as follows: "(Biologos) *poses a major threat to the inerrancy of Scripture…First and foremost, it* (Biologos) *accepts a macroevolutionary view that is inconsistent with a historical-grammatical interpretation of Scripture. Second, by denying a literal understanding of Genesis 1-3, it undermines many important New Testament teachings based on this literal understanding of Genesis, including the depravity of humans (Rom. 5:12), the basis for a monogamous marriage (Matt. 19:4), the divinely appointed order in a family (I Cor. 11:3) and in the church (I Tim. 2:12-13).* **Even more seriously, it undermines the authority and deity of Christ, who understood Genesis 1-3 as literal** (Matt, 19:4; 24: 38-39; Mark 13:19) (present author's emphasis)."

In truth, Biologos is an excellent movement. Its contributors are championing the faith-science quest for truth. As quoted above, Geisler and Roach accused some of their fellow Christians of accepting theistic evolution in order to sustain academic respectability and job security, and/or of undermining *"the authority and deity of Christ."* For me and for Christians known to me who accept theistic evolution, these accusations are false and insulting. We have sought only the truth, as disclosed by God through science.

---

33  www.biologos.org

More importantly, the doctrines of biblical literalism and inerrancy are obscuring a fact that should be obvious to any believer. All the words that have ever been written about creation cannot have given the full story. Only God Himself *knows everything* about how He created everything. Christ was God Incarnate. Christ was God from the very beginning, when God was the Creator of everything.

God cannot be *required* by humans to have explained His entire process of creation in the literal words of a few chapters of Genesis. Moreover, how can anyone assume that Christ's reported use of Genesis 1-3 was an endorsement of its *literal* wording? The wording of Genesis matched the theological and scientific comprehensions of its earliest readers and listeners. It is no insult to God to explore its matches and mismatches with His subsequent truthful disclosures through science.

Apologists who hold that the Bible is literally true and totally inerrant typically hold also that anyone who questions anything about one particular passage must be undermining everything else to which that passage can be linked, however tenuous that link might be. This is a false argument. An inability to believe that the world and the first male and female humans were made exactly as recounted in Genesis does *not* preclude belief in biblical accounts of the Incarnation and the Gospel of Christ or in the writings of the Apostle Paul concerning faith, hope and love etc.

John Stott[34] wrote the following: *"I hope that we are agreed that all Christian preaching is biblical preaching…At the same time, authentic Christian preaching is contemporary. It resonates with the modern world. It wrestles with the realities of our hearers' situations."* Stott quoted from Bishop Stephen Neill as follows*: "Preaching is like weaving. There are the two factors of the warp and the woof* (or weft). *There is the fixed unalterable element, which is for us the Word of God, and there is the variable*

---

34  Stott, J.R.W. 2007. *The Living Church: Convictions of a Lifelong Pastor.* Inter-Varsity Press: Nottingham, UK. 192p.p.104-106. Stott quoted Neill, S.C. 1952.*On the Ministry.* SCM Press: Norwich, UK.p.74.

*element, which enables the weaver to change at vary the pattern at his will. For us that variable element is the constantly changing pattern of people and situations.'"*

Stott called, as follows, for reformation among *"evangelicals"* and *"liberals,"* through building bridges: *"Evangelicals are biblical but not contemporary, while liberals are contemporary but not biblical. Comparatively few are building bridges. But authentic Christian preaching is a bridge-building operation. It relates the text to the context in such a way as to be faithful to the biblical text and sensitive to the modern context. We must not sacrifice either to the other...we must study Scripture until we are really familiar with it. But we must also study the world in which we live."*

I agree with all of the above from Stott and Neill. All preaching is biblical, but the Bible *must* be interpreted with God's ongoing help, so that it what is true for all time and what is His will for our time can be better understood.

Douglas Groothuis[35] put forward 19 theses as his manifesto towards a reformation in Christian apologetics. Here are extracts from the five of his theses that I find most relevant in the context of embracing truthful disclosures through science - retaining here their original numbers, omitting citations to other works, and adding my comments in brackets:

*"3. The fundamental issue for apologetics is not how many apologists one has read or what apologetic method one embraces...Rather, the essential issue is whether or not one has a passion for God's transforming truth - reasonably pursued and courageously communicated... We, like that great apologist* (the Apostle Paul) *should be intellectually equipped and spiritually prepared to enter the marketplace of ideas for the cause of Christ."* (This is the basis on which apologists can contribute to expanding the faith-science quest for truth).

---
35   Groothuis, D. 2007.Postscript - A Manifesto for Christian Apologetics: Nineteen Theses to Shake the World With Truth, p. 401-408. In N. L. Geisler and C. V. Meister (eds.) *Reasons for Faith: Making a Case for the Christian Faith.* Crossway Books: Wheaton IL. 416p.p.402-404.

"4. *The apologist must be convinced of the truth, rationality, pertinence, and knowability* (sic) *of the Christian worldview, which is derived from Holy Scripture as it is logically systematized and rightly harmonized with general revelation (truth knowable outside of Scripture)."* (As part of the *"truth knowable outside of Scripture"* the apologist must include God's truthful disclosures through mathematics and science).

"6. *Any theology, apologetics, ethics, evangelism, or church practice that minimizes or denigrates the concept of objective, absolute, universal, and knowable truth is both irrational and unbiblical... Without a strong, biblical view of truth, apologetics is impossible."* (Again, this *"strong, biblical view of truth"* must be combined with acceptance of God's truthful disclosures through science. Truth cannot contradict truth.)

"9. *Apologetics must be carried out with the utmost intellectual integrity. All propaganda, cheap answers, caricatures of non-Christian views, hectoring, and fallacious reasoning must be avoided."* (Agreed).

"10. *The artificial separation of evangelism from apologetics must end... Therefore, all evangelistic training should include basic apologetic training as well."* (Agreed).

Groothuis'[36] large volume on Christian apologetics has much to recommend it, but its coverage of some topics is disappointing; for example, he gave relatively little weight to religious experiences, which I call subjective evidence from within and regard as the only evidence for *experiencing* God.

Groothuis included a chapter on *"Origins, Design and Darwinism."* He made the usual anti-evolutionist arguments about the fragmentary fossil record before he got to what he saw as the crux of the matter for an apologist: *"Evidence for Intelligent Design."* He then presented

---

36  Groothuis, D. 2011.*Christian Apologetics: A Comprehensive Case for Biblical Faith*. IVP Academic: Downers Grove IL.752p.p.109-110, 329.

## REFORMATIONS AND REVOLUTIONS

the usual flawed material about irreducible complexity and indicators of design and made the correct, though rather bizarre statement that: "*Standing alone, it* [Intelligent Design (ID)] *cannot provide a full apologetic for Christianity.*" Nevertheless, he held that: "*ID should take its rightful place in the overall circle of evidence* (and that) *ID provides strong evidence against the reigning naturalism in the realm of biology, as well as some support for theism as an overarching worldview.*"

'No.' Intelligent Design does nothing like that and merits no place in the faith-science quest for truth. Groothuis' statements illustrate well the confusion that has resulted from the promotion of Intelligent Design. Intelligent Design provides no evidence against any of the wonderful disclosures about nature that God makes available to us through biology. Theistic evolution provides much better support for faith than anything on offer from the Intelligent Design movement. And yet, in much of mainstream Christian apologetics, Intelligent Design is accepted and evolution denied.

Alister McGrath[37] recommended that novice apologists should learn from today's leading apologists. There is much wisdom in that, but it is hard to identify many apologists who can get through to nonbelievers who are scientists and address the specific question about how the truth of faith fits with God's truthful disclosures through mathematics and science. One cannot get around this problem with generalities alone. Where science has settled a specific issue, the apologist must be ready to acknowledge that it is indeed settled. McGrath also recommended that a would-be apologist should develop a personal and "*distinctive approach.*" 'Yes!' Therein lies the path to believers sharing more of their subjective evidence from within.

I prefer the King James Version of the Bible to any other, but the New International Version conveys more clearly the following claim

---

[37] McGrath, A.E. 2012. *Mere Apologetics: How To Help Seekers and Skeptics Find Faith*. BakerBooks: Grand Rapids MI.197p.

## Free Thought, Faith, and Science

by the Apostle Paul about how the Gospel was being preached with truthfulness and integrity: "*We put no stumbling block in anyone's path, so that our ministry is not discredited.*"[38] Dear biblical literalists, anti-evolutionists, Intelligent Design enthusiasts, and all apologists and evangelists who avoid God's truthful disclosures through science, you often put horrible *stumbling blocks* in the path of honest seekers of truth, especially scientists, who would otherwise feel able to explore faith.

For apologists who are biblical literalists, quotations from the Bible are proofs that will clinch any argument. So, off they go proclaiming to their audiences: 'Let me tell you what happened to Jonah, the Hittites, and the woman at the well, and in Second Timothy we find this and in First Thessalonians we find that etc.' Lessons from ancient Middle Eastern battles and early church history are portrayed as absolute truths and behavioural guides for all times and places, albeit from a multi-authored book that has undergone many translations and exists in many versions.

Thus the apologist's pitch often becomes disconnected from present day reality and indeed from honesty about what actually happens in the material realm. The result is an instant turn-off for most nonbelievers, especially scientists, who quite rightly wish to apply their common sense and scepticism to any proposition. Most will not even listen to anything that is put forward as being beyond question. It is pointless to debate with dogma.

If a Christian apologist tells a scientist that Jonah survived inside the digestive system of a fish for three days and nights,[39] then that scientist, like the present author, must point out that this cannot be true, because it is biologically impossible. More importantly, a nonbeliever, especially a scientist, is unlikely to give any credence to claims that a

---

38  II Corinthians 6:3
39  Jonah 1:17

# REFORMATIONS AND REVOLUTIONS

particular book contains no errors and came directly from the mouth of God and is literally true throughout, simply because *it* says so.

The better way to begin sharing one's faith is to say: 'I want to tell you the truth about my experiences, because they are wonderful experiences. I am not asking you to abandon your reason or your scepticism. Ask me whatever you like and let's see where it might lead.'

Christian apologists and evangelists must have the Bible to hand, but they must also be open to all of the revelations that God sends not only through its pages but also through other means, including science. I love the Bible dearly as a major source of revelations from God, but it cannot be the only source. No one can set meaningful limits on how the Living God chooses to send His revelations. He reaches us through the Bible *and* in the form of revelations sent directly to our souls, *and* as truthful disclosures through science.

Christian apologetics needs a science-friendly reformation, after which much greater care must be taken to share only what is known and what is unknown and to speak nothing but the truth. C.S. Lewis[40] advice to apologists in training is worth repeating here and often in apologetics and evangelism: "*One must keep pointing out that Christianity is a statement which, if false, is of* **no** (original author's emphasis) *importance, and, if* ***true*** (present author's emphasis) *of infinite importance.*"

The most important step towards a reformation in Christian apologetics is for apologists to stop insisting that the literal words in a given version of the Bible are totally inerrant and comprise the sum total of all true revelations from the Living God, with no prospect for Him sending any more.

---

40  Lewis, C.S. 1972. Christian Apologetics, p.101, Chapter 10 in W. Hooper (ed.) *God in the Dock: Essays on Theology and Ethics*. William B. Erdman's Publishing Company: Grand Rapids, MI. 346p.

Without that step, I see few prospects for Christian apologists getting through to nonbelievers who might otherwise be prepared to consider that a Living God can send us truthful disclosures, through mathematics and science, as well as other spiritual revelations, directly into our souls, even when we have no Bible to hand or verses memorized.

## THE SECOND VATICAN COUNCIL

The recent history of the Catholic Church can be summed up as pre- and post-Second Vatican Council. Vatican II began a sea change across the Catholic Church, including better relationships with science and increased recognition of the importance of the individual human conscience.

John Quinn[41] emphasized that Vatican II: "*did not come about because of a great groundswell among the people, priests, or bishops of the world.*" Rather, it was a *papal* initiative. It came about because God inspired a single reformer who then enlisted others. That reformer was Pope John XXIII. His Papal Encyclical "*On Establishing Universal Peace in Truth, Justice, Charity and Liberty,*"[42] better known by its short title "*Pacem in Terris,*" is one of the most important documents ever written about order, human rights and morality.

"*Pacem in Terris*" was issued on April 11, 1963, just before the start of Vatican II and was addressed emphatically to: "*all Men of Good Will.*" It includes the following statements, given here with their original numbering:

"*Order in the Universe - 2. That a marvellous order predominates in the world of living beings and in the forces of nature, is the plain lesson which the progress of modern research and the discoveries of technology teach us.*

---

41   Quinn, J.R. 1999. *The Reform of the Papacy; The Costly Call to Christian Unity*. Ut Unum Sint Studies on Papal Primacy. Crossroad Publishing Company: New York.189p.p.181.
42   Available at: www.vatican.va/holy_father/john_xxiii/encyclicals

# REFORMATIONS AND REVOLUTIONS

*And it is part of the greatness of man that he can appreciate that order, and devise the means for harnessing those forces for his own benefit. 3. But what emerges first and foremost from the progress of scientific knowledge and the inventions of technology is the infinite greatness of God Himself, who created both man and the universe."*

*"Order Between Men. Rights. Rights Pertaining to Moral and Cultural Values - 12. (Man) has a right to freedom in investigating the truth, and - within the limits of the moral order and the common good - to freedom of speech and publication, and freedom to pursue whatever profession he may choose... God and the Moral Order - 37. Now the order which prevails in human society is wholly incorporeal in nature. Its foundation is truth, and it must be brought into effect by justice."*

*"Pastoral Exhortations. Scientific Competence, Technical Capacity and Professional Experience. - 148. But in a culture and civilization like our own, which is remarkable for its scientific knowledge and its technical discoveries, clearly no one can insinuate himself into public life unless he be scientifically competent, technically capable, and skilled in the practice of his own profession. 149. And yet even this must be reckoned insufficient to bring the relationships of daily life into conformity with a more human standard, based, as it must be, on truth, tempered by justice, motivated by mutual love, and holding fast to the practice of freedom."*

*"Pacem in Terris"* was a huge step towards reformation. Vatican II followed and was often stormy. Under Pope John XXIII's successor, Pope Paul VI, there were gatherings that totalled over 2,500 persons, exploring and seeking consensus on changes that would open up the Catholic Church to the world. Michael Novak reported on Vatican II in 1964. He published a fuller account of its conduct and importance under the title *"The Open Church."*[43] The *"Transaction Edition"* is required reading for anyone seeking clarity on where the Catholic Church was

---

43  Novak, M. 2002. *The Open Church*. Transaction Publishers: New Brunswick NJ.370p.p.348, 70, xlii-xliii, 308. The original version of this work was published in 1964.

before Vatican II and where it is now striving to be.

Some of Novak's writings are controversial, especially for Catholics who want their old traditional system to last forever. Novak called that system "*non-historical orthodoxy...* (in which) *Catholic thought (is) confined within the conceptual achievements of only a part of the human race.*" This has to be untenable for anyone who seeks truth about the human condition in general and the necessary universality of God's relationships with humans. As Novak put it: "*By its vocation, Catholic thought must be as diverse and extensive as human thought; nothing human is foreign to it.*"

Novak reported that those present at Vatican II included a minority "*School of Fear,*" the members of which wanted the Catholic Church to remain closed and unchanging. Novak wrote: "*The Second Vatican Council is trying to come to grips with the world in which the Church of the twentieth century finds itself. It is trying to insert the Church back into the center of historical life, with respect for the moment in which it acts. There are those who are afraid of this venture; and the story of the second session of the Vatican Council is largely a story of these well-placed, powerful few against the majority. It is a story all the more dramatic because each side believes it is right.*"

Novak continued: "*The stronghold of the closed Church is its theology, complacent in its possession of the absolute truth, full of fears about other ways of pursuing understanding. The men who love the splendor of papal Rome, the clarity of Roman law, the absoluteness of non-historical theology, are not despicable, mean, or uncouth men. They are men who have tried to live outside history. They are good men, victims of the system they faithfully and loyally serve. It is not that they are bad - 'they are only out of touch.' If the opening of the Church means anything, it means displacing their system from the center of the Church.*"

I must note in passing that those statements from Novak could be applied to much of fundamentalist Protestantism today and its sincere but misguided authorities and rank and file members who are afraid of opening up to science in the quest for truth. At Vatican II, those who sought more openness in the Catholic Church won the day. In much of Protestantism today, the trend seems to be in the opposite direction.

In the context of faith-science unity and Free Thought theory, I take great encouragement from Novak's report that Pope Paul VI spoke about "*divine Revelation*" at the closing ceremony of Vatican II and welcomed: "*directives to guide the biblical, patristic, and theological studies which Catholic thought, faithful to ecclesiastical teaching* **and vitalized by every good modern scientific tool** (present author's emphasis), *will want to promote earnestly, prudently and with confidence.*"

I take this as a call for Catholic and other believers to embrace science and to recognize that all truthful disclosures through mathematics and science come from God. The Catholic Church still has far to go in that direction, but the die was cast at Vatican II and there is no going back.

The complete record of Vatican II[44] is a huge document. Much of it is reformist. It reaches out for dialogue with other organized religions, such as Islam and Judaism. It recognizes the importance of the dictates of the individual human conscience to a greater extent than ever before in the history of the Catholic Church. It confirms that truth can be sought and found outside the confines of the Catholic Church.

One cannot do justice to the importance of Vatican II through a few extracts, but here are some more that have particular relevance for this book, with all emphases by the present author.

"*Lumes Gentium*" or "*The Light of the Nations*" (Dogmatic Constitution of the Church) contains the following in its paragraph 8: "*The Church

---

44   Available at: www.vatican.va/archive/hist_councils/ii_vatican_council

(i.e., the Church that was commissioned by Christ to the Apostle Peter) *subsists in the Catholic Church…**although many elements of sanctification and truth are found outside its visible structure**. These elements, as gifts belonging to the Church of Christ, are forces impelling toward catholic unity.*" The lower case "c" in "*catholic unity*" connotes *worldwide* unity.

"*Dei Verbum*" or "*The Word of God*" (Dogmatic Constitution on Divine Revelation) harks back to earlier dogma on biblical infallibility and can be read as supporting almost any position concerning biblical literalism and inerrancy. However, I take heart that it supports clearly the position that divine revelation did not end with the writing of the last of the Apostles. That is a big improvement on the present day position of many Protestant evangelists.

After a discussion on the truth of the preaching of the Apostles, paragraph 8 of "*Dei Verbum*" states: "*This tradition which comes from the Apostles develop* (sic) *in the Church with the help of the Holy Spirit. For there is a growth in the understanding of the realities and the words that have been handed down.*" Paragraph 9 includes the following: "***Consequently, it is not from Sacred Scripture alone that the Church draws her certainty about everything that has been revealed***."

Under the heading "*Handing On Divine Revelation*," it is implied that God's revelations came to the Apostles and have kept on coming ever since to those in the Church who can add to its sacred tradition. This is far from the notion that divine revelations can come to anyone in Free Thought, but it is a big step for a church that for centuries had considered most ordinary folk dependent on its pronouncements and unlikely to be able to receive any personal revelations from God.

"*Gaudium et Spes*" or "*Joy and Hope*" (Pastoral Constitution on the Church in the Modern World), gets into church positions on science, though rather obliquely. Its paragraph 3 states: "*Though mankind is*

# REFORMATIONS AND REVOLUTIONS

*stricken with wonder at its own discoveries and power, it often raises anxious questions about the current trend of the world, about the place and role of man in the world, about the meaning of individual and collective strivings, and about the ultimate destiny of reality and of humanity."*

Paragraph 5 contains the following: *"Today's spiritual agitation and the changing conditions of life are part of a broader and deeper revolution... intellectual formation is ever increasingly based on the mathematical and natural sciences, and on those dealing with man himself, while in the practical order the technology which stems from these sciences is of mounting importance...**Advances in biology, psychology, and the social sciences not only bring man hope of improved self-knowledge**; in conjunction with technical methods, they are helping men exert direct influence on the lives of social groups. At the same time, **the human race is giving steadily-increasing thought to forecasting and regulating its own population growth**."* Maybe a future document will acknowledge how that *"steadily-increasing thought"* has become implemented in the form of increased contraceptive action.

After further review of how humans are changing the world, paragraph 9 states: *"Meanwhile, the conviction grows that humanity can and should increasingly consolidate its control over creation...Still, beneath all these demands lies a deeper and more widespread longing: persons and societies thirst for a full and free life worthy of man."* This a useful summary of the state of our ecologically and socially stressed world.

John Allen[45] reviewed the following 10 so-called *"mega-trends"* across the Catholic Church since Vatican II: a *"World Church;" "Evangelical Catholicism;"* engaging with *"Islam;"* responding to the *"New Demography;" "Expanding Lay Roles;"* responding to the *"Biotech Revolution;"* responding to *"Globalization;"* embracing *"Ecology;"* and *"Multipolarism"* and *"Pentecostalism."* Based on Allen's analysis, I see

---

45  Allen, J.L. Jnr. 2009. *The Future Church: How Ten Trends Are Revolutionizing the Catholic Church.* Doubleday: New York.469p.

two composite mega-trends that will help in the context of reformations and revolutions.

First, the Catholic Church has undoubtedly become more science-friendly since Vatican II. Pope Benedict XVI,[46] now retired, carried on the good work of his two immediate predecessors towards faith-science unity. In his 2006 lecture to representatives of science at the University of Regensburg, he referred to: *"The scientific ethos...the will to be obedient to the truth."*

In March 2009, the Vatican convened an international conference on evolution.[47] It has also appointed a team of 13 astronomers, mostly Jesuits. Brother Guy Consolmagno[48] stated: *They (the team) want the world to know that the Church isn't afraid of science...This is our way of seeing how God created the universe and they want to make as strong a statement as possible that truth doesn't contradict truth; that if you have faith, then you're never going to be afraid of what science is going to come up with...Because it's true."*

The Catechism of the Catholic Church[49] contains the following, with all here emphasized by the present author: **"Since the same God who reveals mysteries and infuses faith has bestowed the light of reason on the human mind, God cannot deny himself, nor can truth ever contradict truth. Consequently, methodical research in all branches of knowledge, provided it is carried out in a truly scientific manner and does not override moral laws, can never conflict with the**

---

46  Pope Benedict XVI. 2006. *Faith, Reason and the University. Memories and Reflections* Lecture at the Meeting With Representatives of Science, 12 September 2006, University of Regensburg, Germany. www.vatican.va/holy_father/benedict_xvi/speeches/2006/september/documents/hf_ben-xvi_spe-2006092_university-regensburg_en.html
47  Auletta, G. Leclerc, M. and R. A. Martinez, Editors. 2010. *Biological Evolution: Facts and Theories. A Critical Appraisal 150 Years After The Origin of Species*. Analecta Gregoriana No. 312. Gregorian and Biblical Press: Rome. 747p.
48  Available at: http://news.bbc.co.uk/2/hi/europe/7021358.stm
49  *Catechism of the Catholic Church.1994. Definitive Version: Based on the Latin Editio Typica*. Episcopal Commission on Catechesis and Catholic Education: Catholic Bishops Conference of the Philippines. Word & Life Publications: Makati City, Philippines. 828p. Para 159.

*faith, because the things of the world and the things of faith derive from the same God. The humble and persevering investigator of the secrets of nature is being led, as it were, by the hand of God in spite of himself, for it is God, the conserver of all things, who made them what they are."*

'Yes indeed!' The creativity of the scientist, as developed and used in her/his Free Thought, is how God leads the march of science and discloses to us the truth about the *"secrets of nature."* I had not expected to find this amount of congruence between Free Thought theory as developed thus far and the products of Vatican II *and* the Catholic Catechism.

Second, Catholic authorities have become more amenable to accepting that people must follow the dictates of their own individual consciences. The current *"Compendium of the Social Doctrine of the Church"*[50] states: *"...we can see the importance of moral values, founded on the natural law (i.e., the Universal Moral Code) written on every human conscience; every human conscience is hence obliged to recognize and respect this law...Contemporary cultural and social issues involve above all the lay faithful, who are called, as the Second Vatican Council reminds us, to deal with temporal affairs and order them according to God's will (cf. Lumen Gentium, 31). We can therefore easily understand the fundamental importance of the formation of the laity, so that the holiness of their lives and the strength of their witness will contribute to human progress."*

Despite all these positive outcomes from Vatican II, it is clear that some rank and file members of the Catholic Church have not yet heard or thought very much about it. Not surprisingly, those who have heard and thought about it differ markedly in their opinions over whether the pace of change over the past 50 years has been too fast or too slow. For many Catholics and watchers of the Catholic Church, some of the

---

50   Pontifical Council for Justice and Peace. 2004. *Compendium of the Social Doctrine of the Church*. Catholic Bishops Conference of the Philippines and Word & Life Publications: Manila, Philippines.557p.p.xviii-xix.

changes agreed at Vatican II are still more like a reformation-in-waiting than a reformation substantially achieved.

Christoph Cardinal Schönborn[51] quoted the following statement from Joseph Ratzinger about creation and evolution, before he became Pope Benedict XVI: *"The Christian idea of the world is that it originated in a very complicated process of evolution but that it nevertheless still comes in its depths from the Logos. It thus bears reason in itself."* Schönborn added: *"The 'Darwinian ladder' has made available to us - thanks also to genetics - a marvelous insight into the way life has ascended, the way it has come into existence and has been shaped and developed."* Those two beautiful statements indicate well the ongoing trend to science-friendly reformation in the Catholic Church.

Pope Benedict XVI's resignation might have been occasioned in part by his frustration at the slow pace of implementation of some of the changes agreed at Vatican II, including those that touch on faith-science relationships. In any case, there is an urgent need to accelerate implementation of more of the openness agreed at Vatican II, including endorsement and support for expansion of the faith-science quest for truth, in research, education and Church-wide dissemination of information.

## Postscript

The Reverend Frank Walters[52] wrote the following, surely with tongue firmly in cheek: *"I do not see why Evangelical Protestants should be so angry with their Catholic brethren for acknowledging the supremacy of*

---

51  Schönborn, C. 2007. *Chance or Purpose? Creation, Evolution and a Rational Faith.* Edited by H.P. Weber and translated by H. Taylor from the original German. Ignatius Press: San Francisco CA. 181p. Opening quotation and p.175. First published as: *Ziel oder Zufall? Schöpfung und Evolution aus der Sicht vernünftigen Glaubens.* Verlag Herder GmBH: Freiburg im Breisgau, Germany.

52  Walters, F. 1890. *Rationalism: What It Is and What It Is Not*, p.2-28. In R.B. Drummond (ed.) *Free Thought and Christian Faith, Four Lectures on Unitarian Principles.* Edited and with Williams and Norgate: Edinburgh. 123p. p.10-11, 14-15, 24. Also published (2005) in Elibron Classics, Elibron: Edinburgh. Available from www.kessinger.net

*the Church, or even of the Pope. I confess that I could quite as easily accept ecclesiastical as I could Biblical infallibility; the dogma of the verbal inspiration of Scripture is not nearly so reasonable as that of the divine authority of a living Church, able to expound its own doctrine, the repository of truth through all ages of the world."*

Walters was making the point that no church, or book, should dictate everything that its members, or readers, must accept. He continued: "*It was authority which would have prevented all reformation where it is, and which has put a barrier against it where it is not...I am told that I must study the Bible for myself, but unless I find certain doctrines in it I am not merely mistaken - no sensible man objects to other people differing from his conclusions, - but I am an enemy of God, and in danger of eternal punishment...I have met with some people who talked as though it were written at the end of their Bibles, 'Here the Holy Spirit spoke for the last time.'*

Walters met those people in the nineteenth century. I am meeting them in the twenty-first. I urge my fellow believers to be honest to God, honest to fellow humans, especially the young, and above all honest to themselves - in the same way that mathematicians and scientists are being honest to God, whether they know it or not, and honest to their fellows and themselves as they practice their professions.

John Polkinghorne[53] wrote: "*Neither science nor religion has access to absolute truth, indubitable beyond the possibility of a challenge.*" He added: "**The kinship thus discerned between science and theology** (present author's emphasis) *rebuts the strident claims made by some atheists that theology is not a proper subject with a rightful place in the academy...* (also) *too many theologians fail to treat what science has to offer with the appropriate degree of seriousness that would enable them to acknowledge adequately its contextual role.*"

---

53   Polkinghorne, J.C. 2009. *Theology in the Context of Science.* Yale University Press: New Haven CT.168p.p.8.

## Free Thought, Faith, and Science

The discerners of that "*kinship*" can be frontline reformers in organized religion *and* revolutionaries in science. In our Free Thought, we are all theologians and scientists. Believers who embrace science will find no conflicts with truths found through faith. Nonbelievers who explore faith will find truths through personal experience that are not accessible through science. Science-friendly reformations in organized religion and faith-friendly revolutions across science will follow. Seeking truth together, through faith and science, is the path to unity.

# CHAPTER 11

# UNITY

*"Outdated models of the relationship of science and theology can be discarded in favor of a joint exploration into a common reality "*[1]

## DRAWING THREADS TOGETHER

Arthur Peacocke's call for *"a joint exploration into a common reality"* is much more than a call for building occasional bridges. It is a vision for expanding the faith-science quest for truth, as threads of truth, disclosed through faith and science, are drawn together by individuals and groups, sharing their findings freely and honestly.

Human life is a complex mixture of daily routine and occasional wonders and horrors. We seek explanations for whatever we experience. Our explanations concerning the higher things of life are outcomes of Free Thought, in which the quest for truth is bedevilled by lies and nonsense. We find unity by seeking and accepting truthful disclosures. Truth unites us. Lies divide us.

Before getting to the big picture of the faith-science quest for truth, I invite believers and nonbelievers to consider and try to explain

---

[1] Peacocke, A.2000. Science and the future of theology - critical issues. *Zygon* 35(1):119-140.

the following horrific happenings. First, in March 2012, 22 Belgian chidren died in a coach crash on their way home from a skiing holiday. Was all that planned long ago by God and then carried out in His perfect timing?

Some believers will say 'Yes,' because they feel sure that God has predestined everything that happens - for His purposes, which are perfect but can be unfathomable to us. Most nonbelievers will say 'No' and I agree with them. Accidents are caused by adverse physical conditions and/or human error and/or mechanical failure. The courses of accidents follow the laws of nature. The laws of physics are non-negotiable. God has limited Himself from intervening in the free process by which He made His material realm to work.

Second, in May 2013, it became clear that three young women had been imprisoned for ten years in a house in Cleveland Ohio, and that their captor had committed rape and terminated some resulting pregnancies. Was all that planned long ago by God and then carried out in His perfect timing?

Again some believers will say 'Yes,' based on their belief that God predestines and dictates the courses of everything that happens. Others will allow that the captor had free will and/or was mentally disordered and/or acted under the influence of the spiritual force for evil. Nonbelievers will offer a range of explanations based on the psychological state and past history of the captor, the vulnerability of young women in modern society, and the ease with which anyone in a modern urban setting can do almost anything behind closed doors and boarded up windows.

My explanation is that the captor's evil behaviour was dictated by his free will and the outcomes of his basic thought and Free Thought. Free process in the material realm contributed to his genome, his experiences and his mental condition. His goals were set by his free will. His basic thought included the pursuit of sexual gratification. The spiritual

# UNITY

force for evil dominated his Free Thought. The results were horrific - to God and to most of humanity.

Matthew Alper[2] described as follows what he saw as the bright consequences of denying the existence of the spiritual realm: *"If it's true that there is no spiritual reality, no God, no soul, and no afterlife, then let's accept ourselves for what we are and make the most of it. Perhaps such a change in our self-perceptions might help us to shift our priorities from the hereafter to the here and now, to deter intolerance, antipathy, and war, thereby minimizing our pain and maximizing our chance of obtaining the greatest amount of happiness. This, more than anything, is what I would hope to gain from a scientific explanation of human spirituality and God. Let the secular revolution begin."*

Faith *per se* does *not* increase pain, diminish happiness, or bring about intolerance, antipathy, and war. The spiritual force for evil is the cause of all lying and disunity and often chooses organized religion as its multiplier. Faith is based on truth. Faith is a unifier. Moreover, believers and nonbelievers alike should welcome Alper's call for a *"scientific explanation of human spirituality and God."* Expansion of the faith-science quest for truth is the way forward.

What Alper called *"the secular revolution"* is already well underway and benefitting millions of people who were formerly constrained by superstition and baseless religious obligations. This revolution is part of the quest for truth. On the basis that all truthful disclosures through mathematics and science come from God, it can also be called sacred. It will remain a partial revolution unless its revolutionaries address the undiminished human need for a spiritual life and spiritual connections.

Our peace of mind depends upon the fulfillment of spiritual needs as well as material needs - drawing together threads of truth from the

---

2   Alper. M. 2006. *The God Part of the Brain: A Scientific Interpretation of Human Spirituality and God.* Sourcebooks Incorporated: Naperville IL.273p.p.246-247.

spiritual and material realms, to satisfy mind and soul. As Brian Davies[3] put it: *"It is time to draw together the various threads. Many scientists are struck by the grandeur of the universe, the simplicity of the laws that govern it, and even its very existence...I look into my soul and wonder why I have one, when all my scientific knowledge tells me that I do not."*

Ian Barbour[4] recognized four main types of relationships between organized religion and science - conflict, independence, dialogue and integration. He saw no clear dividing line between dialogue and integration, which he recommended in combination as the best relationship. The same applies for the faith-science quest for truth, which will expand only if more believers and nonbelievers can agree to widen the scopes of their personal reality checks and appraise all available evidence.

Joining the faith-science quest for truth requires an honest admission that neither science-less faith nor faith-less science is likely to provide sufficient explanations of the human condition. Making progress to unity requires honest seekers of truth to step outside their comfort zones and consider new possibilities. From that honest and humble perspective, more threads can be drawn together to weave a more complete and truthful fabric.

## EXPANDING THE FAITH-SCIENCE QUEST FOR TRUTH

### *PREREQUISITES*

The first prerequisite for the faith-science quest for truth is *total honesty* - to oneself, to one's fellow humans and, as far as His existence is perceived to be possible or not, to God. In other words, all participants must be committed to seek only the *truth* and to share their

---

3 Davies, E.B. 2010. *Why Beliefs Matter: Reflections on the Nature of Science.* Oxford University Press: Oxford.250p.p.239.
4 Barbour, I.G.2000. *When Science Meets Religion: Enemies, Strangers or Partners?* HarperOne: New York.205p.

# UNITY

findings and interpretations freely and honestly. Willful ignorance, baseless generalizations and rejections of inconvenient truths have no place in this quest.

The second prerequisite is to devise and employ *robust methods*. Simon Blackburn[5] quoted from the mathematician William Clifford, who emphasized as follows the importance of method in approaches to difficult and contentious questions: "*In regard then to the sacred tradition of humanity, we learn that it consists, not in propositions or statements which are to be accepted and believed on the authority of the tradition, but in questions rightly asked, in conceptions which enable us to ask further questions, and in* **methods of answering those questions. The value of all things depends on their being tested day by day** (present author's emphasis)."

There is no agreement on theological method comparable to that for scientific method, but this is less of a problem for expanding the faith-science quest for truth than might be imagined. All quests for truth about the spiritual and material realms are rooted at the personal level. Mathematics and science advance through God's truthful disclosures, channeled via human creativity in Free Thought. All choices for faith and unbelief are made through Free Thought. All assessments of spiritual experiences are also made through Free Thought. Free Thought provides anyone with sufficient methods for exploring any experience, as long as the desire for seeking and finding truth prevails over encouragements from others to follow lies and nonsense.

The third prerequisite is *mutual trust*. No joint venture ever gets far without trust around the table. Finding faith-science unity through seeking truth depends on the establishment and continuation of trust on all sides. Mutual trust is a property of persons, not of organizations and institutions. Trust and distrust are always outcomes of Free

---

5   Blackburn, S. 2006. *Truth: A Guide for the Perplexed*. Penguin Books Limited: London.238p.p.6-7, citing p.35 in W.K. Clifford's *The Ethics of Belief*, first published in 1879.

## Free Thought, Faith, and Science

Thought. God move us in our Free Thought towards outcomes of giving trust to trustworthy persons. The spiritual force for evil does the opposite, encouraging distrust of the trustworthy and trust in the non-trustworthy, so that lies and nonsense will prevail over truth.

The fourth prerequisite is *freedom from fear*. Some believers fear that they might lose their faith by honest re-examination of their beliefs in the company of nonbelievers. Some nonbelievers fear that encounters with believers might bring unwanted complications into their lives and/or attract the scorn of their fellows. On both sides, seeking more truth by exploring material *and* alleged spiritual reality can appear dangerous - for believers, as a path to heresy or loss of faith and for nonbelievers as a path to delusion.

For believers, pursuing joint explorations with scientists can never threaten faith based on truth. For nonbelievers, joint explorations with believers will not threaten unbelief from the outset or at all, *unless* unbelief becomes perceived as unlikely to be a *true* worldview and explanation of the human condition. Nonbelievers need not fear allowing a finite, albeit very small, probability that the spiritual realm might exist. Even Richard Dawkins,[6] with his customary scientific correctness, stated only that: "*God almost certainly does not exist.*"

The fifth prerequisite is agreement on *rules of evidence*. Theology and research on the truth or falsity of faith have nothing comparable to the globally agreed rules of objective evidence in science. Nothing in the spiritual realm is discernable through our material realm senses and scientific instruments.

Therefore, from a scientific perspective, the honest seeker must conclude either that the spiritual realm does not exist, or that if it does exist it is totally inaccessible, somewhat like another universe, or is accessible only by spiritual means, as in Free Thought. The faith-science

---
6  Dawkins, R. 2006. *The God Delusion*. Bantam Press: London.406p.p.158.

# UNITY

quest for truth requires agreement on the validity of objective evidence from science *and* rigorously analysed subjective evidence from within concerning faith and unbelief.

The sixth prerequisite is actually a given - our God-given *freedom* to go wherever the search for truth takes us. Free inquiry is the essence of the faith-science quest for truth. The following wise words from E.O. Wilson[7] apply to believers and nonbelievers: "*The search for consilience might seem at first to imprison creativity. The opposite is true. A united system of knowledge is the surest means of identifying the still unexplored domains of reality.*"

## *PERSPECTIVES*

Robert Wright[8] wrote the following on love and truth as unifiers: "*Though we can no more conceive of God than we can conceive of an electron, believers can ascribe properties to God, somewhat as physicists can ascribe properties to electrons. One of the more plausible such properties is love. And maybe, in this light, the argument for God is strengthened by love's organic association with truth - by the fact that these two properties almost blend into one. You might say that love and truth are the two primary manifestations of divinity in which we can partake, and that by partaking of them we become truer manifestations of the divine. Then again, you might not say that. The point is that you wouldn't have to be crazy to say it.*" Truth and love are the basis for faith-science dialogue and faith-science unity. Their pursuit is never crazy.

John Polkinghorne[9] summarized the quest for truth as follows: "*the search for truth through and through is ultimately the search for God... The unity of knowledge is underwritten by the unity of the one true God; the veracity of well-motivated belief is underwritten by the*

---

7    Wilson, E.O. 1998. *Consilience: The Unity of Knowledge.* Vintage Books: New York.367p.p.325, 326.
8    Wright, R. 2009. *The Evolution of God.* Little, Brown and Company: New York.567p.p.459.
9    Polkinghorne, J. 1998. *Belief in God in an Age of Science.* Yale University Press: New Haven CT.133p.p.110, 122.

*reliability of God."*

Polkinghorne[10] also emphasized that: *"Theology, as much as science, must appeal to motivated belief arising from **interpreted experience*** (present author's emphasis)...*A scientist should not be unsympathetic to the notion of an appeal to experience being the best strategy to pursue in the search for understanding."* Progress towards unity in truth depends on individual and collective interpretations, all of which derive from Free Thought. Anyone can participate in and learn from the faith-science quest for truth.

Polkinghorne[11] described as follows the similarity between making contributions in science and theology: *"the feel of doing science is actually one of discovery, rather than pleasing construction. Theologians can claim something similar about the encounter with God. Time and again human pictures of deity prove to be idols that are shattered under the impact of divine reality."* Advances in science and theology are made in similar ways. In science and theology, new theory supplants old theory, providing better explanations for observations and experiences.

Pope Benedict XVI[12] called as follows for theology to belong within the wide gamut of scientific inquiry: *"While we rejoice in the new possibilities* (in science) *open to humanity, we see also the dangers arising from these possibilities and we must ask ourselves how we can overcome them. We will succeed in doing so only if reason and faith come together in a new way, if we overcome the self-limitation of reason to the empirically falsifiable, and if we once more disclose its vast horizons. In this*

---

10 Polkinghorne, J.C. 2007. *Quantum Physics and Theology: An Unexpected Kinship*. Yale University Press: New Haven CT.112p.p.9, 92.
11 Polkinghorne, J. 2005. *Exploring Reality: the Intertwining of Science and Religion*. Yale University Press: New Haven CT.181p.p.4.
12 Pope Benedict XVI. 2006. *Faith, Reason and the University. Memories and Reflections* Lecture at the Meeting With Representatives of Science, 12 September 2006, University of Regensburg, Germany. Available at: www.vatican.va/holy_father/benedict_xvi/speeches/2006/september/documents/hf_ben-xvi_spe-2006092_university-regensburg_en.html

*sense, theology rightly belongs in the university and within the wide-ranging dialogue of sciences, not merely as a historical discipline and one of the human sciences, but precisely as theology, as inquiry into the rationality of faith."*

Alister McGrath[13] called for *"scientific theology,"* through which: *"to encourage and facilitate a respectful and positive dialogue between Christian theology and the natural sciences, without the latter overwhelming the former."* He asked: *"But is this dialogue arbitrary, or a matter of pure convenience? Is it theologically opportunistic? Or are there deeper reasons for seeing this dialogue as a natural element of the theological method?"*

With McGrath, I can shout the answer 'Yes' to that last question. The natural sciences cannot overwhelm the truth of faith, because there is one whole body of truth, which is disclosed in part through faith and in part through science. Truths disclosed through faith and truths disclosed through science will always be mutually supportive. When they are in conflict, something is not true - from either or both sources.

McGrath singled out Christianity as the foundation for scientific theology but qualified his position as follows: *"This is not to say that a scientific theology is unable or unwilling to enter into dialogue with other religious traditions...Instead of offering a 'view from nowhere,' a scientific theology is rooted in the realities and particularities of the Christian tradition, from which it is able to offer an account of the success of the natural sciences, as well as the existence of alternatives to itself."*

The unity of all things is indeed written large in Christianity. The Apostle Paul wrote that God, through His Self-sacrifice as Christ: *"made peace...to reconcile all things unto himself...whether they be*

---

13  McGrath, A.E. 2004. *The Science of God: An Introduction to Scientific Theology*. William B. Eerdman's Publishing Company: Grand Rapids MI.271p.p.21, 33, 173. The translated quotation from Irenaeus was taken from his *"adversus hereses"* IV, xx.

*things in earth or things in heaven.*"[14] McGrath also quoted, as part of Christian theology, the following wonderful words from St. Irenaeus: "*'the glory of God is a living human being, while the life of humanity is the vision of God*'."

*All* humans are potentially God-aware and have the capacity for dialogue with God, regardless of the presence or absence of contact with any particular organized religion, including Christianity. God, the Giver of all souls, is surely open to dialogue with everyone - believers and nonbelievers, Christians and non-Christians. Anyone can use her/his Free Thought to search for truth and indeed to search for God.

McGrath's works[15] proclaim the whole material realm of nature as our "*theological resource.*" God completed us to be good company for Him and gave us our freedom for deciding how to respond. However, beyond our personal experiences of God, *nature* is our largest and most accessible theological resource. The whole of creation is indeed subject matter for ongoing God-human dialogue, through faith *and* science.

Believers who can accept that all truthful disclosures through mathematics and science come from God should have no problems in accepting that science and theology contribute to the one great quest for truth, in which God is asked about nature and about Himself.

Free Thought always includes spiritual revelations and responses. An individual's Free Thought about faith is her/his personal version of theological method, regardless of whether she/he is a believer or a nonbeliever. An individual's Free Thought about science is a combination of her/his personal theological method and her/his personal

---

14 Colossians 1: 20
15 McGrath, A.E. 2006. *The Order of Things: Explorations in Scientific Theology.* Blackwell Publishing: Malden MA.255p. McGrath, A.E.2008. *The Open Secret: A New Vision for Natural Theology.* Blackwell Publishing: Malden MA.372p.

use of scientific method, again regardless of whether she/he is a believer or a nonbeliever.

I propose that prayer has much in common with the framing and testing of hypotheses in science. The Oxford Dictionary defines prayer as *"a solemn request for help"* and defines an experiment as: *"a scientific procedure undertaken to make a discovery...a course of action tentatively adopted without being sure of the eventual outcome."* A believer's prayer is a dialogue with God. The outcome cannot be forecast when the dialogue starts. The same applies to the framing and testing of hypotheses in science.

In prayer, a believer asks God about states, events and future possibilities, according to His will. A scientist framing a hypothesis and testing it by observation and experiment is asking nature and God, Who made it, about states, events and future possibilities, according to the natural laws, which He also made. The scientist applies her/his God-given creativity through her/his God-given capacity for Free Thought.

God makes truthful disclosures to us through prayer and through science. The cumulative results of prayer enhance a believer's relationship with God and can help to advance theology. The cumulative results of scientific research advance an individual's understanding of nature and can advance theory.

David Deutsch[16] advocated a unified worldview of reality, based on four interwoven strands: *"the quantum physics of the multiverse, Popperian epistemology, the Darwin-Dawkins theory of evolution and a strengthened version of Turing's theory of universal computation."* This looks very impressive as a basis for further exploration of the material realm. But is it enough for making progress towards to a theory of everything - a 'final theory'? I think not. What about the possibility of the spiritual realm?

---

16   Deutsch, D. 1998. *The Fabric of Reality.* Penguin Books: London. 390p. p.366.

Steven Weinberg[17] wrote: "*Will we find an interested God in the final laws of nature? There seems something almost absurd in asking this question, not only because we do not yet know the final laws, but much more because it is difficult even to imagine being in the possession of ultimate principles that do not need any explanation in terms of deeper principles. But premature as the question may be, it is hardly possible not to wonder whether we will find any answer to our deepest questions, any sign of the workings of an interested God, in a final theory. I think that we will not.*"

Weinberg considered the concept of God as useful only if it meant: "*an interested God, a creator and lawgiver who has established not only the laws of nature and the universe but also standards of good and evil, some personality that is concerned with our actions, something in short that is appropriate for us to worship.*" He observed that: "*Scientists and others sometimes use the word 'God' to mean something so abstract and unengaged that he is hardly to be distinguished from the laws of nature.*"

Subjective evidence from within believers can indeed indicate some of the "*workings of an interested God.*" The same applies to some of the subjective evidence from within nonbelievers. For example, Roger Penrose's[18] call for a "*new physics*" that frees us from "*the strait-jacket of an entirely computational physics, or of a computational cum random physics*" could have been an outcome of his Free Thought, in which spiritual revelations from God nudged him towards a fundamentally new approach. God works in all Free Thought.

Following the terms used by convention in science to plan, execute and report on research, the *materials* needed for a Free Thought experiment are the individual person and all of her/his experiences,

---

[17] Weinberg, S. 1993. *Dreams of a Final Theory: The Search for the Fundamental Laws of Nature.* Vintage Books: London.260p.p.195-196.
[18] Penrose, R. 2005.*Shadows of the Mind: A Search for the Missing Science of Consciousness.* Vintage: London. 457p.p.420.First published in 1994, by Oxford University Press.

past and present, including her/his current baseline from prior Free Thought. The *methods* are her/his senses and Free Thought itself. The *results* that arise in mind and soul during an episode of Free Thought are subjected to integrated and iterative *discussion* during that same episode. The *conclusions* from an episode of Free Thought are what I call its outcome. They will often contain unresolved questions, some of which will be considered further in subsequent episodes, starting from successive new baselines.

## *Prospects*

The prospects for wider recognition of the true unity of faith and science and for expansion of the faith-science quest for truth are good, despite the fundamental differences between the types of evidence that are available from material realm science and from individual experiences of God through faith. Behind the public faces of organized religion and science, believers and nonbelievers are making Free Thought decisions about faith and science, based on a common desire to seek, find, and tell only the truth.

Everyone is searching not only for material wellbeing but also for spiritual peace of mind. Some nonbelievers are pressured by their peers to deny all possibilities of the existence of the spiritual realm. Some believers are pressured to deny truthful disclosures through science.

On both sides, freedom is there for the taking. Free Thought enables all of us to explore the material and spiritual realms and the human condition. We are all equipped to be theologians and scientists, simply by having the capacity for Free Thought. The challenge is to recognize truth, in the midst of lies and nonsense.

## Research on Free Thought and Further Development of Theory

### *Systematic Gathering and Analysis of Subjective Evidence from Within*

Francisco Varela and Jonathan Shear[19] defined *"first-person events"* as: *"the lived **experience** (original authors' emphasis) associated with cognitive and mental events."* They added: *"What we take to be objective is what can be turned from individual accounts into a body of regulated knowledge."*

The purpose of Varela and Shear's research was: *"to provide the basis for a **science of consciousness which includes first-person, subjective experience as an explicit and active component** (original authors' emphasis)."* Research on subjective evidence from within concerning faith and unbelief can be conducted on the same basis, using rigorous and standardized methods for extracting and combining elements: *"from individual accounts into a body of regulated knowledge."*

Varela and Shear also stated the following: *"the explorations of experience will suffer...from cultural expectations and instrumental bias...whatever descriptions we can produce by first-person methods are not pure, solid 'facts' but potentially valuable intersubjective items of knowledge, quasi-objects of a mental sort. No more, no less."* In other words, diverse personal attitudes and interactions will inevitably constrain and colour the nature and completeness of subjective evidence from within, but it can still be well worth having.

Pierre Vermesch[20] discussed *"introspection,"* by which he meant the

---

19  Varela, F.J. and J. Shear.1999. First-person methodologies: what, why, how? p.1-14. In F.J. Varela and J. Shear (eds.) *The View from Within: First-Person Approaches to the Study of Consciousness.* Imprint Academic: Exeter, UK.313p.p.1-2.First published in 1999 in *the Journal of Consciousness Studies* 6 (2-3): 1-14.
20  Vermesch, P.1999. Introspection as practice, p.17-4.In F.J. Varela and J. Shear (eds.) *The View from Within: First-Person Approaches to the Study of Consciousness.* Imprint Academic: Exeter, UK.313p.p.31, 41.First published in 1999 in *the Journal of Consciousness Studies* 6 (2-3): 17-42.

gaining of access to subjective evidence from within. He emphasized its value as a source of empirical data. Vermesch pointed to the ethical considerations attached to the collection of subjective evidence from within. Free Thought researchers who collect and analyse such evidence will need a code of ethics covering their methods, ownership of data and publication of results.

Vermesch recognized as follows the messy state of this field: "*Are we not under some obligation to add a truly scientific dimension to subjectivity? For my part, I would want to add that what is also at stake is the need to coordinate the innumerable practices which make use of first-person data (teaching, remedial action, re-education, training, coaching, therapy etc.) with the* **present scientific vacuum** (present author's emphasis) *which surrounds all those aspects of cognitive functioning which can only be comprehended at a phenomenological level.*" He concluded the following: "*After all, subjective experience is certainly a fact of personal life. Perhaps we need a more refined theory and practice of intersubjectivity?*"

'Yes,' indeed we do. So, where do we look? Perhaps to psychology, psychiatry and psychotherapy? The diagnosis and treatment of mental disorders and therapies are based substantially on accumulated medical experience and case studies, but the process of getting from bodies of multiple case studies to rigorously established and agreed taxonomies of disorders and therapies is challenging and can be hard to standardize.

Not every case study has the same weight for comparison with another case study or with an attempt to define a general state. The use of precedents in jurisprudence is somewhat analogous. Standard statistical methods must be allied with interpretative guidelines, which can change as experience accumulates. The same will apply in research on subjective evidence from within concerning faith and unbelief.

Martin Brüne [21] commented as follows on diagnosis in psychiatry: *"Current diagnostic systems (Diagnostic and Statistical Manual of Mental Disorders, DSM, 4th revision, and the International Classification of Disorders, ICD, 10th edition) were conceptualized as largely theoretical, purely descriptive frameworks to improve the reliability and validity of the diagnostic process...DSM and ICD are also reductionist in that they constrain the richness of psychopathological signs and symptoms to a compilation of only the most conspicuous clinical symptomatology of disorders."*

Brüne concluded that: *"**current psychiatry lacks a coherent theory of human behaviour*** (present author's emphasis)." I conclude that psychiatry, psychotherapy and works about mental disorders are likely to be of limited use in the development of methods for research on subjective evidence from within concerning faith and unbelief. However, they might provide pointers for addressing the issue of Researcher-Subject relationships and ethical aspects.

Brüne recommended the following with respect to the therapist-client relationship: *"The therapeutic alliance should emulate the safety and stability of a kinship bond, diminish the dominance hierarchy between patient and therapist, and thus help to avoid distant professionalism."* He concluded that: *"a most promising way to improve therapist-client interaction is to emphasize more strongly how minds interact - namely, by making inferences about one another's beliefs, goals, wishes, knowledge and emotions."*

Chris Jaenicke [22] summarized the complexity of analyst-client relationships as follows: *"Understanding the influence of our subjectivity allows us, paradoxically, to unlock our perceptions from the grip of our*

---

21  Brüne, M. 2008. *Textbook of Evolutionary Psychiatry: The Origins of Psychopathology.* Oxford University Press: Oxford.385p.p.315, 305, 317. Brüne cited the Diagnostic and Statistical Manual of Mental Disorders (DSM). Its latest version (DSM 5) was published in May 2013 and is available at www.psych.org. Brüne cited also the International Classification of Diseases (ICD), which is available at: www.who.int/classification/icd10/
22  Jaenicke, C. 2011. *Change in Psychoanalysis: An Analyst's Reflections on the Therapeutic Relationship.* Routledge, Taylor and Francis Group: New York.194p.p.5, 98.

*subjective world and return to the dialogic approach in understanding the 'truth' about our patients. We change what we look at by looking at it and - including now the aspect of mutual influencing - are changed."* Jaenicke recognized *"the primacy of subjectivity within the indissoluble unit* (i.e., the analyst-patient combination), *or the mutual influencing of subjectivities in the intersubjective field"* and concluded: *"In describing the work with a patient, we are implicitly or explicitly describing ourselves."*

The same complexities will apply in research on subjective evidence from within concerning faith and unbelief. Although no one can ever know for sure whether anyone's profession of faith or unbelief is true or false, interviews for research on subjective evidence from within should probably include adequate replication of the following four combinations of professions of belief and unbelief among Researchers and Subjects: Researcher as Believer/Subject as Believer; Researcher as Nonbeliever/Subject as Believer; Researcher as Believer/Subject as Nonbeliever; and Researcher as Nonbeliever/Subject as Nonbeliever. However, even if all four combinations of believer and nonbeliever are included, there will still be mismatches between the positions of Researchers and Subjects across a wide range of issues.

Kenneth Pargament[23] sought: *"a unified field theory…* (meaning) *a unified perspective on human behaviour, one that expands the biopsychosocial perspective to a biopsychosociospiritual perspective."* He lamented the prevailing situation, as follows: *"Unfortunately, many therapists remain uncomfortable about the topic of spirituality, unsure how to deal with spiritual issues, or fearful of intruding in areas too private even for psychotherapy. As a result, they do their best to avoid the spiritual domain…* (but) *Spirituality is part of the psychotherapy process; our choice is to look the other way and proceed with limited vision or to address spirituality directly and knowingly."*

---

23   Pargament, K. 2011. *Spiritually Integrated Psychotherapy: Understanding and Addressing the Sacred.* The Guildford Press: New York.384p.p.x-xi,14-15,9,92,335,336 (Table 16.6),344.

Pargament cited Edward Shafranske,[24] who found that only 24% of clinical and counselling psychologists in the USA believed in a personal God and only 26% considered religion very important to them, compared with 90% and 58% respectively in the whole American population.

Pargament recognized as follows the diversity in how people perceive and choose to relate to *"the sacred:"* *"One person seeks to make the world a sacred place by acts of kindness…Another seeks a personal relationship with Jesus largely through scriptural study, prayer and devotion. One person tries to experience the sense of transcendence in daily life through meditative practices and outdoor experiences. Yet another attempts to discover ultimate truths about the universe through scientific investigation."*

Pargament considered self-awareness to be the professional's *"most effective antidote to the dangers of spiritual imposition on clients."* He devised the following questionnaire as the psychotherapist's guide for writing her/his own *"spiritual autobiography."*

"• *What are my deepest values and what do I strive for in my life?*

• *What do I hold sacred?*

• *How did I discover the sacred?*

• *How has my larger family and institutional religious context shaped my attitudes toward spirituality and religion?*

• *How have I tried to develop and sustain myself spiritually over the years?*

---

24  Shafrankse, E.P. 2001. The religious dimension of patient care within rehabilitation medicine: The role of religious attitudes, beliefs, and personal and professional practices, p. 311-338. In T.G. Plante and A.C. Sherman (eds.) *Faith and Health: Psychological Perspectives*. Guilford Press: New York. Cited on p.9, in Pargament, K. 2011. *Spiritually Integrated Psychotherapy: Understanding and Addressing the Sacred*. The Guildford Press: New York.384p.

- *What kinds of struggles have I encountered in the process of developing and conserving my spirituality?*

- *What kinds of spiritual transformations have I experienced, if any?*

- *Where do I currently stand in the search for the sacred?*

- *What are the areas of spiritual integration and dis-integration in my life?*

- *In what ways has my spirituality affected my life? In what ways has it not affected my life?*

- *What are my areas of spiritual strength and vulnerability in working with clients? Are there clients from particular spiritual or religious backgrounds whom I may not be able to help?"*

Researchers on subjective evidence from within concerning faith and unbelief will need a similar level of self-awareness in order to develop standardized, inclusive instruments for engaging with a wide diversity of Subjects.

Published testimonies of faith seem more numerous than statements of unbelief, but many of those who write about their faith seem to be doing so mainly for audiences of fellow believers. Christian bookshops are stacked with works by Christian authors to be read by fellow Christians. Comparatively fewer agnostics and atheists seem motivated to write about their unbelief, but those who do so tend to write for wide audiences.

Published testimonies of faith and unbelief are highly variable in style and content. Many lack clear and comparable subjective evidence from within. Most are simply statements of belief or unbelief, rather than accounts of experiences and process. Therefore, even the seemingly rich literature on testimonies of faith is not really so rich for present purposes. Edwin Diller Starbuck[25] encountered the same difficulty as he undertook his painstaking research on conversion and other religious experiences.

Starbuck used the term *"autobiographies"* for what I call subjective evidence from within. He came to rely almost entirely on primary data from his own surveys, which he called: *"autobiographies written in response to personal solicitation."* He wrote the following about the secondary data available from published sources: *"Autobiographies in books were usually disappointing and the plan of gathering them from that source was...largely abandoned."* In my limited work so far in this area I have found the same and have concluded that the gathering and analysis of new primary data is by far the best way forward.

Systematic gathering and analysis of subjective evidence from within concerning faith and unbelief requires standard instruments. Starbuck focused on individual experiences of conversion from unbelief to faith in 192 Subjects - 120 females and 72 males. For the parameter of age at conversion, there were 1,265 respondents: 254 females and 1,011 males. William James noted the limitations of Starbuck's reliance on percentages and averages, but concluded that the material gathered by Starbuck contained: *"many acutely interesting individual confessions"* and was *"evidently sincere in its general mass,"* though mostly limited to Protestant Christian respondents *"of the Evangelical sort"* and typified by *"special phraseology* (and) *conceptions."*

L. Shelton Woods[26] studied testimonies of faith at the Protestant Vintar Bible Baptist Church in Ilocos Norte, Philippines and

---

25 Starbuck, E.D. 2010. *The Psychology of Religion: An Empirical Study of the Growth of Religious Consciousness.* General Books: Memphis TN. 226p.p.18, 5-6 (in the Preface by William James) 19,34 (for Table VII). First presented in 1894 and 1895 to the Harvard Religious Union. Published as two articles in the *American Journal of Psychology* in January and October 1897. The date of first publication of the complete study was 1899.

26 Woods. L.S. 2002. *A Broken Mirror: Protestant Fundamentalism in the Philippines.* New Day Publishers: Quezon City, Philippines.297p.p.154.

concluded the following: "(The) *conversion narratives by Vintariñian fundamentalists fit in with the reports of Finney's* (Charles Finney, an architect of American Fundamentalist Protestantism) *revivals and the testimonies from Billy Sunday crusades. The pattern has been passed on through the training of missionaries and Filipino pastors; consequently, the fundamentalist conversion experience was exported beyond the shores of Europe and North America to encompass the Vintariñian farmer, housewife and college student."*

Subjective evidence from within concerning faith and unbelief can include formulaic accounts, based on limited and parroted vocabularies and special phraseology. Moreover, as noted by Matthew Alper,[27] many cultures have a fairly constant vocabulary for describing some allegedly spiritual experiences; for example, assurance, bliss, elation, joy, loss of self, oneness, peace, sadness and sublime happiness. Alper concluded the following: *"That all cultures have described sadness in such a similar way indicates that this sentiment is not learned but an inherent part of our human natures. By the same logic, this should hold true of spiritual experiences."*

In order to illustrate the difficulty of devising questionnaires for research on subjective evidence from within concerning faith and unbelief, I quote here at length from Starbuck's questionnaire concerning conversion to Christianity, omitting his first question, which has no relevance:

*"II. What force and motive led you to seek a higher or better life: - fears, regrets, remorse, conviction for sin, example of others, influence of friends or surroundings, changes in beliefs or ideals, deliberate choice, external pressure, wish for approval of others, sense of duty, feeling of love, spontaneous awakening, divine impulse, etc.? Which of these or other causes were most marked or were present at all?*

---

27  Alper. M. 2006. *The God Part of the Brain: A Scientific Interpretation of Human Spirituality and God.* Sourcebooks Incorporated: Naperville IL.273p.p.135.

*III. Circumstances and experiences preceding conversion: - any sense of depression, smothering, fainting, loss of sleep and appetite, pensiveness, occupation disturbed, feeling of helplessness, prayer, calling for aid, estrangement from God, etc.? How long did it continue? Was there a tendency to resist conviction? How was it shown?*

*IV. How did relief come? Was it attended by unnatural sights, sounds, or feelings? In what did the change consist? - breaking pride, public confession, seeking the approval of others, feeling God's forgiveness, sudden awakening to some great truth, etc.? How sudden was the awakening? Did the change come through, or in spite of, your own thought, deliberation and choice? What part of it was supernatural or miraculous?*

*V. Feelings and experiences after the crisis: - sense of bodily lightness, weeping, laughing, joy, sorrow, disappointment, signs of divine pleasure or displeasure, etc.? How differently did you feel towards persons, nature, ideas, God, etc.? Did you have any unfulfilled expectations or disappointments?*

*VI. What changes did you find that conversion had worked out in your life: - changes in health, habits, motives, conduct, and in your general intellectual and emotional attitude? Did you undertake any private religious acts, as Bible reading, meditation, acts of self-sacrifice, prayer, etc.?*

*VII. Were there any relapses from the first experience? Were they permanent or temporary? Any persistent doubts? What difficulties from habits, pride, ridicule, or opposition of others etc., had you, and what methods did you adopt? Do you still have struggles in your nature? Does that indicate the change was not complete? How have you, and how will you overcome them? What needed helps, if any, were wanting at the time?*

*VIII. Did you always find it easy to follow the new life, and to fit into its customs and requirements? If not, how did you succeed - by habit, pressure and encouragement of friends, a new determination, a sudden fresh awakening, etc.?*

*IX. State a few bottom truths embodying your own deepest feelings. What would you now be and do if you realized all of your own ideals of the higher life?*

*X. What texts, hymns, sermons, deaths, places and objects were connected with your deepest impressions? If your awakening came in a revival meeting, give the circumstances, and the methods used. What do you think of revivals?*

*XI. If you have passed through a series of beliefs and attitudes, mark out the stages of growth and what you feel now to be the trend of your life."*

Free Thought research on subjective evidence from within concerning faith and unbelief will need something much *simpler,* less Christian value-laden, and less shot through with leading questions.

Starbuck's presentation of all his data as percentages is illustrated in the following table about conversion and revival meetings, here condensed and restyled for clarity.

| *Table VII – A comparison of the revival and non-revival cases (both males and females) in regard to motives and forces leading to conversion* | Revival | Non-Revival |
|---|---|---|
| *1. Fear of Death or Hell* | 14 | 13 |
| *2. Other Self-Regarding Motives* | 6 | 5 |
| *3. Altruistic Motives* | 5 | 7 |
| *4. Following out a Moral Idea* | 15 | 19 |
| *5. Remorse, Conviction for Sin, etc.* | 14 | 18 |
| *6. Response to teaching* | 5 | 13 |
| *7. Example, Imitation, etc.* | 15 | 11 |
| *8. Social Pressure, Urging, etc.* | 23 | 14 |

# Free Thought, Faith, and Science

There can be no firm conclusions from such data. Free Thought research will need much better instruments for large-scale surveys and in-depth interviews, well structured for rigorous statistical analysis.

Drawing on various works, Ian Barbour[28] gave the following classification of six major types of religious experience recorded around the world: *Numinous Experience of the Holy; Mystical Experience of Unity; Transformative Experience of Reorientation; Courage in Facing Suffering or Death; Moral Experience of Obligation; and Awe in Response to Order and Creativity in the World."* He continued: *"If the task of the theologian is systematic reflection on the life and thought of the religious community, this will include critical assessment according to particular criteria. I suggest that assessment of beliefs **within a paradigm community** (original author's emphasis) can be undertaken with the same criteria...(as) for scientific theories."*

Free Thought researchers on subjective evidence from within concerning faith and unbelief are likely to encounter all of the above main types of so-called religious experience, among believers and some nonbelievers.

Barbour went on to discuss the factors that affect the interpretation of professed religious experiences - including cultural relativism, conceptual backgrounds and baselines, and expectations of outcomes. Cutting this Gordian knot will require better methods for research on subjective evidence from within, collected from large and diverse populations of believers and nonbelievers and avoiding stereotyping of Subjects by regarding their memberships of organized religions as diverse *"paradigm communities."* There are really only two paradigm communities, or more correctly, two sub-populations - the global sub-populations of

---

28 Barbour, I.G. 1997. *Religion and Science: Historical and Contemporary Issues.* HarperOne: New York.368p.p.111-112 citing Streng, F.J. 1976. *Understanding Religious Life.* Dickenson: Belmont CA., and Smart, N. 1983. *Worldviews.* Charles Scribner's Sons; New York); and 113, 109, 153, citing works by Smart, N., including p.75 and p.79 in Interpretation and Mystical Experience, in *Religious Studies* I, 1965). Barbour's book is: *"A Revised and Expanded Edition of Religion in an Age of Science: The Gifford Lectures, 1989-1991."*

believers and nonbelievers.

Referring to work by Ninian Smart, Barbour stated the following: "*Smart recommends that we use **low-level descriptive terms, with minimal doctrinal ramification** (present author's emphasis), to try to formulate a more phenomenological account on which the mystic and other persons can concur. This would be consistent with my view that the distinction between experience and interpretation, like that between data and theory in science, is never absolute; in both cases the distinction is relative and is drawn at differing points at various times for particular purposes.*"

Now we might be getting somewhere! Working with simple terminology and large, diverse populations and studying faith and unbelief, *not* religious doctrines and practices, can be the way forward for systematic gathering and analysis of subjective evidence from within concerning faith and unbelief.

Victor Stenger[29] stated the following about people who claim to have had spiritual revelations: "*They say that they have been in touch with God or some other form of higher reality. I am convinced that many are sincere in that belief (television evangelists excepted). However, without independent confirmation, the reported experiences could have been all in their heads.*"

Stenger's criteria for distinguishing the "*sincere*" from the rest would be interesting. However, the more important point is that all reported experiences of spiritual revelations and indeed of any experiences related to faith or unbelief come from *within* individuals, as outcomes of their Free Thought. All such evidence must be handled with care and appraised with scepticism.

Antonio Damasio[30] wrote as follows about the self as witness: "*At first glance, after acknowledging the self as our entry into knowledge, it may*

---
29  Stenger, V.J. 2008. *God. The Failed Hypothesis*. Prometheus Books: New York. 302p. p.171.
30  Damasio, A. 2010. *Self Comes to Mind. Constructing the Conscious Brain*. William Heinemann: London.367p.p.13.

*appear paradoxical, not to mention ungrateful, to question its reliability. And yet that is the situation. Except for the direct window that the self opens into our pains and pleasures, the information it provides must be questioned, most certainly when the information pertains to its very nature. The good news, however, is that the self has made reason and scientific observation possible, and reason and science, in turn, have been gradually correcting the misleading intuitions prompted by the unaided self.*"

Teasing out the truth in accounts of spiritual encounters is not as simple as separating the obviously fraudulent from the transparently honest. For example, Thomas Henry Huxley[31] wrote the following about George Fox, founder of the Quakers: "*No one who reads the voluminous autobiography of 'Honest George' can doubt the man's utter truthfulness… It needs no long study of Fox's writings, however, to arrive at the conviction that the distinction between objective and subjective verities had not the same place in his mind as it has in that of an ordinary mortal. When an ordinary mortal would say 'I thought so and so,' or 'I made up my mind to do so and so,' George Fox says, 'It was opened up to me,' or 'at the command of God I did so and so.'*"

The development of methods for collecting and analysing subjective evidence from within concerning faith *and* unbelief will go hand in hand with the research itself. I hope to report some progress in a future book. Meanwhile, I suggest in Appendix V the type of general questionnaire that might be developed further for research surveys among large and diverse populations of believers and nonbelievers.

### BRAIN AND CONSCIOUSNESS RESEARCH

Michael Shermer[32] took what can be called the 'brain is all' position and stated the following, in which all emphases are his: "*The brain is a*

---

31  Huxley, T.H. 1889. The value of witness to the miraculous, p.160-191.In T. H. Huxley. 1899. *Science and the Christian Tradition: Essays*. D. Appleton and Company: New York.419p.
32  Shermer, M. 2011. *The Believing Brain: From Ghosts and Gods to Politics and Conspiracies - How We Construct Beliefs and Reinforce Them as Truths*. Time Books: New York.383p.5, 62, 87, 6.

*belief engine. From sensory data flowing in from the senses the brain naturally begins to look for and find patterns, and then infuses those patterns with meaning. The first process I call* **patternicity: the tendency to find meaningful patterns in both meaningful and meaningless data.** *The second process I call* **agenticity: the tendency to infuse patterns with meaning, intention, and agency.** *We can't help it. Our brains evolved to connect the dots of our world into meaningful patterns that explain why things happen. These meaningful patterns become beliefs, and these beliefs shape our understanding of reality."*

Shermer explained the brain's patternicity as follows: "*we tend to find meaningful patterns whether they are there or not…In this sense, patternicities such as superstition and magical thinking are not so much errors in cognition as they are natural processes of a learning brain. We can no more eliminate superstitious learning than we can eliminate all learning. Although true pattern recognition helps us survive, false pattern recognition does not necessarily get us killed, and so the patternicity phenomenon endured the winnowing process of natural selection. Because we must make associations in order to survive and reproduce, natural selection favored all association-making strategies, even those that resulted in false positives. With this evolutionary perspective we can now understand that* **people believe weird things because of our evolved need to believe nonweird things** (original author's emphasis)."

Shermer acknowledged that his concept of agenticity was derived in part from Daniel Dennett's[33] work on so-called "*intentional stance.*" Shermer explained agenticity as follows: "*we often impart the patterns we find with agency and intention, and believe that these intentional agents control the world, sometimes invisibly from the top down, instead of bottom up causal laws and randomness that makes up much of our world. Souls, spirits, ghosts, gods, demons, angels, aliens, intelligent designers, government conspiracists, and all manner of invisible agents with power and intention are believed to haunt our world and control our lives.*"

---

[33] Dennett, D. 1989. *The Intentional Stance*. MIT Press: Cambridge MA.389p. First published in 1987.

# Free Thought, Faith, and Science

Shermer summarized the importance of patternicity and agenticity together, as follows: *"Combined with our propensity to find meaningful patterns in both meaningful and meaningless noise, patternicity and agenticity form the cognitive basis of shamanism, paganism, animism, polytheism, monotheism, and all modes of Old and New Age spiritualisms."*

Schermer patterned what he called *"belief-dependent realism"* for the human brain and beliefs on *"model-dependent realism,"* as employed by Stephen Hawking and Leonard Mlodinow[34] in their description of where theoretical physics stands concerning a theory of everything - the so-called M-theory (see below). Theoretical physicists often work by devising mathematical models as possible representations of reality and consider them more likely to be correct if they look right in terms of elegance and/or simplicity. I have tried to do the same for Free Thought theory, as developed so far.

According to Shermer, the brain's patternicity and agenticity explain how some of us attribute our circumstances and experiences to external spiritual powers and behave accordingly - even though nothing spiritual actually exists, either 'out there' or inside us. What some of us take to be spiritual entities and influences are the results of the material workings of our brains. Real material realm events and states in the human brain produce fantastic imaginations of a spiritual realm. Although these imaginations reflect nothing that has any real existence or power, they contribute substantially to our beliefs and behaviour because of their *imagined* nature and power.

I disagree and must add to Shermer's mix of *"sensory data flowing in"* the spiritual information that we receive in our souls. Shermer wrote that: *"the brain begins to look for and find patterns, and then infuses those patterns with meaning."* I propose that the complete human self does that, as body-mind and soul, in Free Thought.

---

34  Hawking, S. and L. Mlodinow. 2010. *The Grand Design: New Answers to the Ultimate Questions of Life*. Bantam Press: London.199p.

# UNITY

Antonio Damasio[35] explained as follows the evolution of our brains and consciousness: "*Nervous systems developed as managers of life and curators of biological value, assisted at first by unbrained dispositions but eventually by images, that is, minds...Minded behaviour became very complex in numerous nonhuman species, but it is arguable that the flexibility and creativity that hallmark human performance could not have emerged from a generic mind alone.*"

Damasio continued: "*Once autobiographical selves can operate on the basis of knowledge etched in brain circuits and in external records of stone, clay or paper, humans become capable of hitching their individual biological needs to the accumulated sapience. Thus begins a long process of inquiry, reflection, and response, expressed through human history in myths, religions, the arts, and various structures invented to govern human behaviour - constructed morality, justice systems, economics, politics, science, and technology. The ultimate consequences of consciousness come by way of memory. This is memory acquired through a filter of biological value, animated by reason.*"

Damasio also stated the following: "*Conscious deliberation is largely about decisions taken over extended periods of time, as much as days or weeks in the case of some decisions, and rarely less than minutes or seconds...Conscious deliberation is about **reflection over knowledge*** (original author's emphasis). *We apply reflection and knowledge when we decide on important matters in our lives. We use conscious deliberation to govern our loves and friendships, our education, our professional activities, our relations to others...Decisions pertaining to moral behaviour, narrowly or broadly defined, involve conscious deliberation and take place over extended time periods...such decisions are processed in an offline mental space that overwhelms external perception.*"

Damasio's description of the territory of "*conscious deliberation*" fits

---

35  Damasio, A. 2010. *Self Comes to Mind. Constructing the Conscious Brain*. William Heinemann: London.367p.p. 286-287,290, 271,

well with my description of the territory of Free Thought, including moral choices (explicitly) and creativity (implicitly). He recognized the need for "*offline mental space*" with the capacity to overwhelm external perception. According to Free Thought theory, that "*space*" is the soul, which is "*offline*" to the material realm and online to the rest of the spiritual realm.

I am with Damasio all the way in his descriptions of the evolution of nervous systems, brains and minds. The "*self process*" that he describes looks very like what I call Free Thought. Free Thought takes place in an individual human, in whom "*self* (has) *come to mind*" and the "*autobiographical self*" has been completed by God's gift of a unique soul. Again, a Free Thought research approach might yield more complete explanations of what we can observe among believers and nonbelievers.

Benjamin Libet[36] described his "*cerebral mental field (CMF) theory*" as follows: "*The CMF is not a Cartesian dualistic phenomenon; it is not separable from the brain. Rather, it is proposed to be a localizable system property produced by neuronal activities and it cannot exist without them. Again, it is not a 'ghost' in the machine. But, as a system produced by billions of nerve cell actions, it can have properties not directly predictable from these neuronal activities. It is a non-physical phenomenon, like the subjective experience it represents. The process by which the CMF arises from its contributing elements is not describable. It must simply be regarded as a new fundamental 'given' phenomenon in nature, which is different from other fundamental 'givens,' like gravity or electromagnetism.*"

Libet went back to his earlier work on delays in awareness of sensory events and to the subjective "*referrals*" that we make in our brains, in space and apparently backwardly in time, for what we sense. He concluded that: "*In a broader sense, all of the neuronal activities that lead to*

---

36  Libet, B. 2006. Reflections on the interaction of the mind and brain. *Progress in Neurobiology* 78: 322-326.

*a conscious awareness, as in thinking thoughts, may also be referrals into the CMF."*

CMF theory recognizes only the material realm but somehow incorporates the *"non-physical."* This looks a bit like a step from material realm science towards something like my model for Free Thought. Unfortunately, CMF theory seems untestable without excising parts of brains.

John Searle[37] wrote the following: *"Consciousness is caused by the behaviour of microelements of nervous systems, and is realized in the structure of those nervous systems. Consciousness is not reducible in the way that other biological properties are, because it has a first-person ontology."* This means that brain and consciousness research must address subjectivity while seeking objective evidence for explanations of the human condition. The same applies to research on Free Thought.

We are very similar to some animal species in terms of most of our anatomy, biochemistry and physiology, but we are categorically different from *all* animal species in terms of our level of consciousness and our engaging in what I call Free Thought. It is therefore reasonable to frame the hypothesis that human consciousness is founded on an evolutionary legacy from animals *and* completed by God's gifts to humans alone of souls, mind-soul interfaces, the Universal Moral Code, and Free Thought to ponder all of the higher things of human existence, including the nature and mechanisms of consciousness.

In 1910, James Rorie[38] wrote: *"At the present time there are no questions of greater importance or more pressing than those concerned with the nature of the physical and cerebro-neural changes underlying the mental state, as revealed in consciousness."* Rorie endorsed Sir John Goodsir's views

---

37 Searle, J.R. 1997. *The Mystery of Consciousness: Including Exchanges With Daniel C. Dennett and David J. Chalmers.* A New York Review Book: New York.224p.p.213-214.
38 Rorie, J. 1910. Abstract of a lecture on psychopneumatology; or on the interactions of mind, body and soul. *The British Medical Journal.* June 18, 1910:1477-1478.

that man consisted of three elements: "*a corporeal* (body), *a psychical* (mind), *and a spiritual* (soul)...*treated as the Sarx, Psyche and Pneuma (and that) it was in the last of these, and not in his corporeal element, that his personality resided, and in possession of which he differed so greatly from all lower animals.*" Rorie expressed his conviction that these views: "*will yet receive the attention that they deserve.*" One hundred years later, I echo that conviction.

Drawing from another author, probably Sir Oliver Lodge, Rorie asked: "*May we not regard ether or the medium as not merely a bridge between one portion of the universe and another, but also a bridge between one order of things and another, forming, as it were, a species of cement in virtue of which the various orders of the universe are welded together and made into one?...Nay, is it even necessary to retain the conception of a bridge? May we not at once say that when energy is carried from matter into ether it is carried from the visible into the invisible, and that when it is carried from the ether to matter it is carried from the invisible into the visible?*"

We are way past theories of the ether, but the biggest question of all remains the same for everyone - does the spiritual realm exist and if so how does it interact with the material realm, if at all? Everyone needs to know whether there really is any spiritual stuff or not. Nothing made of spiritual stuff can be detected by material means. According to Free Thought theory so far, only the mind-soul interface enables the sharing of information from the spiritual and material realms.

Those who take the 'brain is all' perspective are challenged to explain how an allegedly soul-less human self, composed only of a material realm brain and its material realm support system, evolved the concept of spiritual meaning and the need to make choices between faith and unbelief. In other words, if humans are made from nothing but material stuff, how have they evolved to become reliant on allegedly non-existent spiritual stuff?

# UNITY

Most brain and consciousness researchers assume that anything and everything to do with human spirituality is entirely material in origin and generated by the brain. Some researchers who are agnostics can allow the possibility that the spiritual realm exists. Researchers who are believers can take human spirituality as indicative of the existence of the spiritual realm and the spiritual stuff of every human self, the soul.

Despite these very different perspectives, Free Thought research and further development of Free Thought theory can be allied progressively with brain and consciousness research, for mutual benefit. Brain and consciousness research can elucidate mental states and events in basic thought and Free Thought. Free Thought research can add information on soul states and events, based on subjective evidence from within. The Free Thought of all concerned will be based on information from diverse perspectives and new findings. Its outcomes will contribute to the faith-science quest for truth.

I anticipate better explanations of human consciousness from the development and testing of a theory of Free Thought than from a 'brain is all' approach. I admit that I can show nothing yet to support this expectation - beyond my Free Thought model, which looks right to me, and my accounts of personal experiences.

### *Towards Combined 'M and S' Theory*

A final 'M' theory of everything in the material realm will probably require every 'something,' from a fundamental particle to a galaxy, to have an actual or potential connection with every other 'something,' through gravitational and other fields. The great minds working towards such a theory will probably succeed. Believers and nonbelievers who are open to science will then be able to agree that to be human is to be a short-lived material assemblage in a wholly interconnected material realm.

Nevertheless, in order to really be a theory of *everything*, any final theory will have to explain *all* aspects of the human condition, including our spiritual longings and experiences. From my own experiences and the experiences that others have shared with me, I see no prospects for any 'M' theory getting there. I believe that a final theory of everything must be a combination of M theory and spiritual realm (S) theory. This will require nonbelievers, especially scientists, to allow the *possible* existence of the spiritual realm.

The prospects might seem limited, but M and S studies are *already* taking place every day in our Free Thought. Free Thought theory, albeit still poorly developed, is already a form of M and S theory. In every episode of Free Thought, an individual is experiencing and assessing the M and the S. The challenge is to devise rigorous methods for sharing and analysing our M-S, material-spiritual, mind-soul interactions and outcomes.

A combined M and S theory could have at its core the individual mind-soul interface, posited as the only link and portal for information exchange between the material and spiritual realms. Quantum uncertainty might be all or part of the mechanism that enables us to be the linkages between the material and spiritual realms - as non-material beables, from menus of material *and* spiritual probabilities, are brought into experienced states and events in mind and soul, by the individual, God, and the spiritual force for evil.

Physics and theology are lacking in explanations that span all scales in the material and spiritual realms. We must think more about how to relate what might be happening at infinitesimally small scales in mind and soul to our scientific and theological appraisals of the big picture.

Nothing spiritual can be detected by material realm means. However, we can continue to note and try to quantify deficiencies in material realm-only explanations of the human condition. We can try investigating how far brain imaging describes and explains our coverage of the higher things of life, in what I call Free Thought. We can study what goes on in the brain during basic thought about choices over practical matters and Free Thought about ethics related to those choices.

Expanding the faith-science quest for truth will facilitate development of M and S theory, including the further development of Free Thought theory. Joining this quest will not require anyone to give up faith, or unbelief, or scientific objectivity. Believers and nonbelievers will be learning together as the quest progresses, putting their sovereign-to-self Free Thought to the task and going wherever the truth leads. This is the faith-science quest for truth. The more that get on board the merrier. There is everything to gain and nothing to lose.

# APPENDIX I: DEFINITIONS

Most of the definitions here are original (O). The remainder are taken, verbatim or modified, from: (A), the New Oxford Dictionary of English;[1] (B), the Merriam-Webster Collegiate Dictionary;[2] and (C), Flew (1984).[3]

## KEY TERMS

*basic thought: the mental process by which an individual makes choices about the practicalities of everyday life - such as what to eat, what to wear and how to keep safe and healthy* (O)

Through basic thought we recognize and pursue our practical needs. We choose actions for their fulfilment, based solely on personal expediency and our assessments of the material risks and benefits. Any consideration of ethics and morality takes us beyond basic thought, into Free Thought.

*free: every individual's internal state of being, which ensures that personal decisions are made independently and can be held in private* (O)

The terms 'free will' and 'Free Thought' are defined fully below. Where the word 'free' is used to mean absence of cost, this is always made clear by the context.

---

[1] A: Pearsall, J., and P. Hanks, Editors. 1998. *The New Oxford Dictionary of English: Thumb Index Edition.* Oxford University Press; Oxford. 2152p.

[2] B: *Merriam-Webster's Collegiate Dictionary.* Tenth Edition. Merriam-Webster Inc., Springfield, MA. 1559p.

[3] D: Flew, A. 1984. *A Dictionary of Philosophy.* Revised Second Edition. St. Martin's Press: New York. 380p.

***Free Thought: the integrated and iterative process that combines soul processing of and responses to information from the spiritual realm with mental reasoning about and responses to information from the material realm*** (O)

Free Thought leads to independent, individual decisions about the higher things in life - faith, moral choices, justice, and the development and use of creativity for making contributions in and/or enjoying the arts, mathematics and science.

Dictionary definitions of the terms "freethinker", "free thought" and "freethought" differ markedly from this definition. For example, "*freethinker*" means "*a person who rejects accepted opinions, especially those concerning religious belief*" (A) or "*one that forms opinions on the basis of reason independently of authority; esp., one who doubts or denies religious dogma*" (B); and free thought is defined as "*unorthodox attitudes or beliefs*" (B).

***free will: every individual's ever-present potential to choose her/his goals privately and to decide freely to what extent those goals are divulged to others*** (O)

Free will enables every individual to choose her/his personal goals independently and to hold them in private. Our free-willed goals might be attained only partially and sometimes not at all. This does not negate the free will that enabled their choice internally. Free Thought is the *process* by which all available information is considered and decisions are made on how to pursue our free-willed goals.

For comparison, the dictionary definitions of free will include: "*the power of acting without the constraint of necessity or fate; the ability to act at one's own discretion*" (A); and "*Voluntary choice or decision; freedom of humans to make choices that are not determined by prior causes or by divine intervention*" (B).

*faith: personal belief and trust in God* (O)

Faith is a private and personal choice. Those who have chosen faith are called believers. The opposite of faith is unbelief. Agnostics and atheists are all nonbelievers. They neither believe in nor trust in God. In this book, faith is never used to mean an organized religion.

*believing (in something): having complete certainty that it is true* (O) and *trusting (in something): having complete confidence in its reliability* (O)

Believing in something and trusting in something are both personal and private choices that were weighed thoroughly and are sustained actively. Belief and trust are never default options. Believing and trusting leave no place for doubt.

*information: everything that can be sensed materially or transmitted and received spiritually* (O)

*material realm: the totality of all existing forms of matter and energy, together with all of their interrelationships, processes and states* (O)

The material realm comprises the entire contents and workings of the physical universe or multiverse, including human body-minds, brains and all of their physical components and processes. The human mind, memory, and all mental maps and images of anything, real and imaginary, are part of the material realm.

*material revelations: information about the material realm, acquired through the physical senses and processed by the mind in basic thought and/or Free Thought* (O)

The composition and workings of the material realm are sensed by the body-mind and become understood by mental processing, thereby guiding actions and improving prospects for survival and material well-being. All of the above takes place in the material realm.

***reasoning: the mental process by which an individual uses information to acquire understanding and/or to form judgments*** (O)

***spiritual realm: the totality of all existing spiritual entities, comprising God, the spiritual force for evil and all human souls, together with all of their interrelationships, processes, and states*** (O)

Nothing in or from the spiritual realm is detectable anywhere in the material realm, except in our minds through our mind-soul interfaces.

***spiritual revelations: information from the spiritual realm, sent by God or by the spiritual force for evil to an individual human soul and processed, through Free Thought*** (O)

Spiritual revelations are the forms of information sent by God and the spiritual force for evil to human souls. Spiritual revelations can be received during prayer or meditation or as guidance sensed at any conscious moment in life.

***science: any intellectual activity, observation or experiment concerning the composition and workings of the material realm and the nature of spirituality in the human condition*** (O)

By convention, the scope of science is restricted to explorations of the material realm; i.e., nature in the broad sense. The scope of science is extended here to include exploration of the spiritual dimensions of the human condition. The systematic gathering and analysis of subjective evidence from within concerning faith and unbelief qualifies as scientific research. Moreover, the creativity in Free Thought that leads to advances in mathematics and science includes spiritual realm revelations and responses.

***soul: the unique spiritual and immortal essence of self that is given by God to each individual human, integrated during life in the material realm with that individual's unique body-mind and taken after death in the material realm to an eternal life in the spiritual realm*** (O)

For comparison, source (C) (Flew 1984; p. 331) contains the following: "*Plato, presumably following Socrates, both identified the soul with the person who reasons, decides and acts, and assumed that this person or soul is not the familiar creature of flesh and blood but rather the incorporeal occupant and director of, even the prisoner in, that corporeal being.*"

## OTHER TERMS

***agnostic: one who believes that nothing is known or can ever be known of the existence or nature of God*** (modified from A)

***altruism: the practice of seemingly selfless concern for the wellbeing of others*** (modified from A)

This definition covers all altruism in animals and humans. Any altruistic action by an animal, while seemingly selfless, always carries prospects of some present and or future pay-offs for the altruist and/or for one or more of her/his fellows having partisan interests in common; for example, the furtherance of genes and the holding of territory.

The same applies in much of human altruism, but humans also practice a special form of altruism, termed *genuine* altruism. Genuine altruism is acting in order to benefit one or more other persons, at one's own disadvantage and without any expectation of or prospect for any present and/or future pay-offs in the material realm to oneself and/or to any of one's fellows having partisan interests in common.

***atheist: one who denies the existence of a God or gods*** (modified from B)

***biblical (or other sacred text) literalist: one who insists that every word in the Bible (or another sacred text) is literally true and is God's unchangeable word for all times*** (O)

Mainstream dictionaries do not define biblical literalism *per se*, though some list it as one of the main uses of the word literalism, which means the acceptance of a given set of words verbatim; i.e., exactly as they are written or otherwise communicated. Many Christians and churches hold that biblical literalism is an absolute requirement for showing obedience to God and specify a particular version of the Bible as authentic and completely inerrant in its every word.

***Christian: one who has faith in Christ as being consubstantial with God, and who strives to be Christ's disciple*** (O)

***church: a distinctive group or organization of believers*** (modified from B)

This term is also used to refer to the total population of all Christians and as a descriptor for particular Christian denominations.

***conscience: an inner feeling or voice acting as a guide to the rightness or wrongness of one's behaviour*** (A)

***consciousness: awareness by an individual of self and surroundings*** (O)

This definition resembles that of Antonio Damasio (2010)[4] who called

---

4   Damasio, A. 2010. *Self Comes to Mind. Constructing the Conscious Brain*. William Heinemann: London. 367p.

human consciousness: *"a state of mind in which there is knowledge of one's own existence and of the existence of surroundings."*

For comparison, source (C) (Flew 1984, p.72) refers to John Locke's definition of consciousness as: *"self-knowledge acquired by virtue of the mind's capacity to reflect upon itself in introspective acts analogous with perception."*

**creed: *a formal statement that expresses individual or shared beliefs*** (O)

**deist: *one who believes in the existence of a God that created the universe, but does not intervene in it, in general and/or on behalf of humans*** (O)

Source (C) (Flew, 1984: p. 87) defined deism as *"the doctrine that belief in God can commend itself to the human mind by its own inherent reasonableness, without either being supported by appeals to alleged divine interventions or imposed by religious institutions."*

**determinism: *see predestination***

**eschatology: *the part of theology concerned with death, judgement, and the final destiny of the soul and of humankind*** (A)

**ethics: *see Universal Moral Code***

**evil: *the quality of actors and acts that pursue lies and/or harm others*** (O)

**God: *the original Spiritual Being Who created everything that has existed and Who sustains His creation through love, inviting every human soul into an eternal relationship with Him, through faith*** (O)

God is Love. Referring to God as a male does not imply male gender in the biological sense. Dictionary definitions emphasize that God is the Creator and Ruler of the material and spiritual realms and the Source of all moral authority.

*good: the quality of actors and acts that pursue truth and/or help others by following the Universal Moral Code* (O)

*heaven: the spiritual realm abode of God and of all the human souls whom He has chosen to share eternal life with Him after their earthly lives have ended* (modified from A)

*hell: the spiritual realm abode of the spiritual force for evil and of all the human souls who have been barred, by God, from sharing eternal life with Him* (O)

*lies: false statements, made in full knowledge of their falsity or repeated in ignorance of their falsity* (O)

*miracle: an event or outcome that is not explicable by natural or scientific laws and is therefore considered to be the work of a divine agency* (modified from A)

Source (C) (Flew, 1984; p. 234) considers that a miracle: "*is most commonly taken to mean an act that manifests divine power through the suspension or alteration of the normal working of the laws of nature.*" He points out that this makes the laws of nature an essential prerequisite for the concept of a miracle.

*organized religion: an institution or organization, the purpose of which is to further specific beliefs and associated practices in recognition of the spiritual realm, spiritual beings and their alleged influences on the material realm, and especially on humans* (O)

This definition includes the institutions and organizations of all varieties of Buddhism, Christianity, Hinduism, Islam, Judaism, and of a host of other belief systems and practices, such as those under various New Age banners.

Most authors do not even attempt to define religion because the word encompasses such wide variations in human activities. Source (C) (Flew, 1984: p.304) contains the following: *"a satisfactory answer to the question 'What is religion?' would be more like an encyclopedia than a one-sentence definition."* The Apostle James' gloriously simple definition of religion is as follows: *"Pure religion and undefiled before God and the Father is this, To visit the fatherless and widows in their affliction, and to keep himself unspotted from the world."*[5]

**postmodernism: a diffuse set of perspectives, aspirations and types of behaviour, varying with identity and culture and denying the possibility of related standards that are universal and lasting** (O)

Defining postmodernism is difficult, because its very essence is flexibility. It began as a concept and style in architecture and the arts, but became a very broad church of philosophy, mainly within the social sciences. Dictionary definitions of postmodernism include the following: *"A late 20$^{th}$-century style and concept in the arts, architecture and criticism, which represents a departure from modernism and has at its heart a general distrust of grand theories as well as a problematical relationship with 'art'"* (A); *"of, related to, or being any of several movements (as in art, architecture, or literature) that are reactions against the philosophy and practices of modern movements and are typically marked by the revival of traditional elements and techniques."* (B)

**prayer: the means by which one's soul communicates with God, in order to offer thanks and praise, to confess faults and seek forgiveness, and to present requests on behalf of oneself and others** (O)

---

5   James 1:27

*predestination:* **divine foreordaining of all that will happen** (modified from A)

The definition published in source (A) restricts the context of predestination to Christian theology and eschatology: "*especially with regard to the salvation of some and not others.*"

*reformation: as pertaining to an organized religion, belief system or church, a major change from an established order to new perspectives and practices* (O)

*revolution: as pertaining to science, a major change, resulting from the establishment and testing of new theory, with radically new perspectives and methods* (O)

*religion: see organized religion*

*self: the totality of an individual human being, comprising a unique spiritual soul, integrated during life on Earth with a unique material body-mind* (O)

*spirit: any inhabitant of the spiritual realm, including God, the spiritual force for evil and all human souls* (O)

*theist: one who believes in the existence of a God that created the universe and intervenes in it: either in general, on behalf of individual humans, or both* (O)

*truth: that which is in accordance with facts and reality* (modified from A)

This simple definition incorporates the common sense and correspondence theory of truth, as discussed in source (C) (Flew, 1984; p.355).

***Universal Moral Code: the standard for moral conduct, given by God for recognition and compliance by all humans*** (O)

The same core elements of moral human conduct - preserving life, telling the truth, and dealing mercifully and justly - are recognized all across humanity. Written expressions of the Universal Moral Code include, *inter alia*, the Ten Commandments[6] and Christ's much abbreviated two commandments.[7] Kent Keith listed 10 principles for avoidance of doing harm and 10 for doing good.[8]

---

6     Exodus 20:2-17
7     Luke 10:27-37
8     www.universalmoralcode.com

# APPENDIX II: CREDO

*"I know whom I have believed"* [1]

### Church Creeds and Personal Creeds

Most Christian churches have official creeds for public affirmations of faith. The Apostles' Creed originated from a shorter text used by the early Christian Church in Rome. The Nicene Creed was adopted at the Council of Nicaea (325 CE) and modified during the First Council of Constantinople (381 CE). The Athanasian Creed, which is used mainly on Trinity Sunday, is attributed to Saint Athanasius (ca. 293-373 CE) but its history is uncertain.

The Catechism of the Catholic Church[2] quotes St. Augustine as follows: *"May your Creed be for you a mirror. Look at yourself in it, to see if you believe everything you say you believe. And rejoice in your faith each day."* Here, the words *"your Creed"* mean the Apostles' Creed, with no omissions or variations.

P. E. Hodgson[3] wrote: *"The Church founded by Christ is a community of believers accepting His teaching in its entirety...Catholic theology is a tightly integrated dynamic unity, a seamless web of essential doctrines. It is not a supermarket, a pick-and-mix stall, wherein we can choose whatever appeals to us and reject what does not."*

Any church can require affirmation of an official creed for membership,

---

[1] II Timothy 1: 12
[2] *Catechism of the Catholic Church.1994.Definitive Version: Based on the Latin Editio Typica.* Episcopal Commission on Catechesis and Catholic Education: Catholic Bishops Conference of the Philippines. Word & Life Publications: Makati City, Philippines. 828p.paragraph1064.
[3] Hodgson, P.E. 1991. Science and the Christian World View, p. 67-77. In N. Mott (ed.) *Can Scientists Believe? Some Examples of the Attitude of Scientists to Religion.* James and James: London.p.74.

but *everyone* has sovereignty of self and Free Thought, with which to craft a personal creed. God made us that way. Only the self, God and the spiritual force for evil know the details of a personal creed.

I am a Christian, which means that I am a disciple of Christ. I am a member of the Union Church of Manila (UCM), Philippines, which uses the Apostles' Creed and the Nicene Creed in its liturgy. I affirm their contents, as worded in the 1662 *"Book of Common Prayer and Administration of the Sacraments and other Rites and Ceremonies of the Church According to the Use of the Church of England."*

John Polkinghorne[4] called the Nicene Creed: *"the loom on which I seek to weave my tapestry."* I follow here his fine example and go through the Apostles' Creed and the Nicene Creed section by section to explain why I can affirm them in their entirety. I explain also why I cannot affirm the Athanasian Creed. I provide brief synopses on my beliefs and opinions about miscellaneous topics: angels; demons; divine interventions; evolution; heaven and hell; miracles; prayer; predestination; souls; the spiritual force for evil; and the Universal Moral Code.

## The Apostles' Creed and the Nicene Creed

***I believe in God the Father Almighty, Maker of heaven and earth:*** (Apostles' Creed)

***I believe in one God the Father Almighty, Maker of heaven and earth, And of all things visible and invisible:*** (Nicene Creed)

I believe in one God and I perceive Him as a Father figure, though I know that He exists in Spirit form and cannot be limited by human concepts of gender, age and familial relationships. I believe that God was the First, Pre-existing Cause and that He created everything in the

---

4  Polkinghorne, J.C. 1996. *The Faith of a Physicist: Reflections of a Bottom-Up Thinker.* Fortress Press: Minneapolis MN.211p.p.6.

spiritual realm, which is invisible to us, and everything in the material realm, including all of its laws and processes. I prefer the Nicene Creed's expression of that truth, because it makes explicit a belief in things that are "*invisible.*" Before God, there was no such thing as before. I believe that existence, as opposed to non-existence, began with God. Everything that ever was, is, and will be began with Him.

I take the word "*earth*" to mean the entire material realm; i.e., the entire observable universe and possibly the multiverse. The fact that God does not fix everything that is wrong does not make Him less than Almighty. He has given us free will and He has made the material realm so that it works through free process. Every human life is lived in the context of all that God-given freedom.

The material realm appears to be finely tuned for the existence of diverse chemical elements and carbon-based life. The Anthropic Principle troubles some nonbelievers because the existence of a finely tuned material realm really does look like the result of a purposeful decision. Some believers hail the Anthropic Principle as clear proof of God's existence. From the observed physics, chemistry and biology of the material realm, I conclude only that it is a very special system to be explored and that we have evolved to become its self-aware and potentially God-aware explorers. I hold that our existence is strong evidence for God's existence, but I recognize that our personal experiences of Him provide the only proof of His existence.

I find no conflict between the theory that the observable universe arose from a singularity by means of a Big Bang and the main message in Genesis, which tells the truth of creation through beautiful poetry. St. John's Gospel expresses the same truth: "*In the beginning was the Word, and the Word was with God, and the Word was God. The same was in the beginning with God. All things were made by him; and without him was not anything made that was made.*"

***And in Jesus Christ his only Son, our Lord, Who was conceived by the Holy Ghost, Born of the Virgin Mary,*** (Apostles' Creed)

***And in one Lord Jesus Christ, the only begotten Son of God, begotten of his Father before all worlds, God of God, Light of Light, Very God of very God, Begotten not made, Being of one substance with the Father, By whom all things were made: Who for us men and for our salvation came down from heaven, And was incarnate by the Holy Ghost of the Virgin Mary, And was made man,*** (Nicene Creed)

I believe all of the above. Both creeds describe Jesus Christ as the Son of the One Father God Who, as the Holy Ghost, implanted Jesus, as a portion of God, in the womb of His earthly mother Mary, while she was still a virgin bride-to-be. The Nicene Creed proclaims at greater length that Jesus always was, is, and will be entirely God, but that He took human form to be visible on Earth, fully God and fully human. God came to Earth as a son. A daughter would not have got very far preaching to a male-dominated society.

I affirm the words in the Apostles' Creed: "*conceived by the Holy Ghost.*" God was God all the way along. Prior to the Incarnation, God had always been in existence as the *Unity* of the Father, Son and Holy Spirit. God then became, in part, a man on Earth: "*the Word was made flesh and dwelt among us.*"[5]

Note that if the Word (i.e., God) had to be "*made flesh*" in order to dwell among us, then God in heaven is *not* made of flesh, original or resurrected. Apart from taking material form as Christ during the Incarnation, God was always entirely Spirit and will remain so, unless He chooses otherwise.

Christ was always consubstantial with God. The Apostle John refers to Jesus as having been present before creation: "*The same was in the*

---

5   John 1:14

*beginning with God."*[6] As a portion of God and therefore a Spirit, Christ became a male human embryo, foetus, child and adult, entirely in dualistic material and spiritual form.

Any mechanism for that divine intervention must have been miraculous. God chose a virgin birth as the mechanism for His Incarnation, which was the most important divine intervention in the material realm throughout its history. Christ became God Incarnate, God sacrificed and God resurrected.

Some scholars hold that the prophesy *"a virgin shall conceive"*[7] has been mistranslated from the original Hebrew text, in which the words *"a young (nubile) woman shall conceive"* could be the truer meaning and need not signify virginity. I believe that Mary was a virgin when she became the human means for the Incarnation of God as Jesus, but I do not regard this belief as being absolutely essential, as an article of faith. The Incarnation is not explicable in terms of human biology and behaviour.

We are integrated assemblages of material and spiritual stuff (body-minds and souls), but both the spiritual *and* material stuff of God Incarnate must have originated entirely from spiritual stuff, whereas our material stuff is derived only from other material stuff. The word *"begotten"* means always coming from. It emphasizes the Incarnate Christ's exclusively *spiritual* origin as a portion of God. It indicates God's capability and choice to enter His material realm creation in human form, on that one very special occasion. .

**Suffered under Pontius Pilate, Was crucified, dead and buried, He descended into hell;** (Apostles' Creed)

**And was crucified also for us under Pontius Pilate, He suffered and was buried** (Nicene Creed)

---

6 John 1:2
7 Isaiah 7:14

I believe that Jesus Christ lived on Earth, as a man and as God Incarnate, in Roman-occupied Palestine. The Gospels have been criticized for lacking sufficient corroboration from other historical sources and for their lack of consistency. I disagree. In particular, I consider the variations and spurious points of detail in the Gospels as typical of eyewitness accounts and indicative of their authenticity. Fabrications could not have produced the convincing big picture that shines from the three synoptic Gospels (Matthew, Mark and Luke).

Both creeds proclaim that Jesus died and was buried. The Apostles' Creed says that He descended into hell. The alternative wording "*descended to the dead*" is gaining wide acceptance, based on scholarly opinions that the oldest available texts are more accurately translated as "*descended to Hades*" (meaning the realm of the dead) and seeking to avoid use of the archaic word "*hell.*"

I prefer the older wording. I take the word "*hell*" to mean the state of a soul's complete and eternal separation from God. The "*dead*" are obviously all separated from the living, but their souls are *not* all separated from God. I believe that the body-mind of Jesus was killed and that His Spirit (His portion of the very Soul of God) became separated from God, albeit temporarily. In other words, He ceased for a while to be a portion of God.

I miss in both creeds an explicit statement that it was through Christ's suffering, death and resurrection that the price for all human failings was paid in full and that we can all be fully forgiven, reconciled with God and given eternal life in His presence. Why did the compilers of these creeds not proclaim this more explicitly, as the greatest ongoing free gift imaginable? It is not mentioned at all in the Apostles' Creed. The Nicene Creed states only that Christ came to Earth "*for us men and for our salvation*" and was crucified "*for us.*"

*The third day he rose again from the dead, He ascended into heaven, And sitteth on the right hand of God the Father Almighty; from thence he shall come to judge the quick and the dead.* (Apostles' Creed)

*And the third day he rose again according to the scriptures, and ascended into heaven, And sitteth on the right hand of the Father. And he shall come again with glory to judge both the quick and the dead: Whose kingdom shall have no end* (Nicene Creed)

I believe that Christ was resurrected and ascended into heaven in spirit form and reigns there as God - having no need of flesh or bones or anything resembling our feeble material realm body parts. No one on Earth can know whether Christ sits down in heaven or whether He and His Father choose to follow right over left superiority. I take those human-conceived images as indicators of Christ's co-equal authority and consubstantial identity with God.

I dissent from the Nicene Creed's implication that God's final judgement will be all at one time and for all of us together - those still alive and those long dead, when He will *"come again with glory."* I believe that God is always with me and with everyone else. I believe that God is *never* separated from His glory. I believe that He comes in His glory repeatedly and often to judge all who die *when* they die and that He will come to Earth again and judge all who are living then.

I take the Apostles' Creed and Nicene Creed's descriptions of God's judgement on human souls as including the possibility of continuous process and not as signifying only a one-time event. I do not believe that the judgment of all who die - irrespective of whether they had sufficient opportunities during life to decide for or against faith - will be delayed until the second coming of Christ.

In common with many of those officiating at funerals and probably with most of the bereaved who are believers, I believe that the soul of a deceased person enters the presence of God immediately after death. Once a blessed soul has left Earth, why would God require that spiritual soul to wait for being fitted out to enter heaven?

I believe that God and the souls of all who have died are beyond time, as time is conceived and measured by humans. I believe that after death our souls enter at once an everlasting and spiritual realm 'now.' I see no basis for postulating any delay, in an earthly and temporal sense, between the death of one's body-mind and the passage of one's soul to its eternal destination in the spiritual realm.

My beliefs here differ from mainstream Christian eschatology, which holds that the souls of all those who died in faith will be resurrected simultaneously - as a New Heaven somehow comes down and becomes as one with a New Earth. I do not believe that, as I explain further below.

I believe that we will all be judged on how we have used God's gifts of life, Free Thought, the Universal Moral Code and His forgiveness of our confessed transgressions. I believe that God accepts not only believers, but also some nonbelievers who, under His influences, have lived their lives as honest seekers and tellers of truth.

**I believe in the Holy Ghost; The Holy Catholick Church; The Communion of Saints; The Forgiveness of sins; The Resurrection of the body, and the life everlasting. Amen.** (Apostles' Creed)

**And I believe in the Holy Ghost, The Lord and giver of life, Who proceedeth from the Father and the Son, Who with the Father and the Son together is worshipped and glorified, Who spake by the Prophets, And I believe in one Catholick and Apostolick Church. I acknowledge one Baptism for the remission of sins, And I look for the Resurrection of the dead, And the life of the world to come. Amen.** (Nicene Creed)

I believe that the Holy Ghost or Holy Spirit is the form of God Who provides guidance, inspiration and comfort to all believers during their earthly lives. I believe that I am part of the church of Christ that is described as "*Catholick*" (worldwide) and "*Apostolick*" (comprised of disciples).

I know that I belong to a communion - a caring and sharing extended family of all believers. I believe that we are all saints in the making. I believe that all our sins can be forgiven through confession and repentance. I believe that salvation leads to an everlasting spiritual life with God.

The resurrection of Christ is the most important doctrine of Christianity. I believe that Christ's adopted material body form on Earth was ultimately interchangeable with His eternal spiritual form. He entered the material realm in spiritual form to become God Incarnate. He was later resurrected and went back to heaven in spiritual form. When I say the words "*The Resurrection of the body,*" I mean the resurrection of Christ's restored body, as Spirit.

Handel set the following words to beautiful music in "*Messiah:*" "*And though worms destroy this body, yet in my flesh shall I see God.*"[8] I recall a UCM Chancel Choir Retreat during which the Leader assured my fellow choristers and I that we would all get completely new flesh and blood bodies in heaven. Some mused that they could finally be slimmer in heaven. I believe that in the "*life everlasting*" or "*life of the world to come*" my fellow believers and I will not need anything like our earthly bodies. We will be spirits, in company with God.

The Nicene Creed affirms the deity of the Holy Ghost and the unity of the Holy Ghost with God as Father and Son and the role of the Holy Ghost in having connected with and spoken through humans, here identified as prophets. I believe all of the above and I am sure that the

---

8    Job 19:25-26

Holy Spirit can use anyone as a spokesperson for the Gospel, as well as for justice, morality and speaking the truth about the complementarity of faith and science.

I believe that all Christians should be baptized as infants or as adults, but I do not believe that God refuses salvation on principle to all the non-baptized. Some will have lacked opportunities for baptism, through no fault of their own. The Catechism of the Catholic Church includes hope of salvation for the following unbaptized persons: a. those "*who suffer death for the sake of the faith* (para 1258)"; b. "*catechumens...who die before baptism but desire it, repent their sins and show charity*" (paragraph 1259); c. those who, though "*ignorant of the Gospel of Christ and of his Church*" seek the truth and do the will of God in accordance with their understanding of it, because it "*may be supposed that such persons would have **desired Baptism explicitly*** (original authors' emphasis) *if they had known of its necessity*" (paragraph 1260); and children who died before they could be baptized (paragraph 1261).

Neither creed mentions explicitly a triune God or Holy Trinity, but both include the three ever-present and consubstantial forms - Father, Son and Holy Ghost. I believe that God is one Spirit, but that He has connected and continues to connect with humans in His different forms which, as best we can conceive and try to describe them, are a loving Father, a Son Who lived for a brief period on Earth as God Incarnate and a Holy Spirit Who guides us.

## THE ATHANASIAN CREED

The Athanasian Creed, also called the "*Quincunque vult*" or "*Whosoever will*," begins by proclaiming the "*Catholick* (i.e., worldwide) *Faith*" and sends out the following dire warning: "*Which Faith except every one do keep whole and undefiled: without doubt he shall perish everlastingly.*" It attempts to describe the interrelationships of the Holy Trinity and the dual God-man identity of the Incarnate Christ and ends with another warning: "*They that have done good shall go into life everlasting: and they that have done evil into everlasting fire. This is the Catholick Faith: which except a man believe faithfully, he cannot be saved."*

I cannot affirm the Athanasian Creed. Its absolutism leaves no scope for Free Thought or for any further revelations from God to change anything that its human compilers might have got even slightly wrong in less than infallible moments. I cannot accept any creed that consigns explicitly to everlasting fire all who fail to believe in its every word. It is God's prerogative alone to decide the fate of every human soul.

I believe that we are given everlasting life with God by His grace and through our faith[9] - as a free gift and not by having "*done good*" rather than having "*done evil."* How then will God deal with the souls of my fine friends and colleagues who are agnostics, atheists, Buddhists, Hindus, Jews and Muslims when they die? I don't know, but I believe that He will deal kindly with anyone who has lived life as an honest seeker and teller of truth and according to the following principles: *"to do justly. and to love mercy, and to walk humbly with thy God."*[10]

## Miscellaneous Topics

### *Angels*

It is possible that beings called angels exist as spirits in the spiritual realm called heaven. I see no basis for equipping them with haloes and wings. Reports of fully-fledged angels coming to Earth have just about ceased, but some children are still taught that that everyone has a God-assigned and named guardian angel, waiting to snatch her/him out of the path of a wayward bus or whatever. Life does not work like that. Free process prevails in the material realm.

---

9   Ephesians 2: 8
10  Micah 6:8

Most Christians who believe in angels do not regard those in heaven as having gender; though some allegedly highly placed angels are named males; for example, Gabriel and Michael. I have no idea what angels in heaven might look like, how many there are, and whether they have gender, rank and names. In truth, nobody knows anything about any of that.

On Easter Sunday 2009, UCM Pastor Emeritus Alex Aronis preached on the following text: "*the angel of the Lord descended from heaven, and came and rolled back the stone from the door, and sat upon it. His countenance was like lightning, and his raiment white as snow.*"[11]

Alex noted that the introduction of an angel at the start of the Resurrection story did not make it easy for us to believe it, because few of us have had definitive experiences of angels. He recalled seeing a man, who had appeared from nowhere, save a young boy caught in a riptide. The hero vanished so quickly that no one had time to thank him. Alex continued: "*Was it an angel; we don't know; angels are debatable; the scriptures cannot compel us to believe in angels; we've got to weigh the evidence and see… We must use our minds.*"

I am with Alex in believing that a resurrection angel appeared on that miraculous day. However, outside the Incarnation of Christ and the immediate preceding period I do not believe that any angels in spirit form have come to Earth. Nor do I believe that any angels are leaving heaven today and assuming human forms to help other humans. I believe that any extant angels are all in heaven. However, I accept that some humans can act angelically to help other humans.

On a winter's day, en route to York University from the Midlands, the engine of my Velocette Venom 500 motorbike stopped suddenly and defied all my attempts to restart it. I was stranded in the cold on the Doncaster by-pass. Along came a BSA 650 Twin, ridden by Nigel Box.

---

[11] Matthew 28:2-3

*"Are you stuck mate?" "Yes, good spark and plenty of petrol, but she won't go." "OK, I'll tow you to our place."* At Nigel's home, we found that my bike's timing pinion had shattered. The best of Yorkshire hospitality flowed and I was invited to stay for the night.

Nigel took me on to York the following morning. Two weeks later he lent me his trials bike for a trip to a breaker's yard to seek a secondhand pinion. When that failed, Nigel's mate Moonbeam lent me a pinion from his collection of old Velocette engines, to be used in mine until I could order a new one. Nigel and Moonbeam spent hours setting up the timing on my bike. I was back on the road. They were delighted. Did they help me in expectation of any future help from anyone? 'No,' they just felt good about helping me. Were they angels? 'No,' they were humans helping a fellow human. Thanks again Nigel and Moonbeam.

## DEMONS

Carl Sagan[12] and others, including Richard Dawkins, Sam Harris and Christopher Hitchens, have ridiculed rightly the belief that the world is populated with demons seeking to ensnare us. That belief has caused immense human suffering and injustice, especially the persecution of women as alleged witches.

Some nonbelievers hold that choosing faith means believing in all of the demons and supernatural happenings that are reported in sacred texts. This is emphatically not true. I am a believer and a Christian. I believe in the spiritual force for evil, but I do not believe in demons. I do not believe in the demonology of the Bible or in any other accounts of alleged demons in the material realm and/or inside humans. I do not believe that Christ Incarnate or any of His contemporary and subsequent disciples healed people by expelling demons.

---

12  Sagan, C. 1997. *The Demon-Haunted World: Science as a Candle in the Dark.* Ballantine Books: New York. 457p.

I believe that Christ healed some mentally disordered persons miraculously. I believe that His first disciples contributed to the healing of others by various medical means, including counseling, and that His disciples up to the present day have continued to do the same. My beliefs do not detract from the power and beauty of sacred texts that tell of healings from demon-possession. Those stories are all about faith and victories over pain and suffering.

Most people today, irrespective of whether they are atheists, agnostics or believers, do not describe mentally and/or physically ill or handicapped persons as demon-possessed. In conventional medicine, doctors are not diagnosing demon possession and prescribing exorcism. Nevertheless, some organized religions still encourage belief in demon-possession and in the need for rituals to ward off the devil and its alleged army of demons.

Matt Baglio[13] described how Catholicism retains rites for exorcism and trains new exorcists. David Kiely and Christina McKenna[14] reported on ten cases of alleged demon possession in Ireland. Having witnessed two exorcisms, M. Scott Peck[15] concluded: *"I had been converted from a belief that the devil did not exist to a belief - a certainty - that the devil does exist and probably demons (under the control of the devil) as well."* I agree that the spiritual force for evil exists, but I do not believe in the existence of any demons.

The so-called demon-possessed are mentally disordered. Their natural states, experiences and surroundings have caused them to accumulate anti-Christ and other mental images and memories from which they can act out convincingly their alleged demon-possession. The exorcism team and the prescribed rituals interact strongly with those images and memories. The result is a battle of wills and outcomes of Free Thought

---

13  Baglio, M. 2009. *The Rite: The Making of a Modern Exorcist*. Doubleday: New York.288p.
14  Kiely, D. and C. McKenna. 2007. *The Dark Sacrament*. HarperOne: New York.398p.
15  Scott Peck, M. 2005.*Glimpses of the Devil: A Psychiatrist's Personal Accounts of Possession, Exorcism, and Redemption*. Free Press: New York.259p.p. 238-239.

- the will and Free Thought outcomes of the patient, who seeks to remain alienated from goodness, are pitted against the wills and Free Thought outcomes of those who seek her/his recovery.

## *EVOLUTION*

Fellow Christians sometimes ask me: 'Do you believe in evolution?' Some ask rather accusingly, implying: 'How can you possibly believe in evolution and be a Christian and/or a member of this church?' or 'Do you believe that my ancestor was an amoeba and/or a monkey?' Others are genuinely interested in how best to explore the truth about evolution.

I try to respond always by asking first what the questioner means by evolution. This can elicit some bizarre answers. Some questioners have done no serious reading about evolutionary theory beyond the demonization of Darwin and Darwinism in Christian literature. Some are adamant that evolution is only a theory and not a fact - the mistake that Kenneth Miller[16] tackled so well.

Richard Dawkins[17] wrote that when asked why he thinks it so important for everyone to 'believe' in evolution he replies as follows: "*I don't like the word belief. I prefer to ask people to look at the evidence.*" I like the word belief, but I can see what Dawkins was getting at. For anyone who has experienced God, He is as real as anything else experienced in life, but experiences of God are entirely subjective. The evidence for gravity, electromagnetism and indeed evolution is objective and can be observed collectively. The word belief seems inappropriate to all of the above and much more in scientific theory. It is clear that these phenomena and mechanisms truly *exist*.

---

16   Miller, K.R. *Only a Theory; Evolution and the Battle for America's Soul.* Viking: New York. 244p.
17   Dawkins, R. 2009. *The Greatest Show On Earth: The Evidence For Evolution.* Free Press: New York. 470p.p.201.

I try to respond as follows to questions about my belief in God: 'Please consider the subjective evidence from me and from other believers for faith and experiencing God.' I try to respond as follows to questioners asking about my belief in evolution: 'Please explore some of the evidence and current theory and then we can talk further.'

I believe that God created our observable universe by engineering a Big Bang, in which a maximally dense singularity exploded and led to the evolution of the observable universe and fixed laws of nature. The conditions in our universe were appropriate for the evolution of carbon-based life. All living things share the same basic chemistry. Our material body-minds are made of recycling *"stardust,"* as stated in Joni Mitchell's[18] lovely song. Our souls are unique and everlasting assemblages of spiritual stuff.

Planet Earth had all the required ingredients, energy sources and environmental conditions for the evolution of proteins, RNA, viruses, living cells, microorganisms, fungi, plants and animals. God made all of the required mechanisms. The theory of evolution is a testament to the creativity of God. Genesis has great beauty and valuable insights, but it cannot be the literally true story of how biodiversity and humans came into being. Truths disclosed through Genesis and through science all come from God.

### Heaven and Hell

I believe that the human soul passes into the spiritual realm immediately after death and is then outside space and time forever. I believe that there are only two possible destinations for the soul after death - an eternal state of communion with God called heaven and an eternal state of separation from God called hell. I do not believe in any gradations of heaven or hell or in any prequalification places of atonement, such as a purgatory.

---

[18] Joni Mitchell. 1970. *Woodstock*.

I believe that the souls of believers go directly to heaven to be with God - by His grace. I believe that the souls of some nonbelievers - whom God deems worthy of sharing an eternal life with Him, considering what was possible or not in their lives - also go directly to heaven. I believe that all of the rest go into a state of eternal separation from God. That state is hell.

Revelation describes heaven as having an abundance of precious earthly substances - such as crystal, gold and jasper - and its occupants as having crowns and white robes, with not only diverse angels for company but also some fantastic vertebrates, including six-winged beasts. I do not believe that heaven or hell can be described reliably in terms of material realm entities.

Will an entrant to heaven be able to enjoy there all of the good things of life on Earth; for example, art, food, music, pets, sport and wine? I believe that heaven will provide anything and everything that God wishes to be there. It will all be better than anything experienced on Earth.

The Apostle John wrote: "*Love not the world, neither the things that are in the world.*"[19] I love some of the things of Earth very much - Manx glens, Welsh mountains, Shropshire hills, coral reefs, music, fishing, Labradors and good guitars. God's message is not to love them too much and to expect much better things in the life to come.

I do not hanker after an Earth-like heaven. I expect something literally out of this world and incomparably better than anything in it. I believe that souls in heaven or in hell retain their individual identities. You will be you and I will be me. I believe that in heaven we will all be equal as the spiritually resurrected children of God - without gender, nationality and religious affiliations.

---

19   I John 2:15

Many Christians believe in a present heaven and a future and final heaven that is going to be a New Earth, which will be the old Earth - allegedly made perfect again and kept perfect by the present heaven coming down to it. John the Divine[20] wrote the following: *"And I saw a new heaven and a new earth; for the first heaven and the first earth were passed away; and there was no more sea. And I John saw the holy city, new Jerusalem, coming down from God out of heaven prepared as a bride adorned for her husband."*

No sea? Why not? In any case, the presently observable universe seems destined to come to an unobservable end as it flies apart. A so-called New Earth would have to be located in the spiritual realm. What would it have to be like in order to qualify as perfect? The Bible foresees the end of all food chains that involve predators and prey. Wolves and lions are supposed to become herbivores.[21] That sounds like the old Earth, but with all killing restricted to the killing of plants. What can anyone make of such fantastic predictions? In common with most imagery in Revelation, they are inevitably Earth-centric.

I prefer not to speculate on possibilities in heaven that are so obviously unknowable from an Earth-bound perspective, but my best guesses are as follows: there is only one heaven; it existed before time began and will last forever; the souls of the faithful departed are already there, with God; Christ *will* come to Earth again and the existence of the entire material realm will then come to an end, with its final inhabitants judged by God; all other human souls will have faced already the judgment of God, immediately after their deaths.

N. T. Wright[22] wrote the following: "(The Apostle) *Paul says that God will give us new bodies; there may well be some bodily continuity, as with*

---

20 Revelation 21:1-2
21 For example, Isaiah 65:25
22 Wright, N.T.2008. Surprised by Hope: Rethinking Heaven, the Resurrection, and the Mission of the Church. HarperOne: New York.332p.p.163. Wright cited John Polkinghorne's entire discussion, especially p.107-112, in *The God of Hope and the End of the World*, published by SPCK: London and by Yale University Press: New Haven CT.154p.

*Jesus himself, but God is well capable of recreating people even if (as with the martyrs of Lyons) their ashes are scattered into a fast-flowing river."* I believe that after death we will exist only as spirits. We can forget bodies, ashes and all mortal remains. I wish to be cremated and to have my ashes scattered in the Santon River, Isle of Man, where it meets the sea. I do not expect God to recombine my swirling ashes into a new body-mind. I expect to be a happy *disembodied* spirit.

I love very much the image of heaven in a prayer by John Donne:[23] "*Bring us, O Lord God, at our last awakening into the house and gate of heaven to enter into that gate and dwell in that house, where there shall be no darkness nor dazzling, but* **one equal light**; *no noise nor silence, but* **one equal music**; *no fears or hopes, but* **one equal possession**; *no ends nor beginnings, but* **one equal eternity**; *in the habitations of thy glory and dominion, world without end* (present author's emphases)." Donne's vision of heaven is worth more than tons of biblical literalist dogma and Earth-centric guesses.

I do not know what happens in hell and neither does anyone else. I neither believe in nor see the need for visions of hell that include devils with tails, horns and pitchforks, everlasting fire, worms that never die[24] and the like. I believe that hell is an everlasting spiritual state of separation from God. I believe that hell is the part of the spiritual realm where God's will does not operate and where there is no hope of any change for the better.

## MIRACLES

I believe that the Big Bang was God's initial mega-miracle in the material realm and that His gifts to humans of souls, higher consciousness and the Universal Moral Code and His personal relationships with

---

[23] www.spck.org.uk/classic-prayers/john-donne/ This form of Donne's prayer is attributed to Eric Milner-White in 1963. Milner-White was Dean of York Minster from 1941 to 1963.
[24] Mark 9: 44, 46 and 48

humans are all ongoing miracles. I believe that the entirety of Christ's life on Earth was a miracle and that the miraculous could and did happen during that time.

I do not believe in any of the alleged Old Testament miracles that would have required suspension of God's established laws of nature. I believe that all of those stories were tailored for audiences that sought supernatural signs and/or colourful metaphors. I believe that all such alleged miracles have natural, material realm explanations. Here are a few examples.

I. *Lot's wife is said to have become an instant pillar of salt.*[25] *Beside saline waters her corpse could have become salt-encrusted with time, but instantaneous conversion to a pillar of salt is not possible.*

II. *The so-called 'Red Sea,' which must have been a reedy coastal lagoon in or close to the Nile delta, was allegedly parted and then reconnected miraculously, to let the Israelites through and drown the pursuing Egyptians.*[26] *An abnormally large ebb and flow of water could have been caused by a tsunami after a volcanic explosion - at Santorini or elsewhere in the eastern Mediterranean.*

III. *Three men are alleged to have survived incarceration in a furnace that could melt metals.*[27] *This must be an exaggerated account of their survival from lesser ordeals.*

IV. *Jonah is alleged to have survived for three days and nights in the digestive system of a big fish.*[28] *This is biologically impossible. It must be a metaphor for temporary separation from God, through exile and/or being cast adrift at sea.*

---

25  Genesis 19:26
26  Exodus 14:21-30
27  Daniel 3:6-28
28  Jonah 1:17 to 2:10

Neither the Apostles' Creed nor the Nicene Creed specifies that Christians *must* believe in anything allegedly miraculous that happened between Christ's birth on Earth and His death on the cross. I believe the biblical accounts of some of Christ's alleged miracles during His Incarnation; for example, His instant healing of mental disorders and His ability to walk on water, for which He could have assumed a spiritual form. I am not sure that He accomplished instant healings over physical disorders, such as leprosy. Nor am I sure that He turned water into wine and/or materialized, from almost nothing, enough bread and fish to feed thousands of people. Pre-existing wine, bread and fish might have arrived or been uncovered.

Many Christians believe that Christ turned water into wine and created piles of instant bread and fish. I believe that He encouraged some of the haves who were present to help the have-nots. With either interpretation, His love and power remedied situations of human need. I believe that the Incarnate Christ made miracles happen in the souls of His disciples and other contemporaries and that the Risen Christ has continued to do the same ever since.

Many Christians believe that God can reconstruct miraculously human organs and tissues that medical science would assess as damaged beyond repair, and that He can eliminate cancers and kill pathogens. I believe that God's spiritual revelations alleviate the suffering of injured and diseased persons and have beneficial healing effects. I do not believe that God brings about direct material change in organs, tissues, pathogens and parasites.

I believe that we must interpret the Bible in the light of God's truthful disclosures through science. I do not believe that God has ever suspended or changed the laws of nature that He created, except during the period of the Incarnation. I agree with the following statement by

Nevill Mott:[29]"*the more we can shed any belief in miracles, while retaining our concept of the supernatural...the more we can learn from Him* (present author's emphasis)."

## PRAYER

I believe that prayer is the way by which an individual contacts God and receives many of His spiritual revelations. Learning how to pray is a lifelong voyage of discovery, guided by God. Among the many books that teach us how to pray, I recommend the one by UCM Pastor Emeritus Alex Aronis.[30]

Despite all the distractions of life, I try to keep reminding myself to be: "*Praying always with all prayer and supplication in the Spirit.*"[31] Christ reportedly told us "*always to pray, and not to faint,*"[32] meaning that we should never give up on prayer. We have been told to pray directly to God and we have been given the best prayer ever written - the Lord's Prayer - to do exactly that.[33] God invites us to ask for outcomes that will accomplish *His* will.

Each of us has been dealt a unique genetic hand that will be expressed in a succession of unique personal environments. Some enjoy good looks, good health and long lives. Others are burdened with genetic defects, abnormalities and chronic sicknesses, which can lead to early death. Some are born into poverty and circumstances where they face high risks of violence and dread diseases. I believe that God knows all human circumstances and needs and the extents to which He can and will help, subject to His Self-imposed

---

29  Mott, N.1991. Christianity Without Miracles, p.17-22. In N. Mott (ed.) *Can Scientists Believe? Some Examples of the Attitude of Scientists to Religion.* James and James: London.p.17.
30  Aronis, A. B. 2002.*Developing Intimacy With God: An Eight-Week Prayer Guide Based On Ignatius' Spiritual Exercises.* Union Church of Manila Philippines Foundation Incorporated: Makati City, Philippines.190p.
31  Ephesians 6:18
32  Luke 18:1
33  Matthew 6:6-13

limitations of free process in the material realm and His gifts of our free will and Free Thought.

I believe that we should be very careful about what we pray for in terms of our personal needs and outcomes that would please or benefit us. No believer should pray for an easy life or for victories over rivals in armed conflicts, business, sports and other competitive fields. I believe that God welcomes intercessory prayers for others, knowing that the results will be limited by the free will of those prayed for and others around them, as well as by free process in the material realm. I believe that God hears all prayers.

I also believe that we should recognize the true nature of the material realm that God has given us and should not pray for it to be divinely manipulated for our benefits. John Polkinghorne[34] put that beautifully as follows: "*God allows the whole universe to be itself. Each created entity is allowed to behave in accordance with its nature, including the due irregularities which may be part of that nature...God is not the puppet master of either men or matter.*"

Where is God during the horror shows of life? Where was He during the Holocaust and other atrocities? He was there all the time. He is everywhere, always - watching, weeping, and keeping records for the vengeance that was, is, and always will be His alone. I believe that there is no contradiction between God loving us and remaining with us but being unable to protect us from the evil acts of humans and from free process in nature.

Those who believe that God plans and executes all good and bad happenings in order to accomplish His mysterious and divine purposes are not only implicating Him in all free-willed human evil but are also making Him the Perpetrator of all natural disasters. I do not believe

---

[34] Polkinghorne, J. 1996. *The Faith of a Physicist: Reflections of a Bottom-Up Thinker.* Fortress Press: Minneapolis MN.211p.p.83.

that a God operating like that could be called a God of Love. Through prayer we can all be blessed in our souls and increase our *peace of mind*, regardless of our material realm circumstances. The Apostle Paul wrote: "*For which cause we faint not; but though our outward man perish, yet the inward man is renewed day by day.*"[35]

### PREDESTINATION

Has our Creator God predestined every human soul for an eternal life in heaven or an eternal life in hell? Surely the only sensible and God-honouring answer must be 'No.' How could a loving God be seeking to bring His willful children into loving relationships with Him, if He had already predetermined those who would accept His invitation and those who would refuse?

I do not believe that any individual's life on Earth and/or the future of her/his soul is completely preprogrammed with no scope for change. I believe that God gave Free Thought to every one of us, together with the obligation to use it and to make a personal decision for or against faith.

The fates of our souls are in *our* hands as well as God's hands. Predestination would take away from us and from God all prospects for choosing to make changes. Predestination would also contradict the following reported words of Christ: [36] "*Come unto me **all** ye that labour and are heavy laden* (present author's emphasis)." I believe that God deals kindly with the souls of *all* honest seekers of truth. I do not believe that He plans all of our good deeds and bad deeds and then redeems only His pre-selected ones, thereby somehow glorifying Himself.

---

35  II Corinthians 4:16
36  Matthew 11: 28

## SOULS

I believe that every human has a unique, immortal soul - the spiritual core of self. The human soul is a spiritual entity that is integrated during life on Earth with a material body-mind. After death, the soul lives forever in the spiritual realm - with God in heaven or separated from Him.

I do not know at what stage of life a human being receives her/his soul and neither does anyone else. According to a recent article,[37] Islamic doctrine has the soul entering the foetus: *"between 40 and 120 days after conception."* This doctrine has allowed research on human embryonic stem cells to proceed in Iran without censure from religious authorities.

I do not believe that a spermatozoon or an ovum has a soul or half a soul. I do not believe that a zygote has a soul. I believe that God gives us our souls when we begin Free Thought and embark on our spiritual lives.

I applaud the work and lasting legacy of Robert Edwards and Patrick Steptoe, who pioneered *In Vitro* Fertilization (IVF) and Assisted Reproductive Technology (ART) and made it possible for infertile couples to know the joy of raising their own biological children. Kevin Coward[38] reported that ART had facilitated more than 12,000 births a year in the UK (1.4 % of all births) and had enabled about 8 million births worldwide. He also mentioned the religious opposition to IVF/ART on the grounds that it involves destruction of human embryos.

In IVF/ART, Pre-implantation Genetic Diagnosis (PGD) and Comparative Genomic Hybridization (CGH) ensure that a healthy embryo is chosen. Those procedures are necessary and moral. Having faced the stresses of infertility, no couple seeking IVF/ART should have

---

37  Anon. 2013. Islam and Science: The Road to Renewal. *The Economist* 406, Number 8820: 50-52.
38  Coward K. 2011. The grandfather of IVF. *The Biologist* 58 (4): 41-45.

to face raising an abnormal child, if that risk can be lessened or avoided.

I believe that artificial abortion is justified in the following circumstances: rape; incest; proven serious risks to the life of the mother and/or the child; and the proven presence of very severe genetic and/or physical defects in the foetus that would reduce her/his future quality of life to an untenably low level. I do not believe that any abortion, artificial or natural, kills a soul.

I believe that suicide is almost always wrong because it is against the Universal Moral Code. One should not kill others and one should not kill oneself. Nevertheless, I believe that God can reveal to someone who is terminally ill and undergoing terrible suffering His divine endorsement for hastening the release that would come with death. I do not believe that anyone can deny God the freedom to endorse that course of events.

### THE SPIRITUAL FORCE FOR EVIL

I believe that the spiritual force for evil is the source of all lies. I believe that it opposes personal communion with God and influences individuals so that their actions will achieve evil outcomes at multiple levels - from one-on-one interpersonal relationships to families, groups, communities, institutions, organizations and nations. I do not believe that the spiritual force for evil engineers any accidents, diseases, mental and physical abnormalities and natural disasters.

Evil in the self is manifested in revelations from the spiritual force for evil to the soul and acquiescent responses. Evil behaviour follows outcomes of individual Free Thought in which revelations from the spiritual force for evil and the individual's acquiescent responses became dominant. Human evil includes betrayal, corruption, cruelty, deceit, discrimination, hatred and above all *injustice*.

God sustains my faith. The spiritual force for evil opposes my faith. As the Apostle Paul put it: *"we wrestle not against flesh and blood, but against principalities, against powers, against the rulers of the darkness of this world, against spiritual wickedness in high places."*[39] Those *"high places"* include our souls, which are part of the spiritual realm.

The spiritual force for evil urges us to say hurtful words and to respond in kind when others say hurtful words to us. It tells the believer that her/his faith is just imagination, because there really is no God. It influences individuals to join forces and to establish institutions and organizations with evil agendas. It encourages people to endorse and participate in evil acts, such as torture, killing and sustaining poverty.

According to Christian doctrine, the spiritual force for evil was present with Adam and Eve in the Garden of Eden and persuaded them to disobey God - thereby, allegedly, causing all other humans to be stained with original sin. I do not believe in original sin. I believe that the earliest humans were choice makers and were endowed with Free Thought, just like us. I do not believe that there were ever any humans or human environments that could be called 'perfect' in any meaningful sense. A Garden of Eden that included the spiritual force for evil was clearly less than perfect.

### THE UNIVERSAL MORAL CODE

I believe that God implants His Universal Moral Code in every soul in order to instruct everyone to follow His example of acting with justice, love and mercy.

If God did not exist, there would be neither need nor basis for any absolute and lasting standards of morality. All behaviour would be seen merely as more or less *expedient* for the time being, not as right or wrong. The human conscience would become a mechanism for

---

[39] Ephesians 6:12

assessing expediency. I find all of this untenable.

Transgressing the Universal Moral Code has consequences. For example, those who perpetrate injustice will be held accountable in this world and/or the next. I believe that God records and ultimately corrects all injustice.

I believe that violence and responding to violence with more violence are always wrong. I believe that no war has ever been just. Christ's reported command "*love your enemies*"[40] is not compatible with any claim to have a God-endorsed right to dispossess, maim, torture and kill anyone. Wars always lead to unjustifiable horrors. I believe that we must always strive to "*overcome evil with good.*"[41] Violence is never good.

---

40    Matthew 5:44
41    Romans 12:21

# APPENDIX III: ABOUT ME

*"But by the grace of God I am what I am..."*[1]

## Origins

My father worked in the furniture trade around High Wycombe, Buckinghamshire. My birth certificate lists his occupation as *"theatre outfitter and woodworker."* His father kept the Prince of Wales pub in the village of Little Kingshill. My mother grew up on a farm at Bratton, Shropshire. Her father suffered financial setbacks in farming and greyhound racing and the farm was lost. The family scattered, looking for work. My mother went into service as a housemaid and nanny.

My parents met and married during World War II. My father was with RAF Bomber Command and my mother was working in a munitions factory. I arrived as their firstborn in 1944, and was christened in the tiny village church of St. John the Baptist, Little Missenden. My parents were not regular churchgoers. Christening a new baby was just the done thing.

After the war, there were good prospects for work in the West Midlands. We settled in Whitmore Reans, Wolverhampton. My father sold insurance - mostly in the poorer areas of town. His clients loved him and the prospect of security that he brought into their hard lives, with great diplomacy and care. He loved the freedom of being on two wheels and progressed from a 98cc Villiers autocycle to a 650cc BSA. He suffered a severe leg injury when a car ran into him. My mother got on her bicycle and did the rounds of his clients, in all weathers.

---

[1] I Corinthians 15:10

We lived in the political constituency of Wolverhampton South-West, where the Rt. Hon. Enoch Powell was Member of Parliament. Enoch was my father's hero, next to Winston Churchill. My father and I fell out over Enoch's predictions that rivers of blood would flow through British streets because of the growing ghettoes of black and brown immigrants.

Most of the adults around me supported Enoch. They spouted a 'monologue' rather like the one described by Doris Lessing[2] among white southern Africans. The Wolverhampton monologue was more xenophobic blah than rabid racism and was not usually aimed at known individuals. One day I found my father working on the accounts of a small factory owned by a Mr. Pal. *"Mr. Pal, Dad? Could he be from Pakistan?"* *"Yes, but he's a really genuine, hardworking sort. I want to help him."* We both saw the funny side of that.

I cannot recall any meaningful talk with my parents about faith. We did not say grace before meals or any other prayers together. I did not see either of my parents open a Bible at home. My father hardly ever went to church, except to attend a wedding or a funeral. My mother attended church occasionally, especially at the major festivals. I am sure that she was a believer. My father used to joke that he would be playing a harp or shoveling coal one day. I overheard him praying when he became terminally ill.

## School and Church

I went to St. Jude's primary school. We said prayers at morning assembly and acted in typically British nativity plays. I was quite irreverent and was caned for laughing when someone had farted during a solemn silence.

The low Anglican parish church of St. Jude's was nearby. I joined

---

2    Lessing, D.1993. *African Laughter: Four Visits to Zimbabwe.* Harper Collins: New York.442p.

sequentially its cub pack, scout troop and rather feeble Youth Club. I sang for seven years with its all-male SATB choir - Decani and Cantores, splendidly attired in cassocks and surplices. The organist and choirmaster, Aubrey Burgess, was superb. He led me to a life-long love of sacred music, especially hymns.

The vicar of St. Jude's, Reverend W. Simmonds, and his wife Lily were gems. They had been missionaries in China and were friends of Gladys Aylward, who came to address our congregation. Billy Graham's 1954-55 campaigns stimulated some growth of evangelism in the UK and one of his London rallies was broadcast live into St. Jude's. Lily Simmonds was the first to walk forward at the altar call. I did not follow. I was not a believer then. I just loved the music.

I took confirmation classes, but felt like backing out from the ceremony until I opened the Bible at random one evening and saw at once the text: "*Who shall also confirm you unto the end that ye may be blameless in the day of our Lord Jesus Christ.*"[3] I prayed there and then that Christ would come into my heart. I went on to be confirmed by the Bishop of Lichfield.

I was lucky enough to gain entrance to Wolverhampton Grammar School - founded in 1512 and steeped in tradition. I became entranced with science, particularly biology. I had superb teachers and made life-long friends. Like my schoolmates, I gloried in the Wolves, who were then one of the best football teams in Europe, and in the advent of rock and roll. When the film "*Rock Around the Clock*" came to town, the ticket seller quizzed all its young viewers about which schools they attended. I answered "*The Grammar School.*" She was horrified and said: "*You surprise me.*" The film's final credit was "*The Living End.*" The establishment got that one very wrong.

My other love was motorbikes. My best mates were all bikers. We

---

3   I Corinthians 1:8

had what we called the 'madness factor' - pushing the limits around bends and laughing at harsh weather. We thought of ourselves as fast but courteous knights of the road. The idea was to look cool around town and then speed off into Shropshire and Wales. A.E Housman's[4] memories of the same beautiful hills and highways speak volumes to me.

## University

Neither my parents nor I really knew what a university was. Nobody in our family had ever been to one. My father was lukewarm about the idea of anyone studying for three or more years after having already spent what to him seemed too long at school. In his world, you either did factory work - which was bad and led nowhere - or you started low in something like a bank and rose to security and respectability, which was good.

Most of my schoolmates applied to six universities. I was not even sure that I wanted to go. I applied to just two - Imperial College and Liverpool. My exam results brought offers of places at both. My headmaster wanted me to stay at school for another year and try for an open exhibition to Oxford or Cambridge. I said no thanks and chose London over Liverpool.

At Imperial College, I was taught in the Huxley laboratory. The madness factor took over and I wanted to get involved in everything. I took up SCUBA diving and was soon elected College Diving Officer in those very exciting early days of scientific diving. I joined the Imperial College Choir, which was allied with more singers and the orchestra from the adjacent Royal College of Music. I began to play folk and blues guitar and made the most of the amazing early '60s music scene

---

4   Housman, A.E. 1987. *A Shropshire Lad*. Poem XL. Palmers Press: Ludlow, U.K.99p.p.59. First published in 1896.

in London.

Money was very tight with a motorbike habit to support. I did factory or farm work during every vacation and some night work as a radio operator for a London doctors' emergency service. My faith was swamped and forgotten. I remember discussing with classmates the discovery that one of our number wanted to leave and become a priest. Someone shouted *"Jesus Christ!"* I wanted to be funny and said: *"No, just a priest."* I had moved so far away from faith.

I graduated with a First Class Honours B.Sc. and was awarded the 1965 Edward Forbes Memorial Medal and Prize. The graduation ceremony was in the Royal Albert Hall, adjacent to Imperial College. None of my family came and I looked and felt a bit of a loner. I sat with the choir and sang syrupy stuff by Borodin and *"Gadeamus Igitur."* I remembered our better gigs - especially a Verdi Requiem, which had shaken me up but failed to rekindle faith. Then I made a mistake.

I was inexperienced with girls. When I met one with whom life seemed fine, I thought that it would be a great idea to get married and she agreed. I had been awarded an Agricultural Research Council Fellowship to work for a Ph.D. at the new University of York. I was anticipating a life in which marital bliss would now be a given and I could get on with science, music and the rest.

Many advised me against the marriage, but I was unstoppable. My father was aghast at the prospect of me spending three more years *"at school"* while married. We were married in her parents' church, *"before God and this congregation."* I was immature and arrogant - neither ready nor fit to be married. We fell apart after a year and divorced. The whole debacle was my fault. I felt bad because I had caused much hurt. I had no faith for consolation.

My years in York from 1965 to 1968 were overfull as usual. I studied the liver fluke, which is a parasite of cows and sheep. I worked on its larval stages in amphibious snails and tried to culture them *in vitro* as a prelude to possible work on vaccines. My supervisor, Alan Wilson, deserves a medal for his patience in keeping me on track. He is still at York University, working in the same field.

In York, I got further into playing guitar and the folk music scene. I rowed for York University's first and only eight in its new Boat Club. I gave no thought to faith. I remember some of my fellow researchers joking that a believer in our laboratory would not know whether an instrument had been reading true or whether God had tweaked its dials to give a good result as a personal reward. I laughed and agreed with them.

When my Ph.D. fellowship came to an end, I could see that jobs in biology were very scarce. I was one of eight unsuccessful applicants interviewed for an Assistant Lectureship in Invertebrate Biology at the University of Wales in Cardiff. Over sixty candidates had applied. I survived on casual work for a while - selling ice cream from a van and welding portable toilets.

Without holding much hope for success, I applied for another Assistant Lectureship - at the University of Liverpool; to teach and to do research on fish reproduction, in support of marine aquaculture. I went for interview prepared to argue that my knowledge of laboratory methods and animal physiology would be transferable from parasites of livestock and snails to farmable fish and their gonads. My interviewers were much more interested in whether I could live and work happily on the Isle of Man. I convinced them and got the job.

## The Isle of Man

The Isle of Man became home. It remains the place that I love most and call my home. The wonders of nature are around me there every day.

On that beautiful Island, I found all of the things that I had sought on escapes to Shropshire and Wales and much more. I could walk through bracken and heather and fish for trout and salmon.

My girlfriend from York joined me on the Island. Three years later we married, in Douglas Registry Office. The Registrar said: *"remember, this is for life."* I really wanted it to be that way. We had two beautiful daughters and took them to be christened in Laxey church, though we were not self-professed Christians or churchgoers. We renovated an old miner's cottage above a beautiful glen and shared it with Beth, the world's best Labrador, and various eccentric cats.

At the Port Erin Marine Biological Station, I became one of a happy band of nine academic staff. A bust of the famous Manx marine biologist, Edward Forbes - in the same likeness as on my Imperial College medal - looked down on us in our well-stocked library, which was named after him. As far as I could tell, none of my colleagues was a believer. I remember a staff room discussion about Sir Alister Hardy. I joined the consensus was that he had gone gaga when he decided to explore spiritual matters. Some of my present day friends and colleagues have come to the same conclusion about me.

The madness factor took over again. I became Founder Secretary of the Manx Wildlife Trust. I played regularly with a succession of folk and rock groups and began to write and record original songs, one of which found its way into the Manx National Songbook.[5] I was a partner in a small trout farm, which took up most of my weekends. It was too much of everything. My wife told me (not unreasonably) that I seemed to love only fish, music and conserving the Isle of Man. I loved her and our children as well, but had failed to show it. Our marriage went cold and something had to change.

---

5   Pullin, R.S.V. 1968. *Song for the Terns*, p.116-118. In Charles Guard. Compiler. 1980. *The Manx National Song Book. Volume Two.* Shearwater Press: Douglas, Isle of Man.126p.

In 1977, the university allowed me to consult on assessment of oil spill damage to some marine fish farms in Hong Kong. I saw that Asian aquaculture had far greater opportunities than those in Europe and began to look for an overseas position. I hoped that a shared adventure might revive my marriage. In 1978, the newly established International Center for Living Aquatic Resources Management (ICLARM) advertised for a scientist experienced in aquaculture and fish reproduction, to be based at its headquarters in the Philippines. I was interviewed in Manila and offered the job.

My father and several colleagues advised me not to leave my university position and beloved Isle of Man, but I was in a make or break frame of mind. I agreed to join ICLARM in April 1979. Before that could happen, the madness factor failed me. Under the stresses of completing commitments on the Isle of Man and a less than harmonious home life, I collapsed with nervous exhaustion and was rushed into hospital on oxygen.

After a week, my tests showed nothing physically wrong. I asked a friend to bring me any letters from work. I tried to draft a reply to a letter from ICLARM and found that I was unable to write a single word. I thought that my working days were over. My doctor scolded me and asked what I did to relax. He commanded me to do nothing for at least two weeks except walk the dog, go fishing and play the guitar. If I did only that, I would be fine. If I tried to do any work, I would be in deep trouble. I obeyed and recovered. I never prayed for help - even when I was in a collapsed state and had assumed that I was dying.

## The Philippines

On my journey out to the Philippines, ICLARM required me to visit some potential research partners in the USA, including universities in New York, Texas, Alabama and Hawaii. I was having a ball and could see a wonderful professional life ahead, but there was also some fear. If I failed to make it in the exciting but very demanding world of international research and development, there would be very few opportunities back home.

In Honolulu, on a bright Sunday morning, I went on a random walk and came to a gleaming white church, surrounded by grass and white frangipani trees. A service was in progress and I went inside. It was the first time that I had been in a church for about 15 years except for christenings, funerals and weddings. Despite my long-lapsed faith, I said a little prayer that all would go well.

About 6 months later, I accompanied my family to Manila. My wife and I went with other newly arrived expatriates to a social gathering at the Union Church of Manila (UCM). I felt uncomfortable there. What would my fellow scientists say if they knew? 'You went to a church! Why?' Our daughters enrolled at the British School, Manila, which had its classrooms in the UCM buildings. So, I was often dodging in and out of UCM.

The world's best job in international aquaculture research had dropped right into my lap. I was tasked to design and lead international strategic research for the development of tropical aquaculture. ICLARM's Director General, Ziad Shehadeh, wanted to prioritize coastal aquaculture of mullets and inland aquaculture of carps and tilapias. We agreed to focus on the tilapias, which proved a very wise choice.

I became a workaholic again and relaxed mainly around the Nomads Sports Club and the Manila bars. Not surprisingly, marital relations failed to improve and my wife and I agreed eventually that she and the girls should return to the Isle of Man, while I continued to earn a good international salary in Manila. The prospects of re-employment for me back home were near zero anyway. So began an awful separation from my daughters and a very difficult life for them and their mother. We divorced in 1983.

I felt numb. I had failed at marriage again. Our daughters received one of the best educations in the world and we were all financially secure, but the damage was great. I knew the airline timetables by heart, so that I could route international work travel through the UK and see my girls. There were some awful goodbyes and wrenching times. We all came through it somehow - scarred, but with abiding love and respect.

Over the next 17 years with ICLARM, I worked with partner organizations in over 30 countries, mostly on genetic improvement of farmed fish and improvement of low cost farming systems, especially the integration of inland aquaculture with agriculture and other sectors. I had exceptionally gifted colleagues who became great friends, including *inter alia*: Rainer Froese, Clive Lightfoot, Jay Maclean, John Munro and Daniel Pauly.

I took refuge from work in marathon running and diving and became Founder Chairman of the Philippines Sub-Aqua Club. That life was very hectic and very good, but completely empty of anything spiritual. I could feel the emptiness in my soul. I was singing the Peggy Lee song: "*Is that all there is?*" I started attending UCM services, where I enjoyed especially the preaching of Pastor Alex Aronis and the music of the UCM Chancel Choir.

I found peace gradually, as God brought me back to Him. I came to realize that He had brought everything into being. Exhausted by the hectic and spiritually empty life during my long separation from God, I searched for Him. My soul battlefield was undergoing territorial shifts. During a Good Friday service, I recommitted my life to Christ. The clincher was His last word from the cross: "*it is finished*" or "*consummatum est*" or "*es ist vollbracht,*" or "*it is paid in full.*"[6]

---

6   John 19:30

What else could anyone need to hear and why did it take me so long to get the point? I kept repeating those words. They meant that my old life was over and had been changed for eternity. There is no greater gift. It is free, just for the believing and the asking. 'Ah, how very convenient,' some will scoff. 'All your sins have been forgiven, so now you can go out and sin some more. Great for you!'

'No,' it is not like that at all. My life was changed through faith by God's grace, which is totally unmerited redemption. The grace of God is sufficient for the forgiveness of all repented sins. All believers sin, but faith transforms the believer's attitudes and intentions.

Up to then, my personal life could have been best described as messy and selfish. Then through mutual friends I was introduced to Tess - well known to Filipinos all around the world as the actress Tessie Tomas. Tess and I obviously had the madness factor in common. We married in 1994. Having two madness factors under one roof worked out well.

After admiring UCM's music ministry for years, I joined its Chancel Choir, under the inspiring and loving musical leaderships of Carminda Regala and Eudenice Palaruan. I was also not used to UCM's American system of formal church membership, but I became a member in 2002 in order to be eligible for service on the Music and Worship Committee.

Work at ICLARM went very well until some senior colleagues and I brought to the attention of the Chairman of the Board of Trustees what we saw as problems with the current management. A bitter conflict ensued, culminating in the resignations of the Director General and the Chairman of the Board and severe reprimands for us. The Board wanted to show the world that *"discipline had been restored,"* though in truth it had never been lost. They fired the senior scientist in the programme that I was directing. My colleagues and I went to court and won his reinstatement.

The Board continued their efforts to get rid of him and all the whistle-blowers and their supporters - despite our pledges of loyalty and the ongoing high impacts of our work. Then the Board decided to move ICLARM's headquarters to Malaysia, which meant terminating most of its wonderful Filipino staff. They also took over a huge and problematic aquaculture research facility in Egypt, on the false premise that it was a good location for centering ICLARM's work to benefit sub-Saharan Africa. This was driven entirely by high politics, not science. It was a huge betrayal of longstanding research partners further south.

Board and Management then applied a 10-year maximum tenure rule - perfect for ditching former rebels, regardless of their merits and the impacts of their work. The programmes that I had led were beginning to benefit poor fish farmers and fish consumers. Those benefits became substantial as the years went by.[7] However, according to the 10-year rule, my time to go was long overdue. From 1996 to 1999, I was given a final part-time contract to build a new programme on Biodiversity and Genetic Resources, which went very well.

When I left ICLARM I wanted to remain in Manila. Leaving UCM would have been especially difficult. Tess had become the First Lady of Philippine TV and a movie star. Her fans were calling me 'Ka Roger.' 'Ka' is the abbreviation for the Tagalog 'kasama,' which means comrade. I liked that name and have used it ever since as my stage name for music projects.

I had also found great comrades in music - Sammy Asuncion and Tom Colvin in Manila, Michael Allen in Italy and Don Johnson in Nashville. With their help, I wrote and recorded new songs and performed regularly in clubs, including the famous Hobbit House in Manila. I signed a 3-year recording contract with Tiger Records, Philippines; through which my first CD was distributed by BMG. During a visit

---

[7] For example: ADB. 2005. *An Impact Evaluation of the Development of Genetically Improved Farmed Tilapia and Their Dissemination in Selected Countries*. Asian Development Bank, Manila, Philippines.124p.

to Memphis, I felt a strong call to write and record Gospel material. I made that change and released three more CDs.[8]

At UCM, I was elected chair of the Music and Worship Committee and served for four years on the Council of Elders. Life was great, but I no longer had regular employment in science and music was not going to pay the bills. Then I saw how God always provides. I resumed consulting on oil spill damage assessments. More work opportunities came in from the Asian Development Bank and the Food and Agriculture Organization of the United Nations. As a freelance consultant, I always got just enough work.

## Lessons Learned and Looking Ahead

I recall terrible sadness when some of my personal relationships broke down and especially when I became separated from my children for long periods. There was also great sadness over what came to pass for the two organizations in which I had worked so happily. ICLARM was replaced by a very different organization called the WorldFish Center. The University of Liverpool closed the Port Erin Marine Biological Station, bringing to an end 114 years of research and teaching in a superb location at the centre of the Irish Sea ecosystem. In both cases, aquatic science and its beneficiaries lost something very precious.

St. Jude's Church Wolverhampton is still well attended every Sunday. Wolverhampton Grammar School has just passed its 500th Anniversary. Imperial College and the University of York have increased their national and international standings. The Manx Wildlife Trust has gone from strength to strength and owns and manages numerous nature reserves. UCM is 'United, Centered and Maturing in Christ,' as it moves

---

8   "*Travellers in Time*"; "*Blessèd*"; and "*Acoustic Hymnal.*" Tracks are available at www.song-smiths.com and CDs from www.isleofman.com

on from its 2014 centenary year.

I have learned that all things in the material realm must change and will ultimately pass away. Some people, places and things delight us. Others let us down and can break our hearts. I have let some people down and caused some heartbreak. We can move on from such failings and be changed for the better only through grace and the love of God. As UCM Pastor Charlie Pridmore put it, in a sermon based on John 3:16 - *"God loves us so much that He will not let our bad choices, our pigheadedness and our sin be the final word."*

## Epilogue: 'The Most Important Thing in Life'

Throughout my life, I have had the joy of being involved in music. In the 1970s on the Isle of Man, I belonged to a folk-rock trio called Triad: Tony Killen, joiner and journeyman; Robin Smith, Chartered Accountant; and me, Lecturer in Marine Biology. We looked as if we had come straight from the Woodstock festival. We had a good following and got radio plays.

Tony and Robin organized most of our repertoire. I was doing far too many other things. I had to call up Tony several times to say that I could not make it to rehearsals. His most memorable response was: *"Come on, Roger, man! Don't you know that the music is the most important thing?"* I felt very guilty. Anyone who has had to beg off from attending a group activity - especially in the arts or in sport - will know that feeling.

So, what is the most important thing in life? Some will answer 'personal relationships with a partner, or with family, or with friends.' Some will answer 'making money and putting it to good use.' Some will answer 'loving and serving God and/or fellow humans.' Some will answer 'music.' Some will answer 'science.' Some will answer 'football.'

We make our most important contributions to the history of humanity when we touch and are touched by something bigger than ourselves - something universal and eternal. Our need to contribute explains our devotions to the arts, mathematics, science and sport. Through Free Thought we pursue whatsoever things become for each of us the most important things in life.

In truth, the most important thing in life, for you and me and everyone, is to discover that God *loves* us, permanently and unconditionally. The Apostle Paul expressed as follows his wish that we would all come to know that love: "*That Christ may dwell in your hearts by faith; that ye being rooted and grounded in love, May be able to comprehend with all saints what is the breadth, and length, and depth, and height; And to know the love of Christ which passes all knowledge, and ye might be filled with all the fullness of God.*"[9]

Loving means setting free, as Sting[10] sings so beautifully. God loves us and He has set us free. He has given us free will, Free Thought and a wonderful world that works through free process. We are free to seek truth and to reject lies and nonsense coming from organized religion, New Atheism and anywhere else. We are free to choose whether or not to recognize and return God's love. It took me a long time to make my choice, but it's made for keeps.

---

9   Ephesians 3:17-19
10  Sting. 1985. *If You Love Somebody Set Them Free.*

# APPENDIX IV: BATTLEFIELD LITERATURE[1]

## PRO-FAITH

The Bible is the pre-eminent pro-faith work for all Christians. I much prefer the Authorized King James Version. Some find its language archaic and difficult, but it is well worth the effort to get over that barrier and enjoy some of the most beautiful English ever written. For access to the Gospel in contemporary English, I recommend Eugene Peterson's version of the Gospel of John.[2]

**Collins, F.S. 2006. *The Language of God: A Scientist Presents Evidence for Belief.* Free Press: New York. 294p.**

Francis Collins writes as follows about his leap to faith after reading C.S. Lewis' exposition of the Moral law in *"Mere Christianity"* (see below): *"I was stunned by its logic. Here, hiding in my heart as familiar as anything in daily experience, but now emerging for the first time as a clarifying principle, this Moral Law shone its bright white light into the recesses of my childish atheism, and demanded a serious consideration of its origin...I had to admit that I had reached the threshold of a spiritual worldview, including the existence of God...It seemed impossible to go forward or to turn back...For a long time I stood trembling on the edge of this yawning gap. Finally, seeing no escape, I leapt (p.29-31)."*

Collins led the Human Genome Project. He calls the genetic code: *"God's instruction book."* He writes as follows about opposition from some fellow scientists, who were also fellow Christians, as he told the

---

[1] Works are listed here in alphabetical order of authorship/editorship.
[2] Peterson, E.H. 2003. *The Gospel of John in Contemporary Language.* NavPress: Colorado Springs CO.95p.

truth about evolution: "*I spoke to a national gathering of Christian physicians, explaining how I had found great joy in being both a scientist studying the genome and a follower of Christ. Warm smiles abounded; there was even an occasional 'Amen.' But then I mentioned how overwhelming the evidence for evolution is, and suggested that in my view evolution might have been God's elegant plan for creating humankind. The warmth left the room. So did some of the attendees, literally walking out, shaking their heads in dismay* (p.145)."

Collins tells believers: "*If you picked up this book because of concerns that science is eroding faith by promoting an atheistic worldview, I hope that you are reassured by the potential for harmony between faith and science.*" He tells nonbelievers, especially scientists: "*if you are one who trusts the methods of science but remains skeptical about faith, this would be a good time to ask yourself what barriers lie in your way toward seeking a harmony between these worldviews* (p.230 - 231)."

**Flew, A. 2007. *There is a God. How the World's Most Notorious Atheist Changed His Mind.* HarperOne: New York. 222p.**

The philosopher Anthony Flew was the son of a preacher. Flew had become an atheist by the time he had finished his secondary schooling and was heading for Oxford. In 1950, he presented a paper on "*Theology and Falsification*" to a meeting of the Oxford University Socratic Club, chaired by C.S. Lewis. Thereafter, Flew took part in public debates against prominent Christian apologists, but in 2004 at New York University he announced that he now believed in the existence of God, because he had realized that a higher Intelligence must lie behind DNA and the genetic code.

In this book Flew confirms that major change: "*I now believe there is a God!* (p.1)." He also states: "*The discovery of phenomena like the laws of nature...has led scientists, philosophers and others to accept the existence of an infinitely intelligent Mind. Some claim to have made contact with this*

*Mind. I have not - yet* (p.158)." Flew became a deist and not specifically a Christian, though he also states the following: "*I think that the Christian religion is the one religion that most clearly deserves to be honoured and respected whether or not its claim to be a divine revelation is true* (p.185)."

**Lewis, C.S. 2001. *Mere Christianity*. Gift Edition, including a Foreword by Douglas Gresham and Appendices on related letters of C.S. Lewis and the history of his work. HarperCollins: New York. 237p.**

In this superb rationale for choosing faith and Christianity, C.S. Lewis explains firmly and clearly that there is one "*Moral Law*" for all of us. He is an honest seeker of truth and recommends that we all follow suit: "*Christ wants every bit of intelligence we have to be alert at its job, and in first-class fighting trim* (p.71)." Lewis tells the honest seeker: "*Give up yourself and you will find your real self...Nothing that you have not given away will be really yours. Nothing in you that has not died will ever be raised from the dead. Look for yourself, and you will find in the long run only hatred, loneliness, despair, rage, ruin, and decay. But look to Christ and you will find Him, and with Him everything else thrown in* (p.177)."

**Polkinghorne, J.C. 1996. *The Faith of a Physicist: Reflections of a Bottom-Up Thinker*. Fortress Press: Minneapolis MN. 211p.**

John Polkinghorne states his conviction that theology must be consonant with disclosures through science; including the Big Bang, the states and workings of the material realm and the evolution of life forms. He writes: "*I see no fundamental incompatibility between any of my bottom-up searchings for the truth, whether pursued in the field of science or in the field of Christian belief. Both domains of inquiry are necessary if we are truly to comprehend the way things are. Many puzzles remain, but the scientist and the theologian can make common cause in the search for understanding...of the great ocean of truth lying undiscovered before us* (p.193 -194)."

**Polkinghorne, J.C. 1998.** *Belief in God in an Age of Science.* **Yale University Press: New Haven CT. 133p.**

John Polkinghorne reinforces here the point that there is no conflict between faith and science. He argues as follows against the idea that that God acts in the material realm only by influencing people: "(restricting) *God's actions* (to) *inspiration and encouragement* (of) *human persons...implies that God has been an inactive spectator of the universe for most of its history to date, since conscious minds seem not to have been available for interaction with divinity until, at most, the past few million years or so of that fifteen-billion year history* (p.55)." He discusses divine action as a top-down agent of causality in the material realm and the search for a *"causal joint."*

No one really knows how God acts in the material realm. I believe that He acts *only* in our souls, which are parts of the spiritual realm. Through our Free Thought, *we* choose how to act in the material realm: on God's behalf, or not. I have no problem with that restriction - as a divinely chosen Self-limitation by God concerning His interactions with His Creation. However, Polkinghorne's position and mine on this are both covered in his following statement: *"Modern science, properly understood, in no way condemns God, at best, to the role of a Deistic Absentee Landlord, but it allows us to conceive of the Creator's continuing providential activity and costly loving care for creation* (p.75)." I regard God's truthful disclosures to us through mathematics and science as part of that beautiful vision.

**Smith, J.M. and J. Quenby. Editors. 2009.** *Intelligent Faith: A Celebration of 150 Years of Darwinian Evolution.* **O Books: Winchester, UK. 330p.**

This compilation of essays by the Editors and 16 other authors was supported by the *"Modern Churchpeoples's Union,"* later renamed the

"*Modern Church.*"³ John MacDonald Smith's Introduction gets us to the battlefield as follows, omitting here his citation of another work: "*a serious and continuous attempt to undermine a developing partnership between science and theology began almost a century ago in America with the founding of the movement known as fundamentalism... What has been called the 'American Scenario,' six day creation of a young earth, dispensationalism, the Rapture, were all part of a rather eccentric doctrinal system which should have had no reasonable expectation of lasting* (p.3)." Unfortunately, it is still with us.

All the essays in this compilation make important contributions to the debate about evolution. All seek unity in truth. Anti-evolution believers who are genuinely interested in seeking the truth about evolution would benefit particularly from the following: Denis Alexander's "*Evolution- Intelligent and Designed?* (p.7-22);" Andrew Robinson and Christopher Southgate's "*Intelligent Design and the Origin of Life* (p.49-59);" and Sam Berry's "*Darwin's Legacy: A Regularly Misunderstood and Misinterpreted Triumph* (p.97-120)." John Quenby celebrates the "*remarkable unity*" of this book and "*the power and universality of Darwin's evolutionary idea (p.330).*"

**Stott, J.R.W. 1999. *Basic Christianity*. 1999. OMF Literature: Manila, Philippines. 222p. First published in 1958.**

John Stott focuses on Christ and how each of us can establish a personal relationship with God. Stott writes about belief, discipleship, and striving to live a Christ-like life. He describes how the practices of some churches and the behavior of some Christians can turn nonbelievers away from faith. Stott explains how God revealed Himself to humans as Christ. He describes in plain and concise language the character and claims of Christ and the nature and consequences of the sinful condition of humans. He explains how we can make the choice for faith and break the barriers that separate us from God. Stott's firm and true

---

[3] www.modernchurch.org.uk

message is: "*To live for ourselves is insanity; to live for God and for man is wisdom and life indeed* (p.182)."

**Wright, N.T. 2006. *Simply Christian: Why Christianity Makes Sense*. HarperOne: New York. 240p.**

Tom Wright notes as follows that everyone is involved in the battle between justice and injustice: "*The line between justice and injustice, between things being right and things not being right, can't be drawn between 'us'* (i.e., believers, including Wright himself as a Bishop) *and 'them'* (nonbelievers). *It runs right down through the middle of each one of us... We all know what we ought to do (give or take a few details); but we all manage, at least some of the time, not to do it* (p.6)." We are all participants in the war between justice and injustice, right and wrong, good and evil - through our Free Thought.

Wright provides a very helpful perspective on Bible, as follows: "*The Bible is full of passages which really do intend to describe things that happened in the real world – and...to forbid various types of actions which occur in the real world...But the Bible...regularly and repeatedly brings out the flavor, the meaning, the proper interpretation of these actual, concrete, space-time events by means of complex, beautiful, and evocative literary forms and figures...it is then open to any reader, commentator and preacher to explore, in a particular passage, which bits are 'meant literally,' which bits are 'meant metaphorically,' and which bits might have been meant both ways - before turning, as a second stage, to ask whether the bits which were 'meant literally' actually happened in concrete reality* (p.194 -195)."

## PRO-UNBELIEF

**Atkins, P. *On Being. A Scientist's Exploration of the Great Questions of Existence*. Oxford University Press: Oxford. 111p.**

Peter Atkins traces the origins and evolution of the universe and of life and moves on to insightful descriptions of the processes of birth and death, the genetic code etc. He skewers examples of nonsense from organized religion and non-science, including Intelligent Design, in ways that honest seekers of truth can only admire.

Atkins explains the universe and the human condition under the banner of a Godless science, which is said to have no limits. He signs off as follows: "*My own faith, my scientific faith…respects the powerful ability of the collective human intelligence, which initially groped for understanding through myth but now gives us the capacity to comprehend and, optimistically, given time and cooperation between brains, will do so without limit* (p. 104)."

**Blackford, R. and U. Schüklenk, Editors. 2009. *50 Voices of Disbelief. Why We Are Atheists.* Wiley-Blackwell: Malden MA. 346p.**

Here, in 50 separate essays, professionals from a wide variety of disciplines explain why they opted for atheism. Many equate belief or faith with adherence to an organized religion from which they had to break away. Another recurrent theme is that man allegedly made (i.e., invented) God and not vice-versa. The authors tend to focus on one or a few issues. Nicholas Everitt focuses on arguments against the existence of a benevolent God on the grounds of the amount of suffering in the world, including the suffering of animals. Victor Stenger emphasizes cosmology.

Michael Shermer writes on "*How to Think About God, Theism, Atheism, and Science*" and states: "*When theists, creationists, and Intelligent Design Theorists invoke miracles and acts of creation **ex nihilo*** (original author's emphasis), *that is the end of the search for them, whereas for scientists the identification of such mysteries and problems is only the beginning. Science picks up where theology leaves off.* (p.77)." The alternative to that dismal scenario is an expansion of the faith-science quest for truth. Perpetuation of the faith-science divide helps no one.

**Dawkins, R. 2006. *The God Delusion*. Bantam Press: London. 336p.**

Richard Dawkins gives his belief concerning the existence of God as follows: "*Very low probability, but short of zero...I live my life on the assumption that he is not there* (p.50-51)." He argues mostly against religion rather than against faith *per se*, though he clearly thinks that belief in God is ridiculous.

Dawkins' coverage of personal experiences of allegedly spiritual things is dismissive and highly inadequate. His main examples are: the cry of a bird, the Manx Shearwater, mistaken as the voice of the devil; a few accounts of religious calls to action, such as the Yorkshire Ripper allegedly ordered by Jesus to kill women and George Bush allegedly ordered by God to invade Iraq; optical illusions, including the Necker Cube and the Mona Lisa's eyes; illusions of ghosts in wind sounds and patterns on windows; and the dancing sun, allegedly seen by 70,000 pilgrims at Fatima, Portugal in 1917 (p.87-92).

Dawkins states: "*if a God really did communicate with humans that fact would emphatically not lie outside science...a God who is capable of sending intelligible signals to millions of people simultaneously, and or receiving messages from all of them simultaneously, cannot be, whatever else he might be, simple. Such bandwidth! God...must have something far more elaborately constructed than the largest brain or computer we know* (p.154)." Science must include the study of spiritual revelations and responses. God is Spirit and is not confined by bandwidth.

I am with Dawkins entirely on his rejection of so-called intellectual proofs for God's existence and his abhorrence of all the bloodletting and sexual violence in the Old Testament and the horrible utterances, homophobia and other cruel behaviour of some present day fundamentalist believers. It is clear that some self-professed Christians continue to participate in torture and war crimes and some self-professed Muslims kill indiscriminately in acts of terrorism. But such religious horror shows say nothing about the true nature of God. Dawkins misses the messages of God's love, mercy and justice in the Old Testament and the message of Christ's (i.e., God's) Gospel of love.

Dawkins refers to A.A. Milne's poem about "*Binker*" and equates a believer's God with the same kind of imaginary friend, as follows: "*I suspect that the Binker phenomenon of childhood may be a good model for understanding theistic belief in adults. I do not know whether psychologists have studied it from this point of view, but it would be a worthwhile piece of research* (p. 349)." I would welcome that kind of research.

Dawkins identifies two types of consolation: Type I, direct physical consolation; and Type II, the "*discovery of a previously unappreciated fact, or a previously undiscovered way of looking at existing facts* (p.353)." He finds science superior to religion for providing Type I consolation, as St. Bernard dogs bring brandy, doctors give medical help etc., but admits that religion can provide Type II, for example after a natural disaster. The benefits of faith far exceed all of the above.

**Dennett, D.C. 2007. *Breaking the Spell: Religion as a Natural Phenomenon*. Penguin Books: New York. 448p.**

According to Daniel Dennett, the spell that must be broken is: "*the taboo against a forthright, scientific, no-holds-barred investigation of religion as one natural phenomenon among many* (p.17)." He continues: "*I want to put religion on the examination table…The only arguments worth attending to will have to demonstrate that (1) religion provides net benefits to humankind, and (2) these benefits would be unlikely to survive such an investigation…if we **don't*** (original author's emphasis) *subject religion to such scrutiny now, and work out what revisions and reforms are called for, we will pass on a legacy of ever more toxic forms of religion to our descendants* (p.39)."

Dennett lists as follows what he sees as the possible explanations for the prevalence of religion among humans: a craving, rather like having a sweet tooth; a meme, as a *"cultural symbiont"* or parasite that becomes an accepted part of us; a competitive asset, rather like the bower made by a bower bird; a kind of money, underpinning societal relationships and structure; and an attractive, but useless addition, like a pearl in an oyster.

Dennett continues: *"Perhaps it seems to you that I am somewhat willfully ignoring the obvious explanation of why* **your** *religion exists and has the features that it does: it exists because it is the inevitable response of enlightened human beings to the obvious fact that God exists! Some would add: we engage in these religious practices* **because God commands us to do so**, *or* **because it pleases us to please God** (original author's emphases)... (but) *Whichever religion is yours, there are more people in the world who don't share it than who do, and it falls to you – to all of us, really – to explain why so many people have gotten it wrong, and to explain how those who know (if there are any) have managed to get it right* (p.92)". The existence of God explains personal faith. The diversity among and within organized religions does not negate God's existence.

Dennett embraces meme theory and writes: *"alongside the domestication of animals and plants, there was a gradual process in which the wild (self-sustaining) memes of folk religion became thoroughly domesticated. They acquired stewards... The wild memes of language and folk religion... are like rats and squirrels, pigeons and cold viruses - magnificently adapted to living with us and exploiting us whether we like them or not. The domesticated memes, in contrast, depend on help from human guardians to keep going* (p. 170-171). He explains this further, citing an earlier work:[4] *"the meme for faith exhibits* **frequency-dependent fitness** (original author's emphasis): *it flourishes particularly in the company of rationalistic memes. In a neighborhood with few skeptics, the meme for faith*

---

4    Dennett, D.C. 1995. *Darwin's Dangerous Idea: Evolution and the Meanings of Life.* Penguin Books: London. 587p. Dennett refers to p.349, in which he used slightly different wording as follows: *"In a* **skeptic-poor world**...*is seldom reintroduced into the* **infosphere** (present author's emphases).*"*

*does not attract much attention, and hence tends to go dormant in minds, and hence is seldom reintroduced into the memosphere (p.231)."*

I cannot grasp all of the above, but I am with Dennett as he states: *"The potential benefits of joining the scientific community on these issues are enormous: getting the authority of science in support of what you say you believe with all your heart and soul* (p. 274)." I believe that we have the authority of God to proceed with science in that broadened way. Dennett pleads for *"more research (p.311)"* and adds: *"If you have to hoodwink your children to ensure that they confirm their faith when they are adults, your faith **ought*** (original author's emphasis) *to go extinct* (p.329)." Let's do that research. That which is not true will then go extinct eventually.

**Everitt, N. 2004. *The Non-Existence of God*. Routledge: London. 326p.**

Nicholas Everitt argues convincingly against the main alleged proofs for God's existence; including the ontological, cosmological, argument from design, arguments from miracles etc. I agree that none of those proofs stands up. The only evidence for the existence of God and for anything made of spiritual stuff is the human condition and the believer's experiences of God.

Everitt states that even if we could fix everything in the universe there could be no answer to the question why it exists because we could not get beyond any explanation for its existence that would not be a *"self-explanation."* To do otherwise, one would have to: *"refer to something that is not part of the universe, i.e., is not spatial or temporal or material...* (and)...*we cannot make sense of anything non-temporal, if this is meant to cause or bring about the existence of the universe (and hence of time itself)* (p.81-82)." I agree that the cosmological arguments for God's existence lack any mechanisms for what I call spiritual realm- material realm interaction.

With scrupulous fairness, Everitt states that the inability of modern cosmology to explain why the universe began, as opposed to its development after it had begun: "*does not in itself prove that science cannot provide such an explanation* (p.84)." Again I agree, but any such explanation would always remain a self-explanation of the material realm, *unless* that science included the study of spiritual revelations.

Everitt rejects arguments that our universe could have induced its own constituents, which comprise in total nothing, to arise from nothing. He states: "*If we indeed start from a position of nothing, then (tautologically) there is nothing to do any inducing of anything* (p.84)." Again I agree, while noting that this discussion is entirely about material realm stuff and excludes the possibility of the existence of any other kind of stuff.

Everitt rejects arguments for God's existence based on the Anthropic Principle. Again I agree. I am awed by the fact that the physical laws and constants that govern our universe are set just right for its existence and our existence. Nevertheless, it is tenable to say that is just the way things are. It does not prove God's existence.

Everitt demolishes Intelligent Design well, citing the failures of Michael Behe and others to demonstrate anything biological that is irreducibly complex. Everitt confronts William Dembski and other proponents of Intelligent Design with the killer question: "*What can Dembski mean by a non-designed object?* (p.109 to p.110)"

Everitt categorizes alleged miracles as follows: those that are simply not credible, because they require God to have violated the laws of nature; those that that do not violate the laws of nature and therefore are simply nature taking its course; and those that are simply coincidences. He concludes that alleged miracles provide no evidence for the existence of God.

I agree, but I regard the creation of the universe at the Big Bang as a miracle and I believe that some miracles contravening the laws of nature took place during the Incarnation of God as Jesus Christ. I also consider God's interventions in our lives through His spiritual revelations as miraculous, because they have a spiritual basis that can change our behavior in the material realm and thereby our impacts on each other and on our world.

Everitt rejects the argument that God's existence is necessary for human morality, though he provides a full and fair discussion of the opposing view and finds that there is: "*much disagreement over what morality is* (p.148)."

Everitt discusses free will at length and finds that a "*compatibilist theist*" faces not only the problem of God's "*divine **foreknowledge** of morally objectionable free actions,*" but also the problem of God's "***involvement in the doing*** *of free actions* (p.251) (original author's emphases)." I am an incompatibilist theist. I hold that God gives us free will to choose our own goals and Free Thought to make our own choices. I hold also that God has no foreknowledge of the evil that is done, because the outcomes of Free Thought are not predictable.

Everitt attempts logical arguments against the existence of the God of theism, as follows: "*the defining attributes of God are either individually self-contradictory (omnipotence) or cannot be coinstantiated (omniscience and omnipotence, omniscience and eternity, eternity and personhood, eternity and creatorship, etc.)* (p.303)." In the same vein, he writes: "*the scale of the universe which modern science reveals…(is)…hugely unlike what theism would lead us to expect…if current Big Bang theory is correct, past time is finite. If God's eternity is construed as endless duration, this scientific finding entails that God does not exist, since it entails nothing that has infinite duration* (p. 304)."

One cannot argue convincingly against God's existence based on His alleged attributes as guessed by humans. God's attributes are largely unknown and unknowable. In common with most believers, I hold only to the certainty of God's love, grace and mercy, based upon the reported life of Christ and my own experiences of God through faith. God resides in the spiritual realm, where finite and infinite durations are not an issue. God is beyond time.

**Harris, S. 2006. *The End of Faith: Religion, Terror, and the Future of Reason.* Free Press: London. 336p.**

Sam Harris spends most of his firepower against organized religion, not against faith *per se*. He begins with this statement of purpose: "*I hope to show that the very ideal of religious tolerance - born of the notion that every human being should be free to believe whatever he wants about God - is one of the principal forces driving us toward the abyss* (p.15)." However, he admits that: "*There is clearly a sacred dimension to our existence, and coming to terms with it could well be the highest purpose of human life* (p.16)."

Harris castigates religious moderates and fundamentalists alike and generalizes to the extent that his criticisms become untenable for anyone who knows the diversity and characteristics of the people and institutions that are his targets. Here are a few examples, with emphases by Harris and my comments in brackets:

I. "*every religion preaches the truth of propositions for which no evidence is even* **conceivable** *(p.23).*"

II. "*for the reliable transformation of fearful, hateful, or indifferent persons into loving ones...There may even be a few biblical passages that would be useful in this regard* (p.24)." (A few? Is Harris kidding? The whole message of Christ's Gospel is about transformation to live in love for God and one's fellow humans.)

III. "*Religious faith represents so uncompromising a misuse of the power of our minds that it forms a kind of perverse singularity - a vanishing point beyond which rational discourse becomes impossible (p.25)."* (Really? I agree that one cannot argue against dogma, but to exclude believers from all rational discourse is not tenable).

IV. *"We must find our way to a time when faith, without evidence, disgraces anyone who would claim it (p.48)."* (I feel no disgrace in professing my faith and recounting my experiences of God.)

V. *"The men who committed the atrocities of September 11... were men of faith - perfect faith, as it turns out - and this, it must finally be acknowledged, is a terrible thing to be (p.67)."* ('No,' they were recruited by the spiritual force for evil and they were outliers from the huge populations of their fellow Muslims and other believers.)

VI. Pacifism is said to be *"flagrantly immoral"* and *"absolutely nothing more than a willingness to die, and to let others die, at the pleasure of the world's thugs* (p.199)." (Mohandas Ghandi's pacifism was moral, brave and effective.)

**Hitchens, C. 2007. *God Is Not Great. The Case Against Religion.* Atlantic Books: London. 307p.**

The late Christopher Hitchens covers the misdeeds of organized religions very well. He predicts that there will be no more impressive people of faith like St. Thomas Aquinas and Moses Maimonides because: *"Faith of that sort – the sort that can stand up at least for a while in a confrontation with reason – is now plainly impossible (p.63)."*

I disagree. Scientists who are believers have always shown that same kind of faith. It comes from an honest quest for truth. Hitchens hopes that every believer's faith will be *"undermined"* so as to allow the newly liberated nonbeliever to: *"feel better...and allow the chainless mind to do its own thinking* (p.153)." In faith or unbelief, Free Thought continues.

Hitchens says much that is valuable on many topics including alleged miracles, the non-science of Intelligent Design and the belligerence, hypocrisy and intolerance within organized religion - for example, Japanese Buddhism and Shintoism in World War II, Hitler's SS having 25% Catholic Christian membership, and Jewish Elders prohibiting the reading of works by Spinoza. Hitchens states: "*it has become necessary to know the enemy* (faith and religion) *and to fight it* (p.283). 'Yes.' The enemy is the spiritual force for evil, which operates in every soul and every institution and organization.

## Huxley, T.H. 1899. *Science and the Hebrew Tradition. Essays.* D. Appleton and Company: New York. 372p.[5]

These essays are mostly about the truths that we can learn from careful observation, deduction and analysis in the natural world and about how to approach situations when those truths do not fit with what are held to be the allegedly unassailable so-called truths revealed in sacred texts. In Huxley's time, as now, the biggest misfit was between the obvious explanatory power of evolutionary theory and the biblical literalists' insistence on verbatim acceptance of the written account of creation in the Book of Genesis.

In his Preface, Huxley describes masterfully how sacred text literalism puts believers into an indefensible mess: "*Age after age, they have held it to be an indispensable truth that, whoever may be the ostensible writers of the Jewish, Christian, and Mahometan scriptures, God Himself is their real author; and, since their conception of the attributes of the Deity excludes the possibility of error and - at least in relation to this particular matter - of willful deception, they have drawn the logical conclusion that the denier of*

---

[5] I give lengthy comments on Huxley's essays because of their historical interest and their special significance for me as I began to write this book, as described in the chapter on Faith.

*any statement, the questioner of the binding force of any command, to be found in these documents is not merely a fool, but a blasphemer. From the point of view of mere reason he grossly blunders; from the point of view of religion he grievously sins* (p.v)".

Huxley continues: "*But, if this dogma of Rabbinical invention is well founded; if, for example, every word in our Bible has been dictated by the Deity; or even, if it be held to be the Divine purpose that every proposition should be understood by the bearer or reader in the plain sense of the words employed (and it seems impossible to reconcile the Divine attribute of truthfulness with any other intention), a serious strain on faith must arise. Moreover, experience has proved that the severity of this strain tends to increase, and in even more rapid ratio, with the growth in intelligence of mankind and with the enlargement of the sphere of assured knowledge among them.*

*It is becoming, if it has not become, impossible for men of clear intellect and adequate instruction to believe, and it has ceased, or is ceasing, to be possible for such men honestly to say they believe, that the universe came into being in the fashion described in the first chapter of Genesis; or to accept, as a literal truth, the story of the making of woman, with the account of the catastrophe that followed hard upon it, in the second chapter; or to admit that the earth was repeopled with terrestrial inhabitants by migration from Armenia or Kurdistan, little more than 4,000 years ago, which is implied in the eighth chapter; or finally, to shape their conduct in accordance with the conviction that the world is haunted by innumerable demons, who take possession of men and may be driven out of them by exorcistic adjurations, which pervades the Gospels* (p. vi-vii)."

Unfortunately, there are millions of believers today who are well educated and rational but who still believe in most or all of the above and who would consider their souls to be in mortal danger if they questioned any of it. The battles for truth that Huxley was fighting are still with us. They gave him no choice other than lifelong agnosticism.

They continue to turn honest seekers away from exploring faith.

His first essay, "*On the Method of Zadig* (1880)," celebrates the gathering of evidence by careful observation. It tells of a certain Zadig, who left the pressures of ancient Babylon for a quiet life beside the river Euphrates - as a naturalist and a maker of deductions, somewhat like Sherlock Holmes. Subsequent essays support evolutionary theory: "*The Rise and Progress of Paleontology* (1881);" "*Lectures on Evolution* (1876)" which contains the clear evidence for horse evolution; "*The Interpreters of Genesis and the Interpreters of Nature* (1885)" and "*Mr. Gladstone and Genesis* (1886)*.*" The other essays are: "*The Light of the Church and the Light of Science* (1890)" "*Hasisadras Adventure* (1891),*" which is about geology and great floods; and "*The Evolution of Theology; an Anthropological Study* (1886)" which is a fine background to later studies on the evolution of organized religions and implications for morality.

Huxley's main target is the doctrine of biblical infallibility. He quotes from a sermon preached by Canon H. P. Liddon at St. Paul's Cathedral, London on December 8, 1889: "*For Christians, it will be enough to know that our Lord Jesus Christ set the seal of His infallible sanction on the whole of the Old Testament. He found the Hebrew Canon as we have it in our hands to-day, and He treated it as an authority which was above discussion...He points to Jonah's being three days and three nights in the whale's belly* (p. 208-209)".

Canon Liddon's message was that we must all believe the story of Jonah and everything else reported in the Bible. Huxley could not do that and neither can I. For me, the story of Jonah is one of the worst examples of literal impossibility in the Bible. Jonah lived, allegedly, for three days and nights in the belly of a big fish.[6] God, allegedly, caused a vine to grow enough in one day to shade Jonah from the sun and then,

---

6   Jonah 1:17

allegedly, allowed a worm to destroy it by the next morning,[7] The story uses natural metaphors for spiritual states of disobedience and obedience to God and their consequences.

Huxley had a near-obsession with the New Testament story of the Gadarene swine.[8] In biblical times, mental disorders were explained as possession by demons, or unclean spirits. The story alleges that demons inside a man, or two men, recognized Jesus approaching. They gave their names as "*Legion*," meaning 'many,' and assumed that Jesus was coming to cast them out of their host(s). They asked Him to send them instead into a large herd of pigs that were feeding nearby. Jesus complied. The pigs went berserk and ran down the hill into the nearby "*waters*" or "*sea*" or "*lake*" and drowned. Note that this story contains both biological and spiritual material to illustrate an alleged healing miracle.

I do not doubt that Christ healed a mental disorder at Gadara. I am entirely with Huxley in disbelieving that mental disorders are caused by demons and that demons can be moved from humans to pigs. Demons do not exist. The biblical account was written for readers who believed in demons as the cause of mental disorders. Those poor pigs figure many times in Huxley's parallel volume, which I cover next.

**Huxley, T.H. 1899. *Science and the Christian Tradition. Essays.* D. Appleton and Company: New York. 419p.**

Huxley's main purpose in these great essays was to define agnosticism as the only tenable position for him and for other honest seekers, especially scientists. His Preface, written in 1893, sets the scene. All emphases in the quotations that follow are Huxley's as original author, unless stated otherwise.

---

7   Jonah 4:6 to 7
8   Matthew 8:28-33, Mark 5:11-16 and Luke 8:26 to 36. Only Matthew's Gospel records two persons.

Huxley rejects the notion that Moses wrote nine tenths of the Pentateuch. He then questions the origins and contents of the three synoptic Gospels (Matthew, Mark and Luke), in which he finds: "*the same general view, or **Synopsis**, of the nature and order of the events narrated (and that)…to a remarkable extent, the very words which they employ coincide* (p. xix)." Huxley concludes that: "*each Gospel is composed of a **threefold tradition**, two **twofold** traditions, and one **peculiar tradition*** (p.xxi)," which he represents as: components of Mark = A+B+C+E; components of Matthew = A+B+D+F; and components of Luke = A+C+D+G.

Huxley states: "*Hence it follows that the Synoptic writers have, mediately or immediately, copied one from the other; or that the three have drawn from a common source; that is to say, from one arrangement of similar traditions (whether oral or written); though that arrangement may have been extant in three or more, somewhat different versions* (p. xxii-xxiii)." I take the mixes of congruence and contradiction in the Gospels as indicators of authenticity.

Huxley continues: "*But it is unnecessary, and would be out of place, for me to attempt to do more than indicate the existence of these complex and difficult questions. My purpose has been to make it clear that the Synoptic problem must force itself upon everyone who studies the Gospels with attention; that the broad facts of the case, and some of the consequences deducible from these facts, are just as plain to the simple English reader as they are to the profoundest scholar* (p. xxix)."

Huxley pursues one of those "*consequences*" as follows: "*the threefold tradition presents us with a narrative believed to be historically true, in all its particulars, by the major part, if not the whole part, of the Christian communities. That narrative is penetrated, from beginning to end, by the demonological beliefs of which the Gadarene story is a specimen; and, if the fourth Gospel indicates the existence of another and, in some respects, irreconcilably divergent narrative, in which the demonology retires into* the background, it is none the less there. Therefore, the demonology is an integral and inseparable component of primitive Christianity."

He continues: *"It further follows that those who accept devils, possession, and exorcism as essential elements of their conception of the spiritual world may consistently consider the Gospels to be unimpeachable in respect of the information they give us respecting other matters which pertain to that world. Those who reject the gospel demonology, on the other hand, would seem to be **completely barred** (present author's emphasis), as I feel myself to be, from professing to take the accuracy of that information for granted. If the threefold tradition is wrong about one fundamental topic, it may be wrong about another, while the authority of the single traditions, often mutually contradictory as they are, becomes a vanishing quantity. It really is unreasonable to ask any rejector of the demonology to say more with respect to those other matters, than that the statements regarding them may be true, or may be false; and that the ultimate decision, if it is to be favourable, must depend on the production of a testimony of a very different character from that of the writers of the four gospels. Until such evidence is brought forward, that refusal of assent, with willingness to re-open the question, on cause shown, which is what I mean by Agnosticism, is, for me, the only course open (p. xxix- xxx)".*

Regardless of who actually wrote the Gospels and the extents to which they might or might not have copied from each other, it is simply not true that to be a Christian requires complete belief in their every word. The Gospels were written before the germ theory of disease was formulated and before the recognition that mental disorders and other pathologies are *not* the results of demonic possession. The writers of the Gospels believed in demons, so they wrote about demons. Their readers expected to read about demons. Demonology was the prevailing theory to explain sickness, crop failure and lots of other adverse happenings. It is not true that who do not believe *today* in all of the demonology in the Gospels are thereby *"completely barred"* from honest belief in God, the spiritual force for evil, the Incarnation, death and

Resurrection of Christ and the gift of the Holy Spirit.

In his "*Prologue (Controverted Questions)* (1892)," Huxley emphasizes the problems of belief in biblical and papal infallibility. His beautiful essays that follow are mostly support for his agnosticism, as can be seen from their titles: "*Scientific and Pseudo-Scientific Realism,*" "*Science and Pseudo-Science,*" and "*An Episcopal Trilogy,*" (all from 1887); "*The Value of Witness to the Miraculous*" (1889);" "*Possibilities and Impossibilities* (1891);" "*Agnosticism,*" "*Agnosticism - A Rejoinder,*" and "*Agnosticism and Christianity*" (all from 1889); "*The Keepers of the Herd of Swine*" (1890);" and another salvo directed at Mr. Gladstone, who believed every word in the Jewish and Christian scriptures, "*Illustrations of Mr. Gladstone's Controversial Methods* (1891)." All are highly entertaining and informative.

In his "*The Value of Witness to the Miraculous,*" Huxley describes allegedly miraculous healings by the powers of saintly relics in the 9th century, mostly as recounted by a certain Eginhard as chief witness. Huxley is fair. He finds "*deliberate and conscious fraud…a rarer thing than is often supposed,*" but he also firm as he states: "*There is no falsity so gross that honest men and, still more, virtuous women, anxious to promote a good cause, will not lend themselves to it without any clear consciousness of the moral bearings of what they are doing.*" The same applies today, though with no basis for apportioning blame by gender - as was surely also the case in Huxley's time.

Huxley takes us back to the swine. In his essay "*Possibilities and Impossibilities*" he restates their importance for his main thesis: "*…when such a story as that about the Gadarene swine is placed before us, the importance of the decision, whether it is to be accepted or rejected, cannot be overestimated. If the demonology part of it is to be accepted, the authority of Jesus is unmistakably pledged to the demonological system current in Judaea in the first century. The belief in devils who possess men and can be transferred from men to pigs, **becomes as much a part**

***of Christian dogma as any article of the creeds*** (present author's emphasis). *If it is to be rejected, there are two alternative conclusions. Supposing the Gospels to be historically accurate, it follows that Jesus shared in the errors, respecting the nature of the spiritual world, prevalent in the age in which he lived and among the people of his nation. If, on the other hand, the Gospel traditions give us only a popular version of the sayings and doings of Jesus, falsely coloured and distorted by the superstitious imaginings of the minds through which it had passed, what guarantee have we that a similar unconscious falsification, in accordance with preconceived ideas, may not have taken place in respect of other sayings and doings? What is to prevent a conscientious inquirer from finding himself at last in a purely agnostic position with respect to the teachings of Jesus, and consequently with respect to the fundamentals of Christianity* (p.193-194)?"

Those swine pushed Huxley over the cliff top and into unsound arguments. The words chosen by Jesus and the writers of the Gospels for their contemporaries do not mean that all Christians and would-be Christians, for all time, *must* believe that a legion of devils went from a man, or two men, into a herd of doomed pigs. For Huxley to suggest that believing in that story has that same importance for a Christian as professing the Apostles' Creed or the Nicene Creed is nothing short of ridiculous.

The Gadarene swine figure greatly in Huxley's profession of agnosticism. In addition to devoting two entire essays to them - "*The Keepers of the Herd of Swine* (p. 366-392)," and "*Illustrations of Mr. Gladstone's Controversial Methods* (p. 393-419)" - he mentions them at least 50 times in other essays. He concludes that Gadarene is the correct descriptor for the location of the story, because it refers to the city of Gadara, which was predominantly Greek, not Jewish. Despite inevitable changes in the landscape at Gadara, it might be worth searching there for dateable skeletal remains of a large herd of mixed sex, mixed age, drowned pigs.

**Loftus, J.W. 2008. *Why I Became an Atheist. A Former Preacher Rejects Christianity*. Prometheus Books: New York. 428p.**

John Loftus' book has been recommended not only by Daniel Dennett and Christopher Hitchens but also by the Christian apologist Norman Geisler, as debated online in 2007.[9] Geisler apologized to Loftus on behalf of the Christians who wronged him. Loftus writes: "*I don't need any sympathy at this point, and I especially don't want any pity. I very much appreciate Norm's concern and compassion though. But with this recognition comes the next question: Why didn't God do something to prevent these particular experiences of mine, especially if he could foreknow that I would eventually write this book and lead others astray (p.31)?*"

Loftus pleads that readers evaluate his book as "*one single argument against Christianity (p.12)*" and makes the following case for viewing the whole thing in his personal context: "*The experiences I've had are not typical of why people leave the Christian faith...For me there were three major circumstances that happened in my life that changed my thinking... These things are associated with three people: a woman I'll call Linda... who brought a major crisis into my life...Larry who brought new information into my life...and Jeff (who) took away my sense of a loving Christian community...In the midst of these things, I felt rejected by the Church of Christ in my local community (p.24).*"

Loftus rails against the Christian apologist William Lane Craig as follows: "*When arguing that he knows Christianity is true based on the self-authenticating witness of the Holy Spirit, Craig stands in the steps of some great Christian thinkers like Augustine, Anselm, and even Barth and Bultmann...Unlike these theologians though, the basis for Craig's faith is in the inner witness of the Holy Spirit alone, who may use the Bible, evidence, or nothing at all to create faith for the believer. Does Craig mean to say he cannot be wrong? Yes! He knows Christianity is true. With this understanding, he has insulated himself from any and*

---

[9] Available at: http://debunkingchristianity.blogspot.com

*all objections to the contrary. He know's he's right because he know's he's right, and that's the end of the matter. Since he know's he's right, Christianity is true* (p.214)."

Loftus continues: "*how did Craig come to believe that there is a Holy Spirit and an inner witness in the first place? Through experience? Hardly. Did this inner witness tell him that his views on the self-authenticating testimony of the Holy Spirit are true?...Whatever it might take to change our minds about our faith, one should never say that she 'knows' Christianity to be true, because just like me, she might later come to believe that she was wrong* (p.219)."

That is the whole nub of the issue of faith. What is the worth of anyone's subjective evidence from within for a relationship with God? On Loftus' soul battlefield, God's spiritual revelations became swamped by the hurt and confusion that were sown by Christians in his church. He came to see everything through the fog of his overall sense of hurt,

Loftus laments the following: lives of monks wasted in "*reading a biblical text that was false*;" self-sacrificial celibacy in the Catholic priesthood; Baptist ministers' lacking of experiences of being drunk; time wasted on reading the Bible instead of great novels and on evangelizing others rather than having fun with family and friends; and money spent on cathedrals and churches that could have been spent on people's needs, with some left for "*a cruise in the Bahamas.*" Loftus final statement is: "*The Christian life is ultimately in vain, because it is built on false hope* (p.414)." 'No,' it is built on the truth that is found through faith.

**Mackie, J.L. 1982. *The Miracle of Theism. Arguments for and against the Existence of God.* Clarendon Press: Oxford. 268p.**

J. L. Mackie covers philosophical arguments for and against God's existence, including the following: miracles and testimonies, with David

Hume's argument against theism; the ontological proof, drawing on Descartes, St. Anselm and Alvin Plantinga; and arguments from cosmology, morality, consciousness and design, with reviews of the problem of evil, religious experiences, belief without reason and religion without belief. Mackie concludes: "*The balance of probabilities, therefore, comes out strongly against the existence of a god.*" He adds: "*we cannot escape the implications of this result by making a voluntary faith intellectually respectable.*"

Mackie states further: "*William James thought that the religious experiences of individuals were the nucleus and root of all religion, and that factual claims going beyond what such experiences themselves contain, all metaphysical theology, and all socially organized and institutionalized religion, are merely secondary outgrowths from this root* (p. 187). He continues: "*The most we could allow was James' experimental approach and…it would be very hard for this to yield a favourable result* (p.253)." I disagree. Subjective evidence from within *is* proof of God's existence for the individual believer. The approaches used by James and others can be improved so as to examine and analyse subjective evidence from within concerning faith and unbelief, in large and diverse populations.

**Mark, J. 2008. *Christian No More: a personal journey of Leaving Christianity and how you can leave it too*. Reasonable Press/ Cogspage Media: Cincinnati OH. 428p.**

Jeffrey Mark was damaged by his church experiences in early life. He was well behaved and studious, but was told every Sunday in church that he was a really bad person and that he must beg God for forgiveness or burn in hell and that he must never judge others, who must therefore be good or at least better than him. Those experiences haunted Mark into adulthood. He states the following: "*Even on those occasions where I felt confident that Jesus still loved me, it ate at me: I was convinced that I am a bad person, a **sinner*** (original author's emphasis) (p.25)."

Those who preached to Mark gave him the wrong impression of the Gospel of Christ. Their message was that we are all equally bad and will all go to hell unless we live every day oscillating between abject guilt and tearful thanks. Christ paid in full the price for all human sin. Christianity is about a life of joy, not a life of guilt. I see in Mark's writing a burning sense of injustice that to be a Christian seems to mean admitting that: *"I am worthless (and so are you),"* which is the title of one of his chapters.

Mark uses many of the same arguments that I use against lies and nonsense, including the following: the notion that all except Christians are damned to hell; the notion that God steps into perilous situations to spare His chosen few while maiming or killing the rest; biblical literalism, as illustrated by Balaam's talking donkey; and the hypocrisy of Christians who send death threats and celebrate the deaths of those whom they consider evil.

Mark continues: *"Now compare my mindset, which is not hurting anyone (except me, perhaps, in the eyes of the Christians who feel I'll fry in hell) to the CEO who creates the policies that really do hurt people. Who are those Christians who confronted me calling me evil? Who are they saying will go to hell? Certainly not the God-fearing CEO who goes to church every week and donates his time and money to the local chapter of the Grand Old Republican Party. No, it is not the CEO who must suffer through persistent threats of eternal damnation and hellfire. I am the one, for simply not believing as the fundamentalist Christians do (p.56-57)."*

Mark concludes by emphasizing, as follows, his lack of belief that Jesus even existed: *"Everything I've spelled out in this book makes the case clear to me...Jesus didn't exist. And the Jesus we have in our minds, the spiritual Jesus, isn't real...I've finally had no choice but to accept this simple fact about myself: I am no longer a Christian (p.257)."* Mark wants others to follow his lead and escape from Christianity. He lists

20 websites[10] to help would-be escapers. Christians who would confront Mark and the like-minded must admit the faults of organized Christianity and churches and stop preaching lies and nonsense to honest seekers of truth.

**Russell, B. 2004. *Why I Am Not a Christian* p.1-19, in *Why I Am Not a Christian and Other Essays on Religion and Related Subjects*. Routledge: London. 223p.**

This famous essay was delivered first as a lecture in 1927 in Battersea Town Hall, London, under the auspices of the South London Branch of the National Secular Society. Russell states: *"There is no reason to believe that the world had a beginning at all. The idea that things must have a beginning is really due to the poverty of our imagination* (p.4)." Really? Surely any scientist must see the need for a cause prior to the coming into being of our universe. Russell finds the case for God as the One who remedies injustice: *"a very curious argument* (p. 10)." I disagree. Without God as the record keeper and corrector of denials of justice, we live in a horror show of injustice.

Russell finds good and bad things to say about the character and teaching of Christ. He finds it unlikely that the allegedly loving and merciful Christ would have singled out as forever unforgivable the sin of speaking against the Holy Ghost.[11] Russell also finds it untenable to regard that particular sin as worse than any others. He notes how this idea of the unforgivable sin has caused *"an unspeakable amount of misery"* for those who assume that they have committed it (p.14).

There is no greater offense to anything than to deny its existence. No sin against God is greater than denying His existence in any of His forms. Nevertheless, I am with Russell in doubting that Christ intended the kind of misery that the literal acceptance of these and some of

---

10   Available through Jeffrey Mark's website: www.escapingchristianity.com
11   Matthew 12:31

His other reported words have brought about.

Russell assaults organized religion, especially Christianity. He concludes that: *"the Christian religion, as organized in its Churches, has been and still is the principal enemy of moral progress in the world* (p.17)." Even allowing for the lies, nonsense and intolerance in the history of Christianity, Christians and their churches have often taken some of the same moral stands that Russell supported, including pro-peace and anti-war.

He winds up with a recipe for "*What We Must Do*," which includes the following rationale for junking belief in God: "*The whole conception of God is a conception derived from ancient Oriental despotism. It is a conception quite unworthy of free men* (p.18). 'No,' it is a "*conception*" worthy of everyone, for exploration in honest faith-science quests for truth, through Free Thought.

**Stenger, V. 2009. *The New Atheism. Taking a Stand for Science and Reason*. Prometheus Books: New York. 282p.**

Victor Stenger begins this work with a brief tour of the atheism-faith battlefield: summarizing the New Atheism movement and the views and works of some of its front line troops, including Dawkins, Dennett, Harris and Hitchens, while denigrating some champions of faith, such as Dinesh D'Souza and Alister McGrath. Stenger's chapter titles speak volumes: "*The Folly of Faith; The Sword of Science; The Design Delusion; Holy Smoke* etc."

Stenger explains well why the argument from design must fail as an alleged proof of God's existence. He cites the Dover Pennsylvania case as a victory over Intelligent Design and presents strong evidence against Michael Behe and William Dembski. I agree with Stenger on all of the above, but his case for calling faith a "*folly*" rests largely on the far less than Christ-like views and behaviour found among the American

Christian Right ("*neocons*" and "*theocons*"). He includes a superficial dismissal of Francis Collins' honest testimony of being led to faith through the writings of C.S. Lewis.

Stenger is more interesting on science, apart from a fairly fruitless discussion about what others have said as to whether science itself is based on a kind of faith. He refers to his own work as follows: "*I randomly varied the electromagnetic force strength, the mass of the proton, and the mass of the electron by ten orders of magnitude around their existing values in our universe... I found that over half of the universes generated had stars with lifetimes of at least ten billion years, long enough for life of some kind to evolve* (p.93)." Good. The more possibilities for the evolution of life, the better! Now consider whether a spiritual realm might also have existed throughout and began to interact with the material realm once we came on the scene.

# APPENDIX V: GATHERING SUBJECTIVE EVIDENCE FROM WITHIN CONCERNING FAITH AND UNBELIEF

Recent research on subjective evidence from within concerning faith includes the following, *inter alia*: a Report from the Fetzer Institute,[1] the surveys of "*Religion Among Academic Scientists (RAAS)*" described by Elaine Ecklund,[2] and the "*Test of Faith*" project.[3]

The type of general questionnaire that might be developed for gathering and analysing subjective evidence from within concerning faith *and* unbelief is suggested as Table 1. Table 2 gives the present author's worked example. This questionnaire has not been peer reviewed or pretested. Readers are invited to comment on its utility and faults, with suggestions for improvements, and to try completing it if they so wish. The further development of this and/or other questionnaires for use in research will require the following:

I.  *Improved design; for example, so that logic checks are strengthened and responses are machine-readable.*

II. *Peer review and pretesting.*

---

[1] Fetzer Institute. 1999. *Multidimensional Measurement of Religiousness/Spirituality for Use in Health Research.* A Report of the Fetzer Institute and the National Institute on Aging Working Group. Fetzer Institute: Kalamazoo MI.103p. Available from: www.fetzer.org

[2] Ecklund, E.H. 2010. *Science vs. Religion: What Scientists Really Think.* Oxford University Press: Oxford. 228p.

[3] Bancewicz, R. Editor. 2009. Test of Faith: Spiritual Journeys With Scientists. Paternoster: Milton Keynes, UK. 119p. Test of Faith: Does Science Threaten God? DVD. Both available from: www.testoffaith.com

III. *Guidelines for respondents including the following: statements of the purpose of the questionnaire and thanks for completing it; definitions of terms; guidance on style and the content and length of free text entries; information on how to contact a 'help' website; explanation of how the data will be filed and analysed and the results published; prospects for follow-up.*

IV. *Preparation of an agreement to be signed by all respondents to certify that the information provided has been given freely and that no legal actions will be taken against any person involved in the research, and by all concerned with data gathering, analysis, storage and publication to certify confidentiality of all personal data.*

V. *Supplementary questions to cover specific circumstances and experiences.*

Table 1. Preliminary Questionnaire for Assessing Personal Histories and Present States of Faith, Atheism and Agnosticism

| |
|---|
| Name: Date/Location/Record No. |
| Date of Birth: |
| Gender: |
| Nationality: |
| Main locations (place, country; not full address): |
| Childhood (<5-10): |
| Youth (11-17): |
| Young-adulthood (18-30): |
| Mid-adulthood (31-50): |
| Late-adulthood (51-): |
| Main occupation(s) as an adult, past and present: |
| Young-adulthood (18-30): |
| Mid-adulthood (31-50): |
| Late-adulthood (>51): |

**A. Your past and present status as a believer or nonbeliever in the existence the spiritual realm**

**Over the course of your whole life up to *and including* the present**

1. Have you ever believed in a spiritual realm? (Yes/No/Don't know)
2. Have you ever believed in God? (Yes/No/Don't know)
3. Have you ever believed in the spiritual force for evil? (Yes/No/Don't know)
4. Have you ever believed that you have a soul? (Yes/No/Don't know)

**At present; i.e., right now**

5. Do you believe in a spiritual realm? (Yes/No/Don't know)
6. Do you believe in God (Yes/No/Don't know)
7. Do you believe in the spiritual force for evil? (Yes/No/Don't know)
8. Do you believe that you have a soul? (Yes/No/Don't know)

**B. Your past and present *registered* or *officially recognized* membership in any organized religion or other organization based on faith or unbelief**

**Past membership(s), if any, in any organized religion or other organization based on faith or unbelief, over your whole life, up to and including the present**

1. Have you ever been a registered or otherwise officially recognized member of any organized religion or organization based on faith or on unbelief? (Yes/No/Don't know)
2. If you answered Yes to B.1, please list here the general name(s) of all the organized religions (not specific branches, sects, churches, mosques, temples etc.) and major organizations based on faith or on unbelief, in which you have had registered or otherwise officially recognized membership - (Buddhism, Christianity, Hinduism, Islam, Judaism; Other - please specify; Atheist Organization; Humanist Organization; Other - please specify)

**Present membership in any organized religion or other organization based on faith or unbelief**

3. If you answered Yes to B.1, are you at present a registered or otherwise officially recognized member of any organized religion or organization based on faith or on unbelief? (Yes/No/Don't know)
4. If you answered Yes to B.3, please list here, in the same form as used for B.2, the name (please enter only one) of the organized religion or other organization in which you are at present a registered or otherwise officially recognized member - (Buddhism, Christianity, Hinduism, Islam, Judaism; Other - please specify; Humanist Organization; Other - please specify)
5. **Please add free text, as you wish, to amplify any answers in B. 3,4**

C. Your *privately held* states of faith or unbelief, irrespective of what anyone told you to believe or required you to profess.

1. For the following stage(s) in your life so far, taking the age ranges here as approximate, please enter your true state(s) of belief or unbelief as follows: F (for faith, defined as personal belief and trust in God); A (for atheism, defined as denial of the existence of God); and X (for agnosticism, defined as neither affirming nor denying the existence of God). Enter only one state (F or A or X) for a life stage throughout which your belief status was constant. Enter a sequence of states, separated by one or more semicolons, for any life stage during which your state changed one or more times.

**Childhood, <5-10** (; ;)
**Youth, 11-17** (; ;)
**Early-adulthood, 18-30** (; ;)
**Mid-adulthood, 31-50** (; ;)
**Late-adulthood, >51** (; ;)

2. For any entries in C.1 that describe changes of status within a life

stage and in the progression from one life stage to the next, please provide free text explanations.

**Childhood, <5-10:**
**Youth, 11-17:**
**Early-adulthood, 18-30:**
**Mid-adulthood, 31-50:**
**Late-adulthood, > 51**

**D. Your experiences of receiving any information, revelations, or communications from spiritual realm sources**

1. During the stages of your life so far, have you experienced any receipt of information, revelations, or communications from God or other spiritual realm source(s)? For all 'Yes' answers, please list the circumstances during which you had your experiences; for example - meditation, prayer, reading, lectures and sermons, conversations, films and plays, art, music, nature, others. For any 'Yes' answers that refer to spiritual realm source(s) other than God, please specify the source(s); for example - angels, demons, saints, the Virgin Mary, departed souls, the spiritual force for evil, others.

**Childhood, <5-10:**

From God (Yes/No/Don't know); (Circumstances)
From other spiritual source(s) (specify) (Yes/No/Don't know); (Circumstances)

**Youth, 11-17:**

From God (Yes/No/Don't know); (Circumstances)
From other spiritual source(s) (specify) (Yes/No/Don't know); (Circumstances)

**Early-adulthood, 18-30:**

From God (Yes/No/Don't know); (Circumstances)

From other spiritual source(s) (specify) (Yes/No/Don't know); (Circumstances)

**Mid-adulthood, 31-50:**

From God (Yes/No/Don't know); (Circumstances)
From other spiritual source(s) (specify) (Yes/No/Don't know); (Circumstances)

**Late-adulthood, > 51:**

From God (Yes/No/Don't know); (Circumstances)
From other spiritual source(s) (specify) (Yes/No/Don't know); (Circumstances)

2. **Please add free text here, as you wish, to amplify any answers in D. 1**
3. **Please re-enter here your present privately held belief status - faith, (F); atheism, (A); or agnosticism (X)** – and assess the proportionate strengths of your present influences from people, material sources of information and spiritual revelations that help you to sustain your belief status or threaten it. Taking your sustaining and threatening influences as each totalling 100%, enter the relative % contributions of those three types of influences.

**Percentages of sustaining influences:**

People (%)
Information from material sources (%)
Spiritual revelations (%)
Total = 100%

**Percentages of threatening influences:**

People (%)
Information from material sources (%)
Spiritual revelations (%)
Total = 100%

4. **Please add free text, as you wish, to amplify any of your assessments of past and present influences in D.3; for example,**

> identify people, material sources of information, and spiritual revelations that have sustained or threatened your belief status, or are doing so at present.

E. **Your personal and private attitudes to science and to the compatibility of faith and science**

1. To what extent do you trust science and scientists? (Strongly/Moderately/Weakly/Not at all)
2. **Please add free text, as you wish.**
3. To what extent do you regard faith (defined as personal belief and trust in God) and science as compatible and complementary paths to truth? (Strongly/Moderately/Weakly/Not at all)
4. **Please add free text, as you wish.**

Table 2. Preliminary Questionnaire for Assessing Personal Histories and Present States of Faith, Atheism and Agnosticism: A Worked Example

**Name:** Roger Pullin
**Date/Location/Record No.** N/A
**Date of Birth:** February 4, 1944
**Gender:** M
**Nationality:** British
**Main locations (place, country; not full address):**
**Childhood (<5-10):** Wolverhampton, UK
**Youth (11-17):** Wolverhampton, UK
**Young-adulthood (18-30):** London and York, UK; Isle of Man
**Mid-adulthood (31-50):** Isle of Man; Makati City, Philippines
**Late-adulthood (>51):** Makati City, Philippines
**Main occupation(s) as an adult, past and present:**
**Young-adulthood (18-30):** Student; University Lecturer (Marine Biology)
**Mid-adulthood (31-50):** Aquatic Biologist
**Late-adulthood (>51):** Aquatic Biologist

### A. Your past and present status as a believer or nonbeliever in the existence the spiritual realm

**Over the course of your whole life up to and including the present**

1. Have you ever believed in a spiritual realm? (Yes)
2. Have you ever believed in God? (Yes)
3. Have you ever believed in the spiritual force for evil? (Yes)
4. Have you ever believed that you have a soul? (Yes)

**At present; i.e., right now**

5. Do you believe in a spiritual realm? (Yes)
6. Do you believe in God (Yes)
7. Do you believe in the spiritual force for evil? (Yes)
8. Do you believe that you have a soul? (Yes)

### B. Your past and present *registered* or *officially recognized* membership in any organized religion or other organization based on faith or unbelief

**Past membership(s), if any, in any organized religion or other organization based on faith or unbelief, over your whole life, up to and including the present**

1. Have you ever been a registered or otherwise officially recognized member of any organized religion or organization based on faith or on unbelief? (Yes)
2. If you answered Yes to B.1, please list here the general name(s) of all the organized religions (not specific branches, sects, churches, mosques, temples etc.) and major organizations based on faith or on unbelief, in which you have had registered or otherwise officially recognized membership - (Christianity)

**Present membership in any organized religion or other organization based on faith or unbelief**

3. If you answered Yes to B.1, are you at present a registered or otherwise officially recognized member of any organized

religion or organization based on faith or on unbelief? (Yes)
4. If you answered Yes to B.3, please list here, in the same form as used for B.2, the name (please enter only one) of the organized religion or other organization in which you are at present a registered or otherwise officially recognized member - (Christianity)
5. **Please add free text, as you wish, to amplify any answers in B.3, 4**
I was baptized as a baby and confirmed as a youth as a member of the Church of England. I am presently a member of an international, interdenominational church, where I have served on the Council of Elders and in the Music Ministry.

**C. Your *privately held* states of faith or unbelief, irrespective of what anyone told you to believe or required you to profess.**

1. For the following stages in your life so far, taking the age ranges here as approximate, please enter your true state(s) of belief and unbelief as follows: F (for faith, defined as personal belief and trust in God); A (for atheism, defined as denial of the existence of God); and X (for agnosticism, defined as neither affirming nor denying the existence of God). Enter only one state (F or A or X) for a life stage throughout which your belief status was constant. Enter a sequence of states, separated by one or more semicolons, for any life stage during which your state changed one or more times.

**Childhood, <5-10 (X)**
**Youth, 11-17 (F)**
**Early-adulthood, 18-30 (X)**
**Mid-adulthood, 31-50 (X; F)**
**Late-adulthood >51 (F)**

2. For any entries in C.1 that describe changes of status within a life stage and in the progression from one life stage to the next, please provide brief, free text explanations.

**Childhood, <5-10:**
**Youth, 11-17:** I became a believer just prior to my confirmation, committing my life to Christ.
**Early-adulthood, 18-30:** At college in London, I lost my faith and became an agnostic.
**Mid-adulthood, 31-50:** From an agnostic life that was empty of anything spiritual, I took the leap back to faith.

**D. Your experiences of receiving any information, revelations, or communications from spiritual realm sources**

1. During the stages of your life so far, have you experienced any receipt of information, revelations, or communications from God or other spiritual realm source(s)? For all 'Yes' answers, please list the circumstances during which you had your experiences; for example - meditation, prayer, reading, lectures and sermons, conversations, films and plays, art, music, nature, others. For any 'Yes' answers that refer to spiritual realm source(s) other than God, please specify the source(s); for example - angels, demons, saints, the Virgin Mary, departed souls, the spiritual force for evil, others.

**Childhood, <5-10:**
From God (Don't know); (Circumstances)
From other spiritual source(s) (specify) (Don't know); (Circumstances)

**Youth, 11-17:**
From God (Yes); (Circumstances - prayer, reading, sermons, conversations, music, in daily life)
From other spiritual source(s) (spiritual force for evil) (Yes); (Circumstances - reading, conversations, in daily life)

**Early-adulthood, 18-30:**
From God (Don't know); (Circumstances)
From other spiritual source(s) (specify) (Don't know); (Circumstances)

**Mid-adulthood, 31-50:**
From God (Yes); (Circumstances - prayer, reading, sermons, conversations, music, in daily life)
From other spiritual source(s) (spiritual force for evil) (Yes); (Circumstances – reading, conversations, in daily life)

**Late-adulthood, >51:**
From God (Yes); (Circumstances - (Circumstances - prayer, reading, sermons, conversations, music, in daily life)
From other spiritual source(s) (spiritual force for evil) (Yes); (Circumstances – reading, conversations, in daily life)

2. **Please add free text here, as you wish, to amplify any answers in D.1.** See the sections about me in chapters 5 and 7 and the entire contents of Appendices II and III.
3. **Please re-enter here your present privately held belief status** - faith, (F); atheism, (A); or agnosticism (X) – and assess the proportionate strengths of your present influences from people, material sources of information and spiritual revelations that help you to sustain your belief status or threaten it. Take all of your influences as 100% and enter the relative % contributions of those three types of influences. F

**Percentages of sustaining influences:**
People (10%)
Information from material sources (20%)
Spiritual revelations (70%)
Total = 100%

**Percentages of threatening influences:**
People (10%)
Information from material sources (15%)

Spiritual revelations (75%)
Total = 100%
4. Please add free text, as you wish, to amplify any of your assessments of past and present influences in D.3; for example, identify people, material sources of information and spiritual revelations that have sustained or threatened your belief status, or are doing so at present. The battlefield for me is mainly in my soul, between revelations from God and opposition from the spiritual force for evil. The church community to which I belong is an important influence for sustaining my faith, but is matched by arguments from nonbelievers against faith. The material sources that I encounter, especially books, are also balanced for and against faith. The Bible and sacred music are important material sources for sustaining my faith. God sends me guidance and other revelations through them, as well as directly in my communication with him in prayer and in daily life.

**E. Your personal and private attitudes to science and to the compatibility of faith and science**

1. To what extent do you trust science and scientists? (Strongly)
2. **Please add free text, as you wish.**
   Science is a path to truth and is the main means for improving the material wellbeing of humans.
3. To what extent do you regard faith (defined as personal belief and trust in God) and science as compatible and complementary paths to truth? (Strongly).
4. **Please add free text, as you wish.**
5. Faith and science are entirely compatible. They are complementary paths to seeking and finding, during our material realm existence, all the truths that are accessible to us about the totality of the spiritual and material realms.

# INDEX[1]

Aaron 239
abortion 223, 226, 372
Abraham 219, 229, 233, 234
absolutism 194, 199, 200, 357
accidents xi, 27, 51, 157, 186, 202, 203, 302, 372
Acts 28, 117, 167
acts of evil xii, 10, 77, 157-159, 191, 226, 342, 369, 373, 397, 402
acts of God; see also divine action 27, 157, 226, 302, 393
Adam and Eve 30-32, 240, 373
agenticity 327, 328
agnosticism 15, 41, 190, 275, 319, 333, 338, 340, 357, 360, 411, 412
Alcorn, Randy 246, 247
Alexander, Denis 129, 169, 177, 394
Alexander, James 192, 193
Allah 226
all-embracing monism; see also dual aspect monism 46
Allen, John 295
aliens 25, 62, 148, 160, 177, 327
Allon, Henry 131
Almighty, the 35, 348, 349, 353
Alper, Matthew 49, 55, 303, 321
altruism; see also genuine altruism 40, 217, 218, 249, 323, 340
Ang, Armando 226
angels 104, 158, 191, 210, 228-244, 327, 357-363
Anglican 128, 162, 174, 251, 376

Angus, Joseph 131
Anselm, St. 413, 415
Anthropic Principle 20, 202, 349, 401
apologetics/apologists 168, 187, 195-201, 253, 280-289, 391, 413
apes 171, 206, 210-212
Apostles' Creed 275, 347-367, 412
Aquinas, St. Thomas 113, 404
Aronis, Alex 358, 368, 384
Aslan, Rezla 265
Atkins, Peter 152, 153, 182, 395, 396
Athanasian Creed 347, 348, 357
atheism 1, 3, 11-41, 60-73, 106, 112, 152-180, 208, 299, 319, 338-357, 391, 396
Augustine, St. 57, 153, 244, 347
Aurelius, Marcus 57
Averroës 61
aware; see also God-aware/self-aware 42-47, 57, 68, 79, 106, 112, 194, 239-331, 341

Bacteria 88, 135, 176, 178, 183
Baglio, Matt 360
Balaam 239, 247, 416
Bancewicz, Ruth 14, 420
Barbour, Ian 86, 266, 279, 304, 324
Baron-Cohen, Simon 159
baselines; see also soul baselines 60, 72, 84, 109-116, 180, 313, 324

---

1 Widely used terms and subjects are listed by the main sections in which they occur, though not on every page; for further guidance, see Table of Contents.

basic thought ix, 3-6, 31-37, 67-85, 115, 120, 184, 333-336
battlefields 2-5, 155-188, 223, 267, 278, 314, 390, 418, 431
BBC 146, 172, 250, 226, 273, 296
beables 23, 26, 86-89, 99, 101, 334
beauty 52, 107, 136, 145, 146, 174, 360, 362
Beck, Friedrich 87
Behe, Michael 182, 183, 278, 401, 418
belief/believer ix-1, 9-49, 57-85, 97-179, 188-415
belief-dependent realism 105, 328
Benedetti, Fabrizio 112
Benson, Herbert 258
Benson, Ophelia 195
Berry, R.J. (Sam) 169, 394
betrayal xii, 161, 372, 386
Bible, the Holy 26, 40, 62, 64, 125-127, 158-188, 205, 223-299, 320, 322, 358-390, 403-414
biblical 75, 121, 151, 178, 212-246, 266-296, 341, 367, 403, 408, 414
biblical/sacred text literalism 13, 179-205, 230-242, 263-294, 341, 365, 405, 416
biblical inerrancy 188, 231, 238, 241, 270, 282-299, 341, 407, 411
Big Bang xii, 7, 21, 110, 169, 180, 202, 278, 349-365, 392-402
biodiversity xiii, 7, 51, 172-176, 362, 386
BioLogos 283
biophilia 214, 221
birds 7, 27, 88, 124, 173, 174, 207, 249, 260
Blackburn, Simon 10, 190-199, 305

Blackford, R. 396
Blackham, Harold J. 119
body x, 25-51, 95, 150, 154, 245, 274, 331-355
body-mind x-7, 24-57, 91-120, 151, 328, 338-345, 352, 354, 362, 371
Bohm, David 87, 100, 101
Borg, Marcus 219
bombardier beetles 178, 179, 281, 28
bonobos 184, 208, 212
brain x-12, 22-57, 70-112, 158, 210, 217, 274, 303-341, 396, 397
brain, damage to 55-57
brain imaging, scans 49, 33
'brain is all' 46, 326, 332, 333
Bricklin, Jonathan 70
Britton, Ronald 165
Brooke, Charles 178
Brosnan, Sarah 206
Browne, Thomas 14, 19, 44
Brown, Mick 197
Brownnutt, Michael 88
Brown, T.E. 19
Brüne, Martin 316
bubonic plague 27
Bucaille, Maurice 232
Buckinghamshire 375
Buckoll, Henry J. 123
Buddhism 21, 40, 109, 197, 219, 254, 344, 357, 405
Burke, Eric 254, 256
Byrd, Randolph 255-259

Calvin, John 106, 113
cancer 140, 245, 367
Carse, James 68, 69, 107
case studies 153, 315

Catechism, the Catholic 111, 236, 227, 242, 251, 296, 297, 356, 397
categorical difference or continuum 207-212, 331
Catholic, Catholick 354, 355, 357
Catholic, Roman 60, 111, 163-186, 226-298, 347-360, 405, 414
causal joint 98, 393
cerebral mental field 330, 331
Chadwick, Owen 164
Cha, K.Y. 259, 260
Chapman, A. 173
chaos 23, 66, 80, 84, 98
charity 15, 152, 213-215, 290, 356
Chesterton, G.K. 66, 264
chimpanzees 25, 206, 208
Chisholm, Roderick 69, 200
Chopra, Deepak 48
Christ, see also Jesus xii, 7, 16-29, 42, 75, 104-131, 161, 185-190, 213-309, 341-398, 402-418
Christians, Christianity xiii, 6, 9, 43-119, 127-131, 156-289, 309, 319-329, 341-419
Christians in Science 14, 129
Churinoff, George 197
Clarke, Peter, G.H. 80, 87, 88
Clark, Stephen R. 167
Clifford, William 305
Collins, Francis 14, 169, 390, 391, 419
Colossians 310
Colson, Charles 169
comfort zones 11, 16, 197, 267
common sense 10, 34, 137, 153, 189, 200, 201, 288, 345
compassion 206, 223, 274, 413
compatibilism 5, 402

complexity; see also irreducible complexity xii, 37, 108, 145-150, 177, 180, 202, 257, 281, 282, 316
Compte-Sponville, André 72, 122
confession and repentance 50, 112, 224, 261, 320, 355, 356, 385
conscience 30, 41, 73, 131, 142, 159, 226, 290-297, 341, 373
consciousness ix, 3, 11, 22-55, 62, 70-104, 121-154, 192, 216, 267-333, 339-341, 365, 411, 415
Consolmagno, Guy 296
consubstantial 341, 350, 353, 356
conversion 49, 77, 117, 120, 121, 218, 280, 320-323
Conway, D.J. 243
Conway, Erik 140
correctness 134, 145, 146, 306
Corinthians 29, 59, 162, 205, 213, 233, 268, 288, 370, 375, 377
corruption xii, 140, 200, 204, 221, 372
cosmology, cosmos 21, 156, 271, 278, 396, 400, 401, 415
Coward, Kevin 371
Cox, Harvey 271
Coyne, Jerry 168
Craig, William Lane 413, 414
create xiii, 14, 20-27, 49, 76, 139, 151-187, 251-296, 342-348, 362-369, 391
creationism 7, 21, 40, 53, 59, 76, 94, 132, 134, 165-186, 203, 218, 263-298, 340-351, 393-405
creativity ix-26, 53-89, 130-176, 202, 230, 266-307, 324-339, 362

Creator 64, 97, 132, 151-182, 284, 312, 343, 370, 393
creeds; see also Apostles', Athanasian and Nicene Creeds ix, 11, 38, 62, 342-357, 412, 418
cruelty xii, 133, 159, 276, 372, 397

Damasio, Antonio 36, 37, 43, 44, 83-85, 325, 329, 330, 341
Daniel 230, 239, 251, 252, 366
Darwin, Charles 14, 43, 107, 126-135, 157-182, 212, 281, 311, 361, 394, 399
Darwinian/Darwinism 134-136, 167-175, 278, 286, 298, 361, 393
David 163, 229
Davies, Brian 304
Dawkins, Richard 1, 12, 39, 107, 124, 215, 311, 361, 366, 397, 398, 418
Day, Vox 163
deceit/deception 140, 233, 268, 372, 405
deism xi, 63, 65, 104, 115, 342, 392, 393
deity 101, 283, 308, 355, 405, 406
Dembski, William 177, 178, 401, 418
demons/demon possession 104, 147, 159, 160, 225, 245, 269, 327, 348, 359, 360, 406-411
Dennett, Daniel 11, 43, 73, 106, 146, 147, 157, 280, 327, 331, 398-400, 413, 418
Descartes, René 44-46, 57, 83, 415
descent with modification 40, 43, 135, 168, 271, 277, 281
design; see also Intelligent Design 6, 78, 140, 150-182, 227, 232, 281-287, 328, 394-401, 415, 418

determinism 4, 5, 69, 70, 79, 80, 87-105, 342
Deuteronomy 222, 223, 229, 230, 231, 254
Deutsch, David 17-20, 138, 139, 311
devil; see also Satan, spiritual force for evil 78, 158, 162, 182, 360, 397
de Pomerai, David, 251
de Waal, Frans 184, 206-210
Dewey, John 200
Dirac, Paul 94, 95
discrimination xii, 110, 161, 372
diseases xi, 7, 41, 55, 106, 228, 251, 261, 367, 372, 410
divine action/interventions; see also acts of God xi, 7, 8, 24, 89-120, 166, 203, 278, 337-351, 393, 402
divine collapse causation (DCC) 278, 279
DNA 25, 52, 72, 75, 105, 146, 164-193, 229-297, 342-373, 391, 407
doctrine ix, 29, 188, 227, 238, 299, 325, 347
dogma ix, 104-109, 163-182, 228, 236, 267-299, 337, 365, 404-412
dogs 49, 50, 137, 184, 206, 363, 381, 398
Donne, John 365
D'Souza, Dinesh 168, 418
doubt 63, 80, 106-128, 140, 151, 171, 180, 192, 204-235, 257, 261, 296, 322-338, 356, 408, 417
Dover, Pennsylvania 167-182, 418
Dowd, Maureen 225
Doyle, Bob 5

dualism x, 6, 37, 44, 46, 53, 56, 97, 101, 330, 351
dual aspect monism 46, 97
Dugatkin, Lee 249
Dulles, Avery Cardinal 280
Dylan, Bob 17, 112, 159

Eagleton, Terry 37, 60, 104, 158
Eccles, John 22, 53, 87
Ecklund, Elaine 420
Einstein, Albert 14, 52, 163
electron 25, 86, 307, 419
elegance 24, 107, 145, 146, 328
Elizabeth 224
Elokiim 26
embryo 33, 180, 259, 260, 351
emotion 44, 45, 83, 105, 112, 200, 208, 316, 322
empathy 158, 184, 206
Endara, Miguel Angel 281
energy x, 23-34, 66-97, 136, 151, 180. 181, 244, 262, 267, 332, 338
Enoch 240, 252
Ephesians 30, 130, 159, 193, 213, 236, 357, 368, 373, 389,
Epicurus 64, 166
Episcopalian 111, 227, 242, 251, 274, 296, 347, 411
Equidistant Letter Spacing 232
Erickson, Millard 199
Ernst, Edzard 148
eschatology 342, 345, 354
eternal/everlasting life 19, 32 49-75, 126, 274, 340-370
ethics 4-24, 140-164, 190-197, 245, 258, 280-289, 305-342
evangelism 29, 187, 196, 218, 225, 246-253, 280-295, 320, 325, 377
Evans, C. S. 56

Everitt, Nicholas 108, 113, 156, 396, 400-402
evidence; see also objective/subjective evidence 2, 10-16, 30-42-116, 131-157, 166-210, 238-287, 304-361, 391-418
evil; see also spiritual force for evil, acts of evil xi, xii, 7, 26-34, 59-89, 107-133, 155-191, 211-235, 281, 302, 312, 339-374, 395-416
evolution ix-7, 36-68, 97-116, 134-140, 157-187, 198-329, 348-362, 391-407, 419
Exodus 26, 222, 223, 232, 239, 274, 346, 366
exorcism 159, 160, 360, 406, 410
expediency 40, 184, 190, 205, 211-219, 373, 371
experiences of God; see also spiritual revelations 23-42, 57-88, 101-153, 196-205, 237, 264-289, 300-328, 349-361, 397-415
explicate/unfolded order 100, 101
Ezekiel 162

Fairness 160, 206, 220, 401
faith ix-16, 29, 40-84, 103-165, 189-213, 226, 235-418
faith, defined xi, 1, 8, 28, 103, 338, 341, 342, 347, 354, 357
faith-friendly revolutions in science xiii, 265, 272, 300
faith healing; see also healing 244, 245
faith, leaps to or from; see also conversion 118-124, 390, 420
faith and religion 27-35, 209, 225-227, 266, 269, 325, 397, 405

faith-science quest for truth/unity xiii-2, 13-18, 185, 188, 206, 207, 263-313, 333, 335, 396, 418
faith-science relationships xii-18, 48-81, 129, 142-201, 263-272, 298-310, 335-369, 391-396
falsehood 52, 165, 189, 193, 196, 203, 205, 306, 343, 411
falsifiability 134-137, 143, 172, 205, 308, 391, 412
Faraday Institute for Science and Religion 129, 169, 173, 202
Farrar, Adam Storey 63
Father, the 28, 111, 161, 213, 344-359
Fergusson, David 130
Fernádez-Armesto, Felipe 268
Ferreira, Jamie M. 118
Fetzer Institute 420
Feynman, Richard 15, 140, 145, 152
fideism 195
first-person 314, 315, 331
fish xiii, 25, 115, 124-127, 148, 173-188, 207, 238, 249, 261, 262, 275, 288, 363-386, 407
Fisher, J. 250
flesh and blood 34, 340, 355, 373
Flew, Anthony 73, 133, 180, 336, 340-345, 391, 392
foetus 33, 53
forgiveness 50, 77, 131, 224, 322, 344, 352-355, 385, 415
Forbes, Edward 379, 381
Fox, George 244, 326
Fox, Matthew 274
Frankfurt, Harry 193-199
fraud 59, 140, 141, 172, 326, 411

freedom 5, 8, 25-78, 138, 158-200, 232, 236, 273-313, 337, 349, 372, 375
free process xi, 7, 8, 27, 76, 77, 157, 158, 202, 218, 235, 236, 302, 349, 357, 369, 389
freethinking, conventional sense ix, 61-66, 337
Free Thought 3-8, 33-89, 105-217, 238, 272, 293-338, 354-389
Free Thought episodes 8, 27, 59, 96-113, 142, 157, 220
Free Thought mechanisms x-16, 24-46, 70-101, 113-119, 203, 218, 271, 279, 310, 313, 325-331, 360
Free Thought scope/territory ix-15, 29-85, 109-120, 130-160, 176-226, 297-311, 330-348, 370-405, 418
free will 4, 5, 35, 64-80, 117, 118, 160, 302, 336, 337, 369, 402
free-willed xi, 5, 60-96, 118, 218, 236, 337, 369
Fromm, Eric 50, 71
fundamentalism 27, 164, 185, 245, 270, 271, 293, 321, 394-403, 416

Gadarene swine 408-412
Gage, Phineas 55
Galatians 68
Garden of Eden 31, 32, 250, 261, 373
Geisler, David 187, 188
Geisler, Norman 179, 195, 196, 281-285, 413
gender 11, 248, 250, 274, 299, 343-363, 411, 421, 426

genes/genomes 25, 33, 72, 74, 146, 177, 211-217, 281, 302, 340, 390, 391
Genesis 13, 25, 31, 149, 172-187, 230-252, 283, 284, 349, 362, 366, 405, 407
genetics x, 25-66, 126, 149-183, 214, 216, 248, 249, 368-39
genuine altruism 40, 212-218, 340
Giberson, Karl 169, 212
Gladstone, J.H. 166, 238, 407, 411
God; see also Allah, Almighty, Christ, Elokiim, Father, Ja/ko/vah, Jesus, Holy Ghost, Holy Spirit, Logos, Lord, Word, Wisdom, Yahweh ix-1, 16-32, 63, 73, 100-306, 338-379, 390-408, 418
God-aware ix, 25, 35, 43, 60, 139, 310, 349,
God, existence of 8-42, 108-114, 130, 151-179, 196-215, 281, 304, 341-350, 390-402, 414-418
God-human relationships x-xii, 5-139, 146, 153-162, 188-245, 254-265, 272-313, 321-416
God, perceived nature/will 3, 9-13, 20-29, 42, 76-133, 151-217, 268-318, 339-356, 368-405, 417
Goetz, Stuart 56, 57
Goldacre, Ben 149
good xi, 20-42, 68-77, 104, 113, 133, 141, 155-190, 208-249, 291, 292, 312, 346-395, 411, 415
Gordon, Matthew 74, 228, 229
Gospel, the 28, 29, 187, 264, 266, 280, 284, 390, 416
Gould, Stephen, J. 12

grace 32, 77, 104-117, 162, 211, 213, 275, 357-388, 403
Grayling, A.C. 156, 203
Greek 62, 219, 238, 245, 412
Groothuis, Douglas 83, 84, 200, 201, 285-287
Grossman, Dave 223

**H**adith 250
Hamer, Dean 249
Hameroff, Stuart 48
Hamilton, William (Bill) 217
Hannay, Alistair 118
happiness iii, 37-40, 118, 125, 131, 303, 321
Hardy, Alister 381
Harman, Oren 217, 218
Harris, Judith Rich 37
Harris, Sam 14, 15, 39, 72, 73, 264, 359, 403, 418
Harris, William 256, 257
Hassan, Ihab 198
hatred 160, 265, 372, 392
Havergal, Frances Ridley 118
Hawking, Stephen 78, 150, 151, 328
healing 78, 159, 160, 245, 256, 360, 367, 408, 411
heaven 9, 32, 75, 111, 122, 158, 213, 231-253, 274, 310, 343-371
Hebrews 42, 103
Heisenberg's Uncertainty Principle 80, 86, 87, 94, 95
hell 29, 75, 187, 274, 323, 343-352, 362-370, 415, 416
heresy 32, 306, 309
higher things ix, x, 3, 6, 16, 53, 66, 84, 85, 146, 331-337
Hinduism 41, 219, 344, 357

Hitchens, Christopher 359, 404, 405, 413, 418
Hodge, David 259, 260
Hodgson, P.E. 347
Holocaust 159, 161,188, 191, 369
Holt, Jim 20, 21
Holy Ghost, Holy Spirit 6, 28, 84, 111, 189, 241, 242, 270, 294, 299, 317, 351-356, 411-417
homeopathy 147, 148
Homo/hominids 25, 166
homophobia 250, 273, 397
*Homo sapiens* 25, 36, 139
homosexuality 224-228, 248-251, 273
honest seekers/tellers of truth xii, 11, 23-30, 168, 188, 224, 247, 248, 272, 288, 304, 354, 370, 396, 407, 408
human condition/human nature x, 3, 10-22, 31, 40, 49, 64, 68, 100, 108, 139-151, 200-214, 251, 276, 292-339, 396, 400
humanist 65, 155, 156, 208, 422
human rights 47, 221, 290, 291
Hume, David 141, 208, 227, 356
Huxley, Thomas Henry 15, 127, 326, 378, 405-412
hymns 104, 118, 123-129, 323, 377
hypocrisy 273, 405, 416
hypothesis 48, 54, 108, 109, 132-147, 202, 283, 311, 325, 331

Ignorance 66, 69, 70, 109, 144, 167, 204, 226, 256, 305
immorality 3, 31, 41, 42, 89, 190, 193, 404
implicate/enfolded order 100. 101
improbability 66, 177
Incarnation xii, 7, 17-32, 110, 119, 160, 231, 274-284, 350-367, 402, 410
incest 40, 224, 372
inconvenient truth xii, 133, 168, 187, 195, 278, 305
incompatibilism 5, 402
infallible 11, 13, 191, 227, 291, 357, 411
information ix, xi, 5-11, 24-101, 114-143, 181, 201, 216, 275, 279, 298, 326-339, 410
injustice; see also unfairness xii, 3, 40, 42, 160, 192, 210, 239, 359-374, 395, 416, 417
Inquisitions 161, 226
Intelligent Design 167, 175-183, 278-288, 327, 394, 396-405, 418
intercessory prayer 214, 254-261, 369
interreligious organizations 185
interpersonal 217, 372, 387
intersubjectivity 314-317
intolerance 185, 269, 303, 405, 418
Irenaeus, St. 244, 309, 310
irrationality 163, 149, 286
irreducible complexity 178-183, 287
Isaiah 84, 123, 126, 220, 229, 268, 351, 364
Islam/Islamic; see also Muslim 35, 41, 61, 62, 74, 103, 156, 185, 196, 219-266, 293, 295, 344, 371
Isle of Man, the xiii, 148, 261, 262, 365, 380-388, 397

Ja/ko/vah 26
Jaenicke, Chris 316, 317
James, the Apostle 19, 162, 213, 344
James, William 59, 70, 94, 108, 109, 120-122, 237, 320, 415

Jeffers, John 138
Jeeves, Malcolm 52
Jeremiah 229, 230, 234
Jesuits 296
Jesus; see also Christ xii, 29, 110, 112, 123, 187, 190, 219, 229-282, 318, 350-379, 397, 402-416
Job 161, 232, 283, 355, 380-383, 392
John, the Apostle 26, 28, 132, 160, 189, 190, 231, 235, 318, 350, 351, 363, 364, 390
Johnson, Phillip 169
Johnstone, Gary 182
Jonah 187, 188, 226, 238, 239, 275, 288, 366, 407, 408
Jones, Judge John E. III 182
Judaism 25, 41, 185, 196, 219, 226, 242, 251, 293, 344
justice ix, 3, 15, 42, 59, 160-226, 274-297, 329-373, 395, 398, 417

Kahneman, Daniel 4
Kane, Robert 5, 69
karma 27, 197
Kaufmann, Stuart 99, 100
Keller, Tim 275
Kenneson, Philip 201
Kent, Keith 222, 396
Kiely, David 360
Kierkegaard, Søren 118, 119
killing 9, 41, 191, 219-226, 265, 266, 364, 373, 416
Kinnaman, David 273
Kitzmiller, Tammy 181
Koran; see Qu'ran
Krauss, Lawrence 21, 22
Krebs, Hans 141
Krucoff, Mitchell 253

Kuhn, Thomas 11, 141
Küng, Hans 266

Lahaye, Tim 172, 252, 253
Landes, David 163
laws of nature xi, xii, 17, 23, 67, 72, 86, 125-129, 150-157, 190, 302-312, 343, 362-367, 391, 401
Lebo, Lauri 182
Lehman, F.M. 123
Lennox, John 151, 153
Leviticus 40, 222, 229, 233
Levy, Neil 210
Lewis, C.S. 68, 188, 197, 217, 289, 319, 390-392, 419
libertarianism 217, 391
Libet, Benjamin 70, 78, 79, 94, 330, 331
lies and nonsense; see also falsehood, lying 2, 9, 26, 29, 71, 114, 133, 157-160, 189-226, 269, 275, 301-313, 342, 343, 372, 389
Lilla, Mark 198
literal impossibilities 238, 275, 407, 411
Liverpool 378-387
Loftus, John 413, 414
Logos, the 242, 298
Lohbeck, Kai 87
London 120, 128, 153-156, 217, 218, 261, 377-387, 407
Lord, the 8, 26-29, 49, 75, 106-134, 161, 162, 218-269, 350-377, 407
Lot 233, 366
Love 9, 28-76, 112-130, 152-194, 207-233, 269-307, 321, 342-403
Lubenow, Martin 172

Luke, the Apostle 7, 42, 161, 187, 222, 234, 254, 346, 352, 368, 408, 409
Luther, Martin 270
lying xii, 161, 165, 303
Lyons, Gabe 273
Lynch, John 202

MacArthur, John 12
MacCullough, Diarmaid 229, 230, 265
MacDougall, Duncan 49, 50
Mackay, Charles 9, 71
Mackie, J.L. 156, 192, 414, 415
Maimonides, Moses 13, 241, 404
M and S theory 152, 333-335
Margenau, Henry 85
Mark, the Apostle 17, 185, 187, 229, 242, 254, 283, 352, 365, 408, 409
Mark, Jeffrey 65, 415, 416
Mary, the Blessed Virgin 32, 234, 350, 351
materiality, material stuff, matter; see also material realm, universe ix-3, 18-66, 85-126, 151-197, 214, 234-236, 267, 303-312, 332-371, 400, 408
material (body-mind) responses x, 6, 60, 70, 71, 82, 89, 93, 327, 420
material realm ix-101, 108-158, 174-218, 235-238, 254-288, 302-313, 330-371, 388-402, 419
material wellbeing 10, 37-42, 190, 205, 211, 212, 313, 339, 340, 431
mathematics 10-23, 34, 93, 104, 139-146, 196, 201, 269, 271, 295-305, 328, 337, 339, 389

Matthew, the Apostle 27, 40, 77, 111-127, 155, 185, 213, 214, 223-270, 352-374, 408, 409, 417
McCaffrey, Anne 254
McGrath, Alister 14, 156, 270, 287, 309, 310, 418
McKemmish, Laura 48
McMaster, Joseph 182
McNamara, Patrick 34, 35
Medawar, Peter 261
meditation 318, 322, 339, 424
memes 138, 139, 214, 217, 399
Memphis 125, 387
mental 22-51, 80, 95, 129, 158, 159, 182-245, 314-338, 360-372, 408, 410
mental disorders 55, 56, 158, 158, 302, 315, 316, 360, 367, 408, 410
mental processing; see also reason x, 3, 6, 15, 36, 44, 50, 82, 83, 93, 117, 119, 330, 336, 337, 339
Merton, Thomas 35, 37
meta-analysis 259, 260
meta-paradigm 267, 272
metaphysics 36, 97, 134-136, 168, 203, 277, 415
Meyer, Stephen 177
Micah 28, 233, 357
Miller, Kenneth 167, 361
Mill, John Stuart 38, 39, 165
mind ix, x, 4-7, 17-109, 125, 142-228, 269-358, 391-414
mind-soul interface x, xi, 4-13, 24, 51-60, 88-101, 138, 139, 203, 208, 331-339
miracles xi, xii, 21, 66, 85-87, 104, 120, 151-164, 217, 234, 245, 278, 279, 322-368, 396-414

misery 39, 40, 75, 417
Mitchell, Jerry 226
model-dependent realism 328
monkeys 180, 206
morality ix-15, 30-64, 80-122, 140-228, 265, 290-297, 323-371, 390, 402-418
Mosaic Law 222, 231
Moses 229, 409
Mott, Nevill 126, 347, 368
M-theory; see also theory of everything / M and S theory 150-152, 328, 334
Muhammad, the Prophet 229, 232, 250, 265
multiverse 21, 202, 244, 311, 338, 349
murder 40, 51, 72, 101, 102, 161, 191, 223
music xiii, 66, 124-126, 145, 248, 261, 355-388
Muslim 43, 62, 125, 167, 226, 232, 250-254, 266, 357, 397, 404
mysticism 35, 122, 145, 266, 274, 324, 325

Nahmanides 240, 241
natural disasters xi, 7, 27, 41, 157, 226, 251, 369, 372, 398
naturalism 46, 171, 197, 200, 276, 277, 281, 287, 407
natural selection 40, 43, 135, 168, 174, 271, 277, 281, 327
Natural Theology 14, 166, 309
Needham, Joseph 44
Neill, Stephen 284, 285
neural events and states 87-95, 99
neural networks and processes 36, 37, 46, 48, 88

neural probability fields 88, 92
nervous system/neurons 87, 88, 99, 329-330
neurophysiology/neuroscience 11, 57, 70, 91, 112
New Age 328, 344
New Atheism 73, 246, 389, 418
New Earth 354, 364
New Heaven 354, 364
Newman, Cardinal John Henry 164, 165
New Testament 230, 233, 241, 275, 283, 408
Nicene Creed 167, 347-355, 367, 412
Nicholas of Cusa 69
Noah 172, 229, 240
nonbeliever ix-1, 9-25, 49-66, 97-133, 154-174, 193-219, 247-289, 300-363, 385-395
non-material/non-materiality xi, 7, 23-26, 34, 38, 88, 96, 100, 334
Non-Overlapping Magisteria 12
Novak, Michael 219-293
Numbers 223, 239

**O**bjective evidence x, 138, 205, 269, 306, 307, 331, 361
objectivity 11, 12, 143, 259, 335
observeds 23, 26, 88, 89, 101, 108
Occam, William of 108, 145
offline mental space 329, 330
Old Testament 28, 166, 222, 223, 229-233, 266, 282, 397, 398, 407
ontological proof 156, 400, 415
Oreskes, Naomi 140
original sin 30, 32, 33, 274, 373
orthodoxy ix, 3, 66, 264
Osler, William 244

Pacem in Terris 15, 290, 291
paradigm/paradigm community/ paradigm shift 10, 11, 145, 264-272, 324
Parfit, Derek 20, 192, 193
Pargament, Kenneth 317, 318
partisan interest 205-214, 340, 341
passion 45, 208, 237, 285
patients 159, 160, 244, 255-262, 316-318, 361
Patriarchs 229, 233, 239, 240
patternicity 327, 328
Paul, the Apostle 1, 28, 29, 59, 75, 117-129, 162, 167, 193, 205, 213, 214, 233-252, 268, 284-288, 309, 364-373, 389
Pauly, Daniel 115, 384
Payuto, Bhikku 109
peace of mind 37, 42, 123, 184, 213, 215, 303, 313, 370
Peacocke, Arthur 301
Pearcey, Nancy 169-171
Pennock, Robert 182
Penrose, Roger 20, 22, 23, 46-48, 98, 312
perfect xi, xii, 26, 27, 51, 119, 121, 124, 145, 166, 176, 179, 224, 278, 302, 364, 373, 404
personal experiences of God 9, 34, 103-124, 138, 151, 208, 210, 271, 294, 300, 310, 333, 349, 365, 397
Peter, the Apostle 161, 294
Peterson, Eugene 390
Philip, the Apostle 28
Philippians 1, 122, 123, 127, 129
Philippines, the xiii, 148, 227, 228, 262, 320, 382, 384, 386, 426
Phy-Olsen, Allene 175

Pigliucci, Massimo 136, 137
Pilate, Pontius 190, 351
Pinker, Steven 191
placebo 112, 148, 245, 261
Plantinga, Alvin 106, 113, 200, 276-279, 415
Polanyi, Michael 11, 17, 138, 142
Polkinghorne, John 7, 14, 51, 76, 97, 98, 136, 141, 202, 237, 299, 307, 308, 348, 364, 369, 392, 393
Pollard, Kathrin 25
Pollard, William 86
Pontifical Council for Justice and Peace 297
Pope/papal 222, 290-299, 411
Pope Benedict XVI 106, 296, 298, 308
Pope Gregory I 226
Pope John XXIII 15, 290, 291
Pope Paul VI 185, 291, 293
Popper, Karl 22, 53, 54, 134-137, 143, 311
Posner, Gary 256
postmodernism 83, 197-201, 280, 344
prayer; see also intercessory prayer 15, 27, 49-68, 123-131, 174, 214, 232-272, 311-348, 365-383
preachers/preaching 28, 29, 63, 64, 75, 196, 109, 128, 186-187, 245-294, 350, 358, 384-417
predestination xi, 70-76, 105, 117, 118, 342, 345, 348, 370
Price, George 217, 218
Pridmore, Charlie 388
primates 25, 40, 43, 51, 206-212

private ix, 5, 8, 34-37, 50-79, 105-108, 116, 266, 317-338
probabilities 23, 86-104, 255, 257, 306, 397, 415
process theology 86, 279
prophets 123, 163, 229, 230, 354, 355
Protestant 167, 185, 186, 228, 252-298, 320, 321
Proverbs 49, 125, 132, 222
Psalms 82, 163, 240
pseudoscience 85, 134, 136, 146-148, 269, 411
psyche 50, 332
psychoanalysis/psychotherapy 50, 71, 165, 315-318, 360
psychology 22, 39-106, 121, 153, 202, 244, 269-295, 315, 318, 320
psychopathology 158, 159

**Q**uantum theory 20-28, 48, 67, 80-100, 150, 279, 308, 311
quasi-morality in animals 40, 192, 205-210, 219, 278
quest for truth; see also faith-science quest xii-23, 76, 96, 104-200, 264, 269, 291-361, 396-406
Quinn, John 290
Qu'ran, the Holy 228-232

**R**acism 83, 376
random 67-99, 170, 179, 180, 281, 312, 327, 377, 383
rape 27, 161, 191, 224, 302, 372
Rapture 172, 231, 251-253, 394
Ratzinger, Joseph 298
rationality 37, 62-71, 105, 113, 133-137, 192-210, 263, 280-406

Rawls, John 220
Readiness Potential 78, 79
realism 10, 200, 203, 411
reality x-3, 11, 16-117, 134, 174, 191, 193, 267-327, 345, 395
reason x, 15, 30-45, 60-131, 156, 157, 208-211, 264-269, 286-308, 326-340, 403-406, 415, 418
reasonable 82, 120, 150, 193, 204, 224, 253, 285, 299, 331, 342, 394
reciprocity/indebtedness 206-216
Rees, Martin 202
Reformation, The Protestant 270
reformations/reformers xiii, 2, 263-300, 345
Reformed Epistemology (RE) 113, 276, 279
relativism 194, 195, 198-201, 324
religion, organized religion ix-1, 8-136, 155-168, 184-225, 244-280, 293-329, 338-368, 389-407, 415
Researcher-Subject/therapist-client relationships 316-324
Resurrection 239, 352-364, 411
Revelation, the Book of 32, 158, 231, 246, 252, 363, 364
revolutions/revolutionaries xiii-16, 104, 144, 145, 172, 177, 263-272, 295-303, 345
Ridley, Matt 215, 216
right 14, 30-47, 72-81, 103-114, 145-221, 290-292, 338-341, 373, 374, 395, 399, 414
Ritchie, Robert 250
RNA 180, 181, 186, 362
Roach, William 179, 282, 283
Robertson, David 156
Robertson, John 61, 62

Roche, Patrick 276
Romans 75, 127, 219, 374
Rome 292, 347
Rorie, James 331, 332
Rorty, Richard 199-201
Royal Society 61, 87, 238
Rowe, Dorothy 28
Rozema, Lee 95
Russell, Bertrand 14, 109, 141, 145, 417, 418
Russell, Peter 3, 154, 267

Sacks, Jonathan 130, 19
sacred, the 8, 99, 303-319, 403
sacred texts 13-23, 64, 74, 104, 156, 195, 222-294, 341-360, 403
Sagan, Carl 147, 225, 269, 359
Saints 104, 245, 354, 355, 389
salvation 77, 234, 270, 345-356
Satan 158, 159, 161, 167
Schroeder, Gerald, L. 14, 26, 180, 240, 241
Schönborn, Cardinal Christoph 263, 298
science x-2, 10-22, 27-44, 61-78, 97-114, 130-225, 264-345, 377-419
science-friendly reformations xiii, 263-276, 289, 300
science, philosophy of 133-136, 143, 189, 272
scientific method 11, 47, 133, 134, 141-145, 152, 260, 272, 305, 311
science-religion relationships 14, 21, 48-61, 86, 129, 152-195, 232-304, 347, 367, 371, 391-408

scientists x-xii, 10-23, 49-69, 106-115, 129-188, 205, 241-312, 334, 347, 383-408, 417
Science and Religion Forum 14, 129
science and theology 14, 115, 144, 152, 166, 266-311, 334, 392-396
scientism 136
Scott Peck, M. 78, 159, 160, 360
Searle, John 11, 46, 331
Second Vatican Council (Vatican II) 185, 236, 290-298
Segerstrale, Ullica 217
self x, 2-6, 22, 24, 30-140, 185, 199, 218, 236, 244, 280, 321-348, 371, 372, 392
self-aware/-conscious/-knowledge/ -processing ix-79, 101, 125, 139, 171, 199, 209, 295, 323-342
selfish 9, 52, 210-216, 385
selfless/self-sacrificial 190, 206-217, 322, 340, 414
self-professed 218, 226, 244, 324, 381, 397
Sennett, James F. 106
sense/sensory x, 8, 25-101, 222, 228, 264, 306-339, 413, 414
*sensu divinitatis* 113
sermons 128, 186, 187, 219, 323, 388, 407
sexuality/sexual orientation 208, 222, 224, 240-251, 274, 302, 397
Shafranske, Edward 318
Shanley, John Patrick
Sharp, John 147
Shear, Jonathan 314
Shermer, Michael 29, 105, 140, 326-328, 396

Shintoism 405
Shorto, Russell
Shropshire 363, 375, 378, 381
signs and wonders xii, 16, 17, 202
Silva, Ignacio 97
simplicity xii, 20, 21, 24, 53, 145, 146, 304, 328
sin 32, 33, 159, 385, 406, 415, 417
Singh, Simon 202
Sire, James 197, 198
Sloane Manuscript 32
Smart, Ninian 325
Smith, John MacDonald 175, 393, 394
Society of Ordained Scientists 14, 129, 251
Sola Scriptura 270
soul, as the spiritual core of self ix, x, 2-6, 24-58, 73, 90-93, 101, 115, 119, 120, 151, 209, 290, 328-333, 339-351, 362, 371, 373
soul, as God's gift to humans xi, 8, 28-40, 64-98, 113-124, 131, 139, 158, 176, 201-216, 242, 269-281, 330-344, 365-372
soul, alleged weight of 49, 50
soul baselines and battlefields, processing and shifts x, 5-7, 15, 26-99, 109-124, 155-162, 236, 313, 333-337, 352, 384, 405, 414
soul, destiny of 28-33, 51, 75, 77, 89, 115, 138, 188, 193, 203, 231-236, 253, 279, 328, 342-370
soul/spiritual events and states 73, 88-95, 311, 333, 334
soul, containing the Universal Moral Code 41, 160, 192, 210, 222
soul-less 54, 95, 100, 303, 332
soul territory 47-62, 80, 83, 118, 151, 204, 227, 245, 274, 304, 313, 327, 372, 400, 406
sovereignty of self 5, 29, 71, 99, 105, 114, 208, 266, 335, 348
Spieth, Hermann 249
spirits 25, 44-68, 84, 189, 262, 270, 327, 342-371, 397, 408
spiritual experiences 23, 57, 144, 321, 326, 334
spiritual force for evil x, xii, 5-10, 26-41, 50-220, 269, 302-306, 334-373, 404, 405
spirituality/spiritual life x-58, 85-133, 160, 185, 208, 228-319, 332-351, 371, 381-416
spiritual longing/needs ix, 3, 22, 34
spiritual revelations/responses x-xii, 8, 15, 21-49, 63-98, 113-139, 203-225, 264-290, 310-339, 367, 368, 397-402, 414
spiritual nudges 96
spiritual probabilities 88, 92
spiritual realm x-16, 24, 32-155, 174, 193-214, 263-277, 303-313, 328-373, 393, 403-419
spiritual stuff 22-34, 53, 58, 73, 152, 156, 332-371, 400, 408
spiritual warfare i, 32, 56, 62, 127, 373
spiritual wellbeing 104, 213, 261
Spong, John Shelby 274
Stannard, Russell 78
Stapp, Henry 94, 95
Starbuck, Edwin Diller 121, 320, 321, 323

Stenger, Victor 73, 132, 325, 396, 418, 419
Streeter, Burnett Hillman 219
S-theory 152, 334,
stimuli 6, 70, 71, 90, 93
Stott, John 284, 285, 394
string theory xii, 150, 151
stumbling blocks 16, 195, 288
subjectivity 11, 40, 54, 79, 83, 105, 138, 142, 201, 314-331, 361
subjective evidence from within x, xiii, 10-17, 34, 39, 57, 101-109, 152-157, 197-205, 263-287, 307-333, 362, 414, 415
suicide 9, 51, 218, 223, 372
supernatural xii, 16, 53-64, 104, 106, 131, 202, 210, 251, 253, 270, 278, 322-368
superstition 65, 164, 165, 261, 276, 327
systematic gathering, analysis and clustering x, 10, 102, 152, 153, 204, 207, 314, 320, 325, 339

Taliaferro, Charles 56, 57
Taliban 39
Tanakh 241
Taoism 41
Taylor, Herbert J. 222
TED 14, 206
Ten Commandments 221-223, 232, 346
Teresa of Calcutta, Mother 72, 106
testability 134, 135, 143, 175
Test of Faith Project 14, 420
testimonies 2, 29, 105, 120-129, 182, 204, 319, 321, 410, 414, 419
theism xi, 21, 65, 104, 113, 114, 136, 156, 175, 196, 197, 274-287, 328, 345, 396-402, 414, 415
theistic evolution 283, 287
theology/theologians 14, 35, 61, 97, 115, 144-167, 197, 213, 219, 229, 253-313, 324-347, 391-396, 407, 413, 415
theory of everything/final theory; see also M-theory/M and S theory 150-154, 271, 311, 312, 328-334
Thessalonians 242, 246, 252
Tillich, Paul 106, 280
Timothy 125, 233, 347
tolerance 163, 185, 250, 251, 265, 403
torture 161, 373, 374, 397
Toumey, Christopher 149
Tribulation 251-253
Trinity, the Holy 174, 347, 356
Trivers, Robert 249
trust ix, 1, 8, 24, 28, 40-57, 103-107, 119, 172, 196, 254, 268, 305, 306, 338, 391
truth x-28, 52-84, 104-123, 132-242, 264-362, 389-406, 414
truth, as unifier 16, 141, 187, 195, 301, 303, 307, 308, 394
truthful disclosures from God through mathematics and science x, xii, 11-26, 43-69, 108-165, 189-217, 238-311, 367, 393
Turl, John 6, 56

Unaffirmability 196
unbelief ix-3, 9-17, 29, 50-84, 103-156, 193-205, 236-281, 305-339, 362, 390-405
unbroken wholeness 100, 101

unconscious 4, 22, 73, 78, 79, 83
undeniability 4, 196, 208
unfairness 160, 206, 210
unified field theory 317
unified whole 15, 16, 311
Union Church of Manila (UCM) 125, 129, 186-188, 275, 348-388
Unitarian 62, 63, 298
unity 2, 18, 51, 106, 187, 229, 265, 290-313, 324, 350, 355, 394, 397
untestable 16, 134, 176, 331
Universal Moral Code 14, 32, 41, 85, 139-160, 193-224, 297, 331-374, 390-396
universe; see also multiverse xii, 17-22, 48-89, 150-152, 163-202, 290-318, 332-349, 362-369, 393-406, 417, 419
Ussery, David 182

**V**an Deventer, J. 118
Varela, Francisco 314
Varghese, Roy 73
Vedral, Vlatko 67
Vermes, Geza 223
Vermesch, Pierre 314, 315
Vintar Bible Baptist Church 320
violence 27, 185, 191, 368, 374, 397
Virgin Birth 274, 350, 351
viruses 7, 173, 176, 181, 362, 399
von Balthasar, Hans Urs 265
von Liebig, Justus 141

**W**ade, Nicholas 40
Walters, Frank 62, 63, 298, 299
Walsch, Neale Donald 124
war xi, 41, 113, 122, 155, 158, 184, 191, 208-224, 303, 374-405
Ward, Keith 52
Warfield, Benjamin B. 270
Warraq, Ibn 232, 250
weak nudges 95-97
Weatherhead, Leslie 190, 275
Wedge Document 175
Webster, Douglas 280, 281
Webster, Richard 242
Wegner, Daniel 80
Weinberg, Steven 154
Wesley, John 120, 244
Wesley, Charles 123, 128
Westermarck, Edward 208
White, R.S. (Bob) 173
Willimon, William 83
Wilson, E.O. 18, 249, 307
wisdom 109, 130, 132, 168, 231, 240, 268, 287, 395
Wisdom, the 174, 242
witches 9, 262, 359
Wolverhampton 123, 375-377, 387
Woods, L. Shelton 320
Word, the 132, 174, 231, 349, 350
Wright, N.T. (Tom) 112, 364, 395
Wright, Robert 307
wrong xii, 26-44, 68-81, 164-193, 204-221, 282, 341-373, 395

**Y**ahweh 230
Yancey, Philip 225
York 359, 379, 380, 387, 397

**Z**en 35
Zoroastrianism 242, 251
zygote 33, 371

Lightning Source UK Ltd.
Milton Keynes UK
UKHW020625290419
341788UK00015B/1299/P